# Innovative Technologies for Food Preservation and Processing

# Innovative Technologies for Food Preservation and Processing

Guest Editors

**Hristo Kalaydzhiev**
**Zhivka Goranova**
**Jesus Blesa**

MDPI

Basel • Beijing • Wuhan • Barcelona • Belgrade • Novi Sad • Cluj • Manchester

*Guest Editors*
Hristo Kalaydzhiev
Department of Analytical Chemistry and Physical Chemistry
University of Food Technologies—Plovdiv
Plovdiv
Bulgaria

Zhivka Goranova
Institute of Food Preservation and Quality
Agricultural Academy
Plovdiv
Bulgaria

Jesus Blesa
Department of Food Science and Nutrition
University of Valencia
Burjassot
Spain

*Editorial Office*
MDPI AG
Grosspeteranlage 5
4052 Basel, Switzerland

This is a reprint of the Special Issue, published open access by the journal *Applied Sciences* (ISSN 2076-3417), freely accessible at: https://www.mdpi.com/journal/applsci/special_issues/J1W60ME1EX.

For citation purposes, cite each article independently as indicated on the article page online and as indicated below:

Lastname, A.A.; Lastname, B.B. Article Title. *Journal Name* **Year**, *Volume Number*, Page Range.

**ISBN 978-3-7258-4841-6 (Hbk)**
**ISBN 978-3-7258-4842-3 (PDF)**
**https://doi.org/10.3390/books978-3-7258-4842-3**

# Contents

# Preface

The following Reprint, "Innovative Technologies for Food Preservation and Processing", brings together a collection of eight peer-reviewed research articles originally published in the journal *Applied Sciences*. The focus of this Special Issue lies at the intersection of food science, engineering, and technology, focusing on cutting-edge approaches for enhancing the safety, quality, and longevity of food products.

The scope of the Reprint covers a wide array of contemporary challenges and solutions in food preservation and processing. It addresses topics such as novel preservation mechanisms, innovative processing techniques, food spoilage prevention, and the optimization of health-promoting qualities in food. Contributions also examine the potential for scaling laboratory-validated technologies to industrial applications.

The primary aim of this Reprint is to disseminate scientific advancements that contribute to more sustainable, efficient, and consumer-oriented food systems. Our goal is to offer researchers, engineers, technologists, and industry professionals a consolidated resource that supports innovation and informed decision-making in food production and preservation. The motivation behind compiling this Special Issue was to highlight the increasing need for technological solutions in response to global food system pressures, such as food waste reduction, consumer demand for minimally processed products, and the drive toward clean-label and functional foods. By assembling state-of-the-art contributions from diverse research groups, we hope to encourage further exploration and interdisciplinary collaboration in this dynamic field. This Reprint is intended for academic researchers, graduate students, industry specialists, and policy-makers engaged in food technology, safety, and sustainability. It provides valuable insights for both foundational study and applied research.

We are grateful to the contributing authors for their high-quality submissions and the reviewers for their constructive feedback. Special thanks are also due to the editorial team at *Applied Sciences* for their support and dedication throughout the publication process.

**Hristo Kalaydzhiev, Zhivka Goranova, and Jesus Blesa**
*Guest Editors*

applied sciences

MDPI

Article

# High-Pressure Processing of Reduced Salt Pangasius Catfish (*Pangasianodon hypophthalmus*) Minced Muscle: The Effects on Selected Quality Properties of Its Gels

**Binh Q. Truong [1], Roman Buckow [2], Kha N. H. Nguyen [1], Linh T. Nguyen [1], Tuan N. A. Huynh [3] and Van Chuyen Hoang [3,*]**

1 Faculty of Fisheries, Nong Lam University, Ho Chi Minh City 720100, Vietnam; tqbinh@hcmuaf.edu.vn (B.Q.T.); kha.nguyen@hcmuaf.edu.vn (K.N.H.N.); ntlinh@hcmuaf.edu.vn (L.T.N.)
2 School of Agriculture, Biomedicine and Environment, La Trobe Institute for Sustainable Agriculture & Food (LISAF), La Trobe University, Bundoora, VIC 3086, Australia; r.buckow@latrobe.edu.au
3 Faculty of Chemical and Food Technology, HCMC University of Technology and Education, Ho Chi Minh City 720100, Vietnam; tuanhna@hcmute.edu.vn
* Correspondence: chuyenhv@hcmute.edu.vn; Tel.: +84-948012849

**Featured Application: High-pressure processing (HPP) is a promising technology for developing healthier fish products with reduced sodium levels and improved or similar gel quality compared to conventional heat-induced products.**

**Abstract:** This study investigated the effects of high-pressure processing (HPP) on selected physicochemical properties of pangasius gels at reduced salt concentrations. Minced pangasius muscle was treated at different HPP conditions (300–500 MPa/10 °C/10 min) with 1% or 2% added salt, then cooking (90 °C/30 min) and compared to heat-induced gels (HIGs) with 2% added salt and cooking (90 °C/30 min). The results showed that HPP significantly improved the texture, whiteness, and water-holding capacity of gels added salt 2% as compared to HIGs. At a reduced salt concentration of 1%, the gel texture, whiteness, and water-holding capacity of pangasius gels are similar to HIGs. SDS-PAGE showed a reduction in myosin band intensity of both SDS-soluble and sarcoplasmic proteins with pressure treatment. Sarcoplasmic actin protein was undetectable in all treatments. Fourier transform infrared spectroscopy (FTIR) analysis revealed higher α-helix content in HPP-treated samples, while SEM images confirmed the formation of a denser, more uniform gel network, particularly at 2% salt. These findings suggested that HPP improved the quality of gel with 2% salt added, while producing gels with similar quality compared to HIGs at reduced salt concentration.

**Keywords:** high-pressure processing; pangasius minced muscle; reduced salt; Fourier-transform infrared spectroscopy; gelation

## 1. Introduction

The use of high-pressure processing (HPP) for the gelation of fish muscle has been studied in various fish species such as barramundi and golden threadfin bream [1,2]. HPP followed by heating resulted in gels with improved texture, glossier, and smoother appearance compared to those formed by heat-only gelation [3]. In addition, HPP has been applied to produce reduced salt surimi/gels from fish muscle with comparable quality to heat-treated gels containing 2% added salt [4,5].

Conventionally, thermal gels are prepared with 2% added salt to enhance functionalities such as improved texture, better water-holding capacity and fat retention, and inhibition of microbial growth during storage [6]. Under salt-added conditions, more proteins responsible for the physical stability and functionality of the product are extracted and solubilized, leading to an increase in the connection between protein and protein-lipid structures. Upon heating, these structures denature and aggregate, forming a gel matrix [7]. The reduction in salt concentration in gels may lead to lower quality due to inadequate solubilization of gel proteins. However, the addition of a high salt concentration of 2% could cause many adverse effects on the gel product, such as oxidative reactions in the gel product or high blood pressure in susceptible individuals [8]. Thus, there is a high demand to find alternative methods to reduce salt in gel products.

HPP has recently emerged as a promising method for producing high-quality gels under low salt conditions. Several mechanisms have been proposed to explain the improved quality of reduced salt gelation induced by HPP. One explanation is that high-pressure processing (HPP) strengthens interactions between myofibrillar proteins, enhancing the solubilization of muscle proteins for gel formation under low-salt conditions [9,10]. Iwasaki et al. [11] also suggested that HPP disrupts the myofibrillar structure and depolymerizes the thin filament, contributing to increased gel strength in low salt conditions. Additionally, HPP is believed to promote the formation of ionic, hydrogen, and hydrophobic interactions, which reinforce the gel structure [6]. Several studies have also reported that hydrogen bonding occurs at pressures below 150 MPa, while ionic and hydrophobic interactions dominate at pressures above 200 MPa [12,13].

Pangasius catfish (*Pangasianodon hypophthalmus*) is one of the commercially important aquaculture species in several Asian countries and is commonly farmed in Vietnam, Thailand, Indonesia, India, and Bangladesh [14]. It is traded globally, particularly in the United States, European Union, China, Australia, and other markets, primarily in the form of frozen white filets [14,15]. Pangasius fish muscle has also been utilized to produce various value-added products, such as fish fingers, fish cakes, fish sausages, surimi, and fish patties [16]. However, the application of novel processing technologies like HPP on pangasius fish muscle products remains limited. Therefore, the primary objective of this study is to investigate the effects of HPP on the quality of pangasius gels, particularly in the salt-reduced condition.

## 2. Materials and Methods

### 2.1. Materials and Fish Gel Preparation

Pangasius fish weighing around 2–3 kg was purchased from a local farm in Chiang Mai, Thailand. After filleting, each filet's weight is approximately 500–700 g. The filet was stored in an air blast freezer for 4 h, then minced with 1% and 2% NaCl (*w*/*w*) for about 3 min. During the mincing step, the temperature of the fish paste was kept below 4 °C. The moisture of the fish paste was adjusted to 78%. After mincing, the fish paste was stuffed into collagen cases (24 mm in diameter × 100 mm in length). Six sausage samples were prepared for each treatment. Salt concentrations of 1% and 2% were chosen because 2% is considered a conventional level in surimi and gel-based products for achieving optimal gel characteristics [5,17]. Meanwhile, several studies have also used 2% salt as the control and reduced it to 1% for low-salt concentration [18–20].

### 2.2. High-Pressure and Thermal Treatment of Barramundi Paste

High-pressure treatment of pangasius minced muscle was conducted in a 5 L chamber HPP system (Baotou Kefa High Pressure Technology Co., Ltd., Baotou, China) with a maximum pressure of 600 MPa. Pangasius samples (1% and 2% added salt) were subjected to high-pressure treatments with the following conditions: 300 MPa/10 °C/10 min (PG300), 400 MPa/10 °C/10 min (PG400), and 500 MPa/10 °C/10 min (PG500). Pressure levels of 300, 400, and 500 MPa were selected based on preliminary trials and previous studies, as they are commonly applied in the gelation of fish proteins while maintaining structural integrity. After pressurization, the pressure-induced pangasius gels were cooked at 90 °C for 30 min. Heat-induced pangasius gels (HIGs) were prepared by adding 2% salt to the minced muscle, followed by cooking at 90 °C for 30 min, and these were used as control samples.

### 2.3. Proximate Analysis

The composition of the pangasius muscle, including ash, protein, and fat content, was determined before salt addition. Crude protein was analyzed using the Kjeldahl method, crude fat was determined by the Soxhlet method, and ash content was measured following AOAC (2000) procedures.

### 2.4. Total Plate Count (TPC)

TPC of Pangasius gel was conducted by homogenizing 10 g of fish muscle with 90 mL of sterile diluent solution for 60 s with a homogenizer (HG-15 A, Daihan, Wonju, Republic of Korea). To prepare a series of decimal dilutions, 1 mL of the homogenized solution was transferred into a test tube containing 9 mL of sterile diluent solution, and this process was repeated to achieve dilutions up to $10^{-5}$. Microbial enumeration was performed by spreading 0.1 mL of each dilution onto the surface of a sterilized Petri dish containing Plate Count Agar (PCA, HiMedia Laboratories Pvt. Ltd., Mumbai, India). The total plate count was determined after incubation for 48 h at 37 °C [21].

### 2.5. Color Analysis

Color is an important sensory attribute for seafood products that greatly influences consumer perception and marketability of the products. In surimi or fish gels, whiteness is often associated with freshness, purity, and high quality. The color of the pagasius gels was measured by a Konica Minolta Chroma meter (CR-400, Konica Minolta, Tokyo, Japan). L (brightness), a (+a, red; −a, green), and b (+b, yellow; −b, blue) values were recorded, then whiteness and ΔE were calculated from the following formula [22,23].

$$\text{Whitenes} = 100 - \left[(100 - \text{L})^2 + \text{a}^2 + \text{b}^2\right]^{1/2}$$

$$\Delta\text{E} = \left[(\triangle\text{L})^2 + (\triangle\text{a})^2 + (\triangle\text{b})^2\right]$$

where △L, △a, and △b are the differences in the L, a, and b values between pressure and heat-induced gels.

### 2.6. Mechanical Properties

Hardness, springiness, and gel strength are among the most important parameters for assessing the structural quality of fish gels. For analysis of gel strength, pangasisus gels (2.4 cm × 3 cm) were removed from their cases, and then tempered at 25 °C for 2 h before measuring. Gel strength was investigated by a cylindrical plunger (TA-XT plus, Stable Micro Systems Ltd., Godalming, Surrey, UK) with a diameter = 40 mm at a speed of 50 mm/min with a 200 N load cell. Gel strength was calculated using breaking force × breaking deformation [24].

For analysis of hardness and springiness, the case of pangasisus gels (2.4 cm × 3 cm) was also removed and tempered at 25 °C for 2 h. Hardness and springiness were measured with a texture profile analyzer (TA-XT plus, Stable Micro Systems Ltd., Godalming, Surrey,

UK) using a cylindrical plunger (diameter = 40 mm) at a deformation rate of 2 mm/min and compressing 50% of the gel's height.

### 2.7. Water-Holding Capacity

WHC is an important parameter of fish gels, reflecting the quality of the formed gel and affecting both sensory properties and the economic value of the final product. About 2 g of pagasius gel was wrapped in filter paper and then placed inside a centrifuge tube. The water-holding capacity was determined by calculating the amount of water retained per 100 g of the initial water content in the sample. Centrifugation was performed using a universal centrifuge (Universal 320 R, Hettich GmbH & Co. KG, Beverly, MA, USA) at 9000× *g* for 20 min at 20 °C [25].

### 2.8. Protein Solubility

Pangasius gel (about 4 g) was ground with 40 mL of a 30 g/L NaCl solution for 1 min. The mixture was then centrifuged at 1500× *g* for 15 min at 5 °C. A 1 mL aliquot of the resulting supernatant was homogenized with 10 mL of the same 30 g/L NaCl solution. Protein concentration in the supernatant was measured using the Lowry method [26] at an absorbance of 550 nm, using bovine serum albumin as the protein standard. The results were expressed as protein solubility [4].

### 2.9. SDS-Polyacrylamide Gel Eminlectrophoresis

Isolation of Total SDS-Soluble and Sarcoplasmic Protein Fractions:

The extraction process was performed following the method described by Pazos et al. [27]. Two fractions of proteins were prepared: total SDS-soluble proteins and sarcoplasmic proteins. For the total SDS-soluble proteins, 0.5 g of pangasius gels was homogenized with eight volumes of Tris buffer (10 mM Tris-HCl, pH 7.2) containing 2% SDS as a denaturing agent and 5 mM PMSF as a protease inhibitor. The homogenate was boiled, then processed with an Ultra-Turrax homogenizer, and centrifuged at 40,000× *g* for 12 min at 4 °C. The supernatant was collected as the total SDS-soluble protein fraction. Sarcoplasmic proteins were extracted by homogenizing 0.5 g of the fish gels in eight volumes of non-denaturing Tris buffer (10 mM Tris-HCl, pH 7.2) containing 5 mM PMSF for 2 min. After centrifugation at 40,000× *g* for 12 min at °C, the supernatant was collected and labeled as the sarcoplasmic protein fraction. Both protein fractions were stored at −80 °C until electrophoretic analysis.

SDS-Polyacrylamide gel electrophoresis:

SDS-PAGE was performed using laboratory-prepared 10% (*v*/*v*) polyacrylamide gels with an acrylamide:N,N′-ethylene bis-acrylamide ratio of 200:1. The stacking gel (4% polyacrylamide) was run at 60 V for 30 min, followed by the separating gel at 100 V for 90 min in a Mini-PROTEAN 3 cell system (Bio-Rad, Hercules, CA, USA). Each well was loaded with 30 μg of protein. The running buffer contained 1.44% (*w*/*v*) glycine, 0.67% Tris-base, and 0.1% SDS. After electrophoresis, gels were stained overnight with Coomassie PhastGel Blue R-350 dye (GE Healthcare, Uppsala, Sweden). The stained gels were scanned and analyzed by the LabImage 1 D software (Kapelan Bio-Imaging Solutions GmbH, Halle, Germany).

SDS-PAGE provides information on molecular weight distribution and solubility of proteins; it does not reveal changes in secondary structure. Therefore, FTIR analysis was used to investigate changes in the secondary structure of proteins, providing complementary insights into pressure-induced protein denaturation and gelation mechanisms.

### 2.10. Fourier Transform Infrared Spectroscopy (FTIR)

FTIR was used to investigate conformational changes in protein, particularly within the amide I region (1600–1700 cm$^{-1}$), which is sensitive to the secondary structure elements such as α-helix, β-sheet, β-turn, and random coil. These structural components determine the functional behavior of proteins during gelation.

Approximately 1 mg of the pangasius gel sample was analyzed at room temperature by placing it on the ATR crystal surface and applying pressure with a flat-tip plunger until spectra with clear peaks were achieved. Infrared spectra ranging from 4000 to 650 cm$^{-1}$ were obtained with an FT-IR 4700 (JASCO Corporation, Tokyo, Japan) equipped with an ATR prism crystal accessory. The spectral resolution was set to 4 cm$^{-1}$. All measurements were conducted in triplicate. To enhance the resolution of the spectra, a second-derivative analysis was applied, where the sharp minima corresponded to the regions of maximum intensity in the original spectrum. A second-derivative spectrum was obtained to enhance the resolution of the spectral data. All absorbance versus wavelength spectra were analyzed after background subtraction using OriginPro 2024.

### 2.11. Scanning Electron Microscope

The scanning electron microscopy procedure was carried out following the method of Iwasaki et al. [28] with minor modifications. The pangasius gel was cut from the center of the fish gel and cut into 1–2 mm square blocks. These blocks were fixed in 2.5% glutaraldehyde containing 0.1 M Na phosphate buffer (pH 7.3) and then dehydrated through a graded ethanol series (50%, 70%, 90%, and 100%). The dehydrated samples were immersed in 2-methyl-2-propanol to replace ethanol and subsequently freeze-dried. The specimens were coated under vacuum with platinum-palladium using a sputter coater. Observations were conducted using a Hitachi scanning electron microscope (TM 4000 plus, Hitachi High-Tech Corporation, Tokyo, Japan) at an accelerating voltage of 10 kV.

### 2.12. Statistical Analysis

All experiments were performed in quadruplicate, and the results are presented as mean values ± standard deviation. Data were evaluated using one-way analysis of variance (ANOVA) followed by the least significant difference (LSD) test to identify variations in gel properties across different treatments. Statistical analysis was conducted using SPSS software version 21.0 (IBM Corp., Armonk, NY, USA), with the significance level set at $p < 0.05$.

## 3. Results and Discussion

### 3.1. Proximate Analysis

Pangasius minced muscle was analyzed for protein, fat, ash, and moisture content in triplicate. Pangasius minced muscle contained crude protein: 16.46 ± 0.28%, crude fat: 1.15 ± 0.02%, moisture 80.28 ± 0.3%, and ash 1.73 ± 0.6%.

### 3.2. Total Plate Count (TPC)

The results of TPC (Figure 1) show that the TPC of high-pressure-treated samples is significantly lower than HIGs ($p < 0.05$). TPC of untreated raw pangasius minced muscle was 5.24 ± 0.21 log CFU/g. All treatments resulted in a reduction in microbial load from 3.02 to 4 logs CFU/g, depending on the treatment conditions. Studies on the effects of high pressure on microorganisms have shown that, at pressure levels of 300 MPa or higher for a few minutes and room temperature (RT), it generally inactivates viable microorganisms in various food products [29]. HPP was known to effectively inactivate heat-resistant fungal spores, such as *Talaromyces flavus* and *Byssochlamys nivea*, which can survive conventional thermal processing [30]. Moreover, the combination of HPP and cooking has been shown to improve the inactivation of spore-forming bacteria such as *Bacillus cereus* and *Bacillus subtilis* [31]. Thus, these findings are in agreement with previous research on the antimicrobial effects of high-pressure processing. The lowest TPC was found at a pressure level of 500 MPa, ranging from 1.38 to 1.45 log/mL, and significantly lower than samples treated at 300 MPa and 400 MPa ($p < 0.05$). This result is also similar to the findings of Kunnath et al. [32], in which TPC decreased as the pressure level increased. TPC results show that high-pressure processing effectively reduces microbial numbers in pangasius gel, particularly at 500 MPa.

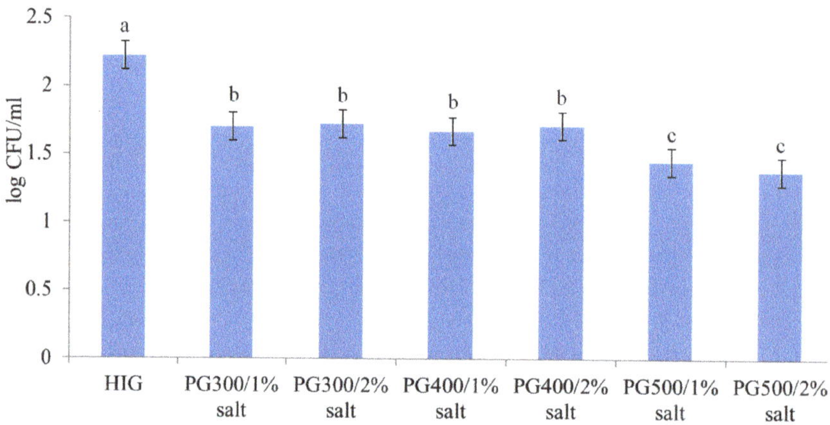


**Figure 1.** TPC of pangasius gels at 1% and 2% salt concentrations at different high-pressure processing conditions. HIGs: conventional heat-induced gels; PG300/1%: salt concentration 1% and 300 MPa/10 °C/10 min; PG300/2%: salt concentration 1% and 300 MPa/10 °C/10 min; PG400/1%: salt concentration 1% and 400 MPa/10 °C/10 min; PG400/2%: salt concentration 1% and 400 MPa/10 °C/10 min; PG500/1%: salt concentration 1% and 500 MPa/10 °C/10 min; PG500/2%: salt concentration 1% and 500 MPa/10 °C/10 min. Different letters (a–c) indicate significant differences ($p < 0.05$).

### 3.3. Color Analysis

The whiteness of pangasius gel at different pressure treatments and salt concentrations is shown in Table 1. PG gels exhibited whiteness values ranging from 63.17 ± 0.26 to 64.56 ± 2.04 and were higher than those of HIGs (60.52 ± 0.10) ($p < 0.05$). This result corresponds with the study by Cando et al. (2015) [5] on Alaska Pollock gel in which brightness increased with higher pressure levels.

It is hypothesized that HPP can induce greater protein denaturation, aggregation, and cross-linking, resulting in structural changes in a way that increases whiteness [4,5]. In addition, HPP may disrupt the heme group and break down globin in fish muscle proteins, reducing pigment intensity and enhancing the whiteness of fish gels [33]. It has been suggested that the increase in light scattering due to compact protein structures and the formation of a homogenous gel matrix also enhances the brightness [34]. In this study, all HPP-treated samples showed significantly increased whiteness regardless of salt concentration and pressure level, as compared to HIGs, which could be caused by protein structure changes and pigment disruption. This result also aligns with other research reporting that pressures above 200 MPa significantly altered the color of white fish meat [35].

**Table 1.** Color of pangasius gel at 1% and 2% salt concentrations under different high-pressure processing conditions.

| Treatment | Whiteness | ΔE |
|---|---|---|
| HIGs | $61.21 \pm 0.55^{a}$ | |
| PG300/1% salt | $64.56 \pm 2.04^{b}$ | $6.53 \pm 0.88$ |
| PG300/2% salt | $63.17 \pm 0.26^{b}$ | $6.15 \pm 0.99$ |
| PG400/1% salt | $63.45 \pm 0.91^{b}$ | $6.59 \pm 0.47$ |
| PG400/2% salt | $64.06 \pm 0.50^{b}$ | $6.18 \pm 0.71$ |
| PG500/1% salt | $63.81 \pm 1.43^{b}$ | $6.42 \pm 0.59$ |
| PG500/2% salt | $63.26 \pm 0.67^{b}$ | $6.54 \pm 0.49$ |

HIGs: conventional heat-induced gels; PG300/1%: salt concentration 1% and 300 MPa/10 °C/10 min; PG300/2%: salt concentration 1% and 300 MPa/10 °C/10 min; PG400/1%: salt concen-tration 1% and 400 MPa/10 °C/10 min; PG400/2%: salt concentration 1% and 400 MPa/10 °C/10 min; PG500/1%: salt concentration 1% and 500 MPa/10 °C/10 min; PG500/2%: salt concen-tration 1% and 500 MPa/10 °C/10 min. Different letters (a, b) in the same column indicate significant differences ($p < 0.05$).

ΔE represents the perceptual difference in color between fish samples, with higher ΔE values indicating a greater color difference [23]. ΔE values ranging from 3.0 to 6.0 correspond to very distinctive differences, while ΔE values ranging from 6.0 to 12 indicate strong differences [36]. In this study, the ΔE value ranging from 6.15 to 6.59 indicates a strong color difference between the pressure-induced gels and HIGs (Figure 2). These results confirm that HPP changed the appearance of pangasius gels compared to HIGs in both statistically and visually significant ways. The effect was apparent at both 1% and 2% salt concentrations, suggesting that pressure treatment resulted in dominant effects on gel color as compared to salt concentration.

**Figure 2.** Visual observation of pangasius fish gel color under varying salt concentrations and HPP treatments. HIGs: conventional heat-induced gels; PG300/1%: salt concentration 1% and 300 MPa/10 °C/10 min; PG300/2%: salt concentration 1% and 300 MPa/10 °C/10 min; PG400/1%: salt concen-tration 1% and 400 MPa/10 °C/10 min; PG400/2%: salt concentration 1% and 400 MPa/10 °C/10 min; PG500/1%: salt concentration 1% and 500 MPa/10 °C/10 min; PG500/2%: salt concen-tration 1% and 500 MPa/10 °C/10 min.

### 3.4. Mechanical Properties

The results of the mechanical properties of pangasius gel under different salt concentrations and HPP conditions are presented in Table 2. The hardness of the pangasius gel ranged from 7.93 N to 11.56 N. HPP increased the hardness of the pangasius gel with 2% added salt compared to HIGs ($p < 0.05$). Notably, HPP pangasius gels with 1% added salt showed comparable hardness to the HIGs with 2% added salt ($p \geq 0.05$). Thus, the hardness of the pangasius gel under reduced salt conditions remained equivalent to that of the HIG control with 2% added salt. Furthermore, treatment at 500 MPa/10 °C/10 min with 1% added salt exhibited hardness comparable to treatments at 300 MPa/10 °C/10 min with 2% salt and 400 MPa/10 °C/10 min with 2% salt. These findings align with previous studies demonstrating that HPP can reduce the required salt concentrations in barramundi [4] and Alaska Pollock surimi [5].

**Table 2.** Changes in the mechanical properties of Pegasus gels added 1 and 2% salt under different HPP treatments.

| Treatment | Hardness (N) | Springiness (mm) | Gel Strength (N.mm) |
|---|---|---|---|
| HIGs | $7.93 \pm 0.64^{a}$ | $0.88 \pm 0.02^{a}$ | $346.21 \pm 16.96^{a}$ |
| PG300/1% salt | $8.34 \pm 0.52^{a}$ | $0.87 \pm 0.01^{a}$ | $417.65 \pm 50.95^{b}$ |
| PG300/2% salt | $10.39 \pm 1.68^{bc}$ | $0.89 \pm 0.02^{a}$ | $424.85 \pm 25.29^{b}$ |
| PG400/1% salt | $8.29 \pm 0.42^{a}$ | $0.87 \pm 0.01^{a}$ | $408.38 \pm 12.43^{b}$ |
| PG400/2% salt | $10.32 \pm 0.87^{bc}$ | $0.88 \pm 0.02^{a}$ | $436.92 \pm 42.25^{b}$ |
| PG500/1% salt | $8.96 \pm 0.59^{ac}$ | $0.89 \pm 0.01^{a}$ | $425.30 \pm 23.09^{b}$ |
| PG500/2% salt | $11.56 \pm 1.76^{b}$ | $0.89 \pm 0.02^{a}$ | $464.92 \pm 13.56^{b}$ |

HIGs: conventional heat induced gels; PG300/1%: salt concentration 1% and 300 MPa/10 °C/10 min; PG300/2%: salt concentration 1% and 300 MPa/10 °C/10 min; PG400/1%: salt concen-tration 1% and 400 MPa/10 °C/10 min; PG400/2%: salt concentration 1% and 400 MPa/10 °C/10 min; PG500/1%: salt concentration 1% and 500 MPa/10 °C/10 min; PG500/2%: salt concen-tration 1% and 500 MPa/10 °C/10 min. Different letters (a–c) in the same column indicate significant differences ($p < 0.05$).

The increase in hardness of PC gels as compared to HIGs could be explained by the combined effects of pressure followed by heat treatment on protein denaturation and aggregation. HPP causes the partial unfolding of myofibrillar proteins, exposing hydrophobic groups and reactive sites for gelation, resulting in stronger protein–protein interactions after heating and the formation of a denser and more compact gel network [5]. HPP also dissociates actomyosin, increasing the content of myosin monomer for better gelation [11]. Furthermore, the formation of non-covalent bonds such as hydrophobic bonds, hydrogen bonds, and electrostatic interactions is also facilitated under high pressure. This could contribute to the improved hardness of the gel matrix [9]. In certain conditions, HPP also enhances the formation of disulfide bonds, especially with conditions of heat following pressure treatment [34]. The compacted network formed through these interactions could exhibit higher hardness and gel strength. The increase in hardness of PC gels as compared to HIGs could be explained by the combined effects of pressure followed by heat treatment on protein denaturation and aggregation. HPP causes the partial unfolding of myofibrillar proteins, exposing hydrophobic groups and reactive sites for gelation, resulting in stronger protein–protein interactions after heating and the formation of a denser and more compact gel network [5]. HPP also dissociates actomyosin, increasing the content of myosin monomer for better gelation [11]. Furthermore, the formation of non-covalent bonds such as hydrophobic bonds, hydrogen bonds, and electrostatic interactions is also facilitated under high pressure. This could contribute to the improved hardness of the gel matrix [9]. In certain conditions, HPP also enhances the formation of disulfide bonds, especially with conditions of heat following pressure treatment [34]. The compacted network formed through these interactions could exhibit higher hardness and gel strength.

The springiness of pangasius gels did not show statistically significant differences in all treatments ($p \geq 0.05$). This finding is consistent with previous studies on the effects of high-pressure treatments on *Nemipterus japonicus* gels [32] or barramundi gels [2].

In contrast, the gel strength of all HPP samples was significantly higher than that of HIGs ($p < 0.05$), ranging from 346.21 N.mm to 464.92 N.mm (Table 3). Among the HPP samples, the gel strength was comparable in all treatments ($p \geq 0.05$), with the highest average value (464.92 ± 13.56 N.mm) observed in samples with 2% added salt subjected to 500 MPa/10 °C/10 min, followed by cooking at 90 °C for 30 min.

**Table 3.** The secondary structures (%) determined by FTIR self-deconvolution of HIGs and pressure-treated pangasius gel added 1 and 2% salt.

| Treatment | α-Helix (%) | β-Sheet (%) | β-Turn (%) | Random Structure (%) | Aromatic Side Chain |
|---|---|---|---|---|---|
| HIGs | $19.41 \pm 2.27^{a}$ | $31.18 \pm 0.73^{a}$ | $19.64 \pm 1.33^{a}$ | $10.57 \pm 0.51^{a}$ | $19.17 \pm 0.37^{a}$ |
| PG300/1% salt | $24.74 \pm 1.99^{b}$ | $29.55 \pm 0.96^{a}$ | $20.14 \pm 0.81^{a}$ | $8.49 \pm 0.66^{b}$ | $17.05 \pm 3.76^{a}$ |
| PG300/2% salt | $24.14 \pm 1.07^{b}$ | $29.54 \pm 1.32^{a}$ | $19.84 \pm 0.84^{a}$ | $8.95 \pm 1.36^{ab}$ | $17.52 \pm 3.89^{a}$ |
| PG400/1% salt | $24.66 \pm 1.87^{b}$ | $29.22 \pm 2.86^{a}$ | $19.85 \pm 1.08^{a}$ | $8.02 \pm 1.26^{b}$ | $18.21 \pm 1.34^{a}$ |
| PG400/2% salt | $23.53 \pm 1.11^{b}$ | $30.72 \pm 1.72^{a}$ | $19.89 \pm 1.33^{a}$ | $9.06 \pm 0.45^{ab}$ | $16.77 \pm 3.73^{a}$ |
| PG500/1% salt | $23.65 \pm 0.98^{b}$ | $30.18 \pm 1.73^{a}$ | $20.11 \pm 1.03^{a}$ | $7.69 \pm 1.38^{b}$ | $18.32 \pm 3.16^{a}$ |
| PG500/2% salt | $23.82 \pm 1.76^{b}$ | $29.43 \pm 0.89^{a}$ | $20.16 \pm 1.96^{a}$ | $9.19 \pm 1.12^{ab}$ | $17.37 \pm 3.81^{a}$ |

HIGs: conventional heat-induced gels; PG300/1%: salt concentration 1% and 300 MPa/10 °C/10 min; PG300/2%: salt concentration 1% and 300 MPa/10 °C/10 min; PG400/1%: salt concen-tration 1% and 400 MPa/10 °C/10 min; PG400/2%: salt concentration 1% and 400 MPa/10 °C/10 min; PG500/1%: salt concentration 1% and 500 MPa/10 °C/10 min; PG500/2%: salt concen-tration 1% and 500 MPa/10 °C/10 min. Different letters (a, b) in the same column indicate significant differences ($p < 0.05$).

The comparable hardness and higher gel strength of high-pressure-induced pangasius gel with 1% added salt, compared to heat-induced gels (HIGs) with 2% salt, may be attributed to the ability of HPP to enhance interactions among myofibrillar proteins, promoting protein solubility at low salt concentrations [9]. Additionally, pressure causes the disruption of myofibrillar structures, accompanied by the depolymerization of protein polymers, resulting in higher gel strength of myofibrillar proteins under low salt conditions [11]. Furthermore, Ramirez-Suarez and Morrissey [22] reported that under reduced salt conditions, pressure can promote the formation of high molecular weight polypeptides, potentially through disulfide bonding, thereby increasing gel strength. HPP enhances protein bonding reactions and induces the formation of dense gel network structures, contributing to increased gel strength under reduced salt conditions [4]. As a result, HPP has been extensively studied for reducing the amount of salt added to protein gels. Studies have shown that the presence of salt improves protein solubility, enhances bonding, and maintains the denaturation/aggregation pattern of major proteins under thermal conditions [37]. However, using high salt concentrations in protein gels is undesirable due to their negative impact on product quality and consumer health. Therefore, HPP can be beneficial for producing high-quality pangasius gel at low salt concentrations. In this experiment, high-pressure treatment resulted in pangasius gel at a lower salt concentration of 1% with comparable mechanical properties to HIGs.

### *3.5. Water-Holding Capacity*

Figure 3 presents the water-holding capacity (WHC) of pangasius gel at different salt concentrations. HPP improved the WHC of all pangasius gel treatments compared to HIGs ($p < 0.05$). Pangasius gels with 2% added salt and treated at 500 MPa/10 °C/10 min achieved the highest WHC at 88.62%, while HIGs had the lowest WHC at 76.28%. HPP pangasius gels with 1% added salt exhibited similar WHC to those with 2% added salt

($p \geq 0.05$). The improved mechanical properties of pangasius gel under low salt conditions could be attributed to the enhanced interactions among myofibrillar proteins by HPP, which promote protein solubility and gel formation at low salt concentrations, resulting in a gel with better WHC [9]. Additionally, the myofibrillar structure is disrupted and degraded into smaller protein fragments, leading to an increased solubility of proteins under low-salt conditions. This also facilitates the binding of free water and fish proteins, resulting in gels with high strength and water-holding capacity under low-salt conditions [38]. Furthermore, HPP creates a stable and compact spatial structure of the gel, which helps immobilize and retain water within the gel structure [1]. HPP also exposes hydrophobic groups, enhancing hydrophobic interactions and stabilizing the water/protein system [5]. These findings suggested that the improvement in WHC by HPP treatment could be caused by several mechanisms, including protein unfolding by HPP, which can promote protein–water interactions; exposing polar and hydrophobic groups that help to bind water molecules more effectively; and better formation of gel structure, which better holds water within the gel matrix as reported by Ma et al. [1]. These mechanisms also explain HPP HPP-treated pangasius gels at 1% salt concentration have comparable WHC to those with 2% salt, thereby supporting the potential of HPP in reducing salt without reducing WHC.

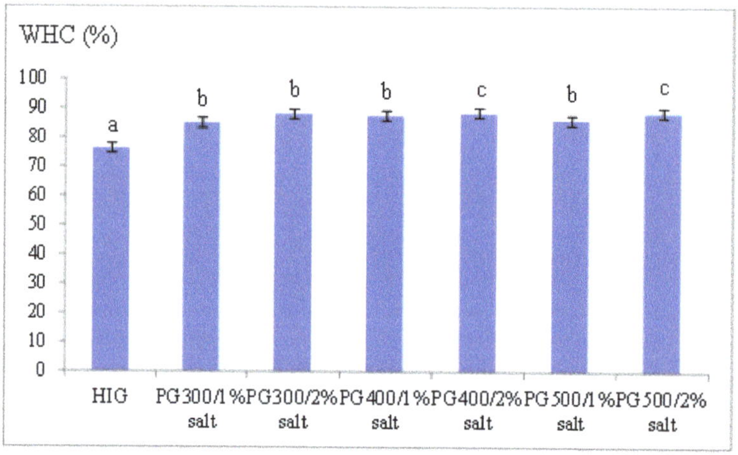


**Figure 3.** WHC of pangasius gel was added with 1 and 2% salt under different HPP treatments. HIGs: conventional heat-induced gels; PG300/1%: salt concentration 1% and 300 MPa/10 °C/10 min; PG300/2%: salt concentration 1% and 300 MPa/10 °C/10 min; PG400/1%: salt concen-tration 1% and 400 MPa/10 °C/10 min; PG400/2%: salt concentration 1% and 400 MPa/10 °C/10 min; PG500/1%: salt concentration 1% and 500 MPa/10 °C/10 min; PG500/2%: salt concen-tration 1% and 500 MPa/10 °C/10 min. Different letters (a–c) indicate significant differences ($p < 0.05$).

### 3.6. *Protein Solubility*

Protein solubility is an important parameter of the functional structure of muscle proteins, particularly myofibrillar proteins, which are responsible for in-gel formation. Protein solubility exhibits the extent of protein denaturation and aggregation induced by processing treatments such as heat or HPP. The decrease in solubility shows the formation of insoluble protein aggregates or network structures, which may contribute to better gelation. In contrast, high solubility indicates that proteins are insufficiently denatured and therefore, less able to form a well-structured gel matrix. Therefore, protein solubility provides important mechanistic insight into how HPP influences the gelation behavior and structural quality of fish protein gels. The protein solubility of HIGs and HPP gels ranged from 7.75 mg/L to 8.52 mg/L and showed no significant differences ($p \geq 0.05$).

The reduction in protein solubility indicates protein denaturation and the precipitation of myofibrillar proteins, particularly myosin, following the denaturation process [39]. These findings show that both heat treatment and high-pressure processing result in complete protein denaturation, facilitating the formation of a well-structured gel network at both 1% and 2% salt conditions.

### *3.7. SDS-Polyacrylamide Gel Electrophoresis*

For SDS-soluble proteins, the electrophoresis profile (Figure 4S) showed that HMW proteins, including myosin heavy chain (MHC), were present at 250 kDa and actin at 45 kDa. As pressure increased, the intensity of the 250 kDa band decreased, and this effect was more remarkable at 2% salt compared to 1% salt. At PG500/2%, the 250 kDa band was nearly undetectable. The reduction in high molecular weight proteins, particularly with 2% salt, is mainly attributed to pressure-induced aggregation, which decreases their SDS solubility. Salt could increase this effect by promoting aggregation and reducing the extractable fraction of HMW proteins in the gel [40]. In contrast, no significant differences were observed in the actin bands among the treatments.

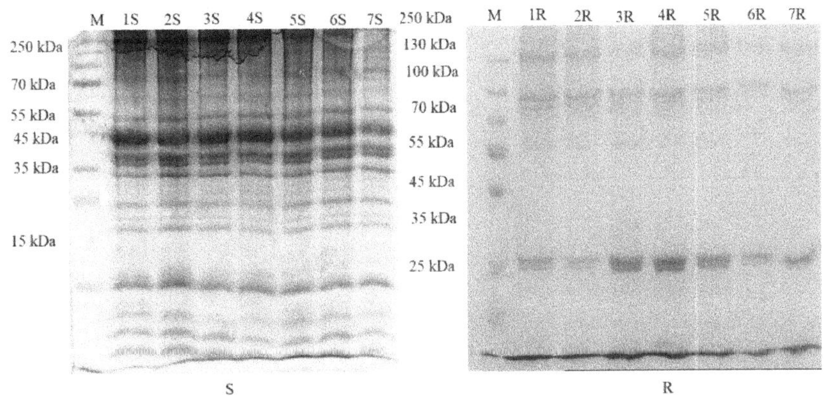


**Figure 4.** SDS PAGE pattern of SDS soluble protein (S) and sarcoplasmic protein (R) of pangasius gel obtained under various pressure treatment conditions. M: molecular ladder; 1S/1R: HIGs; 2S/2R: PG300/1%; 3S/3R: PG400/1%; 4S/4R: PG500/1%; 5S/5R: PG300/2%; 6S/6R: PG400/2%; 7S/7R: PG500/2%. HIGs: conventional heat-induced gels; PG300/1%: salt concentration 1% and 300 MPa/10 °C/10 min; PG300/2%: salt concentration 1% and 300 MPa/10 °C/10 min; PG400/1%: salt concen-tration 1% and 400 MPa/10 °C/10 min; PG400/2%: salt concentration 1% and 400 MPa/10 °C/10 min; PG500/1%: salt concentration 1% and 500 MPa/10 °C/10 min; PG500/2%: salt concen-tration 1% and 500 MPa/10 °C/10 min.

In general, the electrophoresis pattern of sarcoplasmic proteins exhibited a similar trend to SDS-soluble proteins, with a reduction in high molecular weight bands, including myosin heavy chain (MHC), as pressure increased (Figure 4R). Particularly, the band intensity in pressure-induced gels with 2% salt was lower than in those with 1% salt, suggesting that higher salt concentration can promote more protein aggregation. A similar observation was found in arrowtooth flounder gels, where the MHC band intensity decreased with increasing pressure [34]. The actin band of sarcoplasmic proteins disappeared in all treatments, indicating that actin proteins were extensively denatured and aggregated.

### *3.8. FTIR Analysis*

The vibrational spectra of PG and HIGs at different salt concentrations were analyzed by FTIR to gain insights into their secondary structure. Spectral analysis within the frequency range of 4000–500 $cm^{-1}$ revealed peaks at 435 ± 8.6 $cm^{-1}$, 1391 ± 9.4 $cm^{-1}$, 1455 ± 2.2 $cm^{-1}$, 1549 ± 12.1 $cm^{-1}$, 1624 ± 0 $cm^{-1}$, 2921 ± 3.37 $cm^{-1}$, and 3267 ± 7.37 $cm^{-1}$. Among these bands, the amide I band was observed at 1624 ± 0 $cm^{-1}$, amide II at 1549 ± 12.1 $cm^{-1}$, and amide A at 3267 ± 7.37 $cm^{-1}$ (Figure 5).

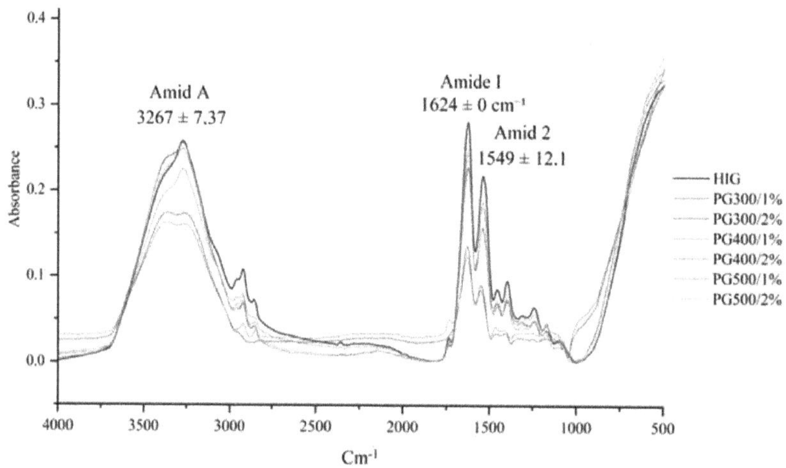


**Figure 5.** FTIR spectra of pangasius gels with 1 and 2% salt under different HPP treatments over the frequency range 4000–500 $cm^{-1}$. HIGs: conventional heat-induced gels; PG300/1%: salt concentration 1% and 300 MPa/10 °C/10 min; PG300/2%: salt concentration 1% and 300 MPa/10 °C/10 min; PG400/1%: salt concen-tration 1% and 400 MPa/10 °C/10 min; PG400/2%: salt concentration 1% and 400 MPa/10 °C/10 min; PG500/1%: salt concentration 1% and 500 MPa/10 °C/10 min; PG500/2%: salt concen-tration 1% and 500 MPa/10 °C/10 min.

The amide I region, characterized by C=O stretching vibrations, typically consists of overlapping component bands derived from structural elements such as α-helices, β-sheets, turns, and disordered or irregular conformations [41]. Amide II (1480–1575 $cm^{-1}$) primarily corresponds to CN stretching and NH bending, while the amide A band (3100–3600 $cm^{-1}$) is mainly linked to OH and NH stretching vibrations, with a broadened absorbance reflecting the muscle's high water content [42,43].

Within the myofiber spectrum, the amide I band (ranging from 1700 to 1600 $cm^{-1}$) is particularly significant due to its high sensitivity to hydrogen-bonding, dipole–dipole interactions, and the structural configuration of the protein's polypeptide backbone [5]. Like the study of Herranz et al. [41], the application of HPP resulted in a reduction in the absorbance of the amid I band in pangasius protein as compared to HIGs. This reduction indicates structure alterations, due to the disruption or formation of protein interactions. However, the amide I frequencies were similar in all treatments, suggesting that high pressure caused only minor changes in the protein structure.

To enhance spectral resolution and investigate the changes in the secondary structure of myofibrillar proteins, the amide I region (1700–1600 $cm^{-1}$) was analyzed using second-derivative spectra (Figure 6). This analysis found nine different bands including 1604 ± 0.58 $cm^{-1}$, 1625 ± 0.73 $cm^{-1}$, 1637 ± 0.35 $cm^{-1}$, 1648 ± 1.3 $cm^{-1}$, 1654 ± 0.84 $cm^{-1}$, 1665 ± 1.32 $cm^{-1}$, 1675 ± 0.26 $cm^{-1}$, 1680 ± 0.66 $cm^{-1}$ and 1696 ± 0.66 $cm^{-1}$. The bands at 1625 ± 0.73 $cm^{-1}$, 1637 ± 0.35 $cm^{-1}$, and 1696 ± 0.66 $cm^{-1}$ were attributed to β-sheet structures; the band at 1654 ± 0.84 $cm^{-1}$ and 1665 ± 1.32 $cm^{-1}$ corresponded to α-helices; the bands at 1675 ± 0.26 and 1680 ± 0.66 $cm^{-1}$ were assigned to β-turns [43]. On the other hand, the band at 1648 ± 1.3 $cm^{-1}$ was attributed to random structures, and the band at

1604 ± 0.58 $cm^{-1}$ could be aromatic side chain vibrations of amino acids such as tyrosine, phenylalanine, or tryptophan [44–46].

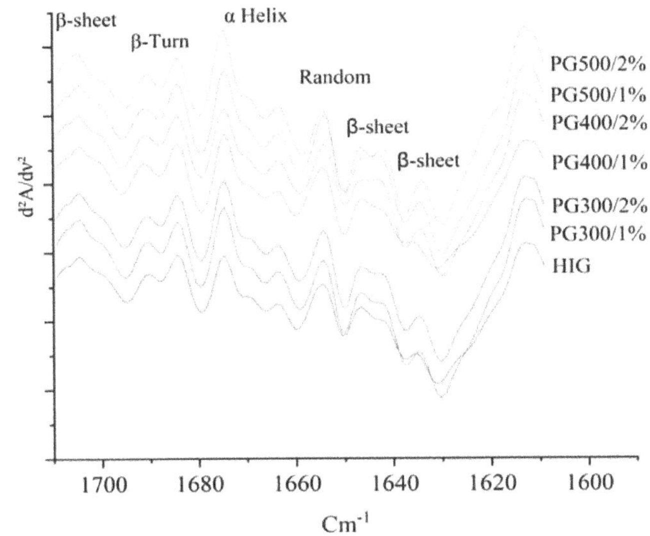


**Figure 6.** Second-derivative of Amide I band (1700–1600 $cm^{-1}$) of infrared spectra of pangasius gels added 1 and 2% salt under different HPP treatments over the frequency range 1700–1600 $cm^{-1}$. Peaks corresponding to α-helix (~1654 $cm^{-1}$), β-sheet (~1625, ~1696 $cm^{-1}$), β-turn (~1680 $cm^{-1}$), and random coil (~1648 $cm^{-1}$) indicate structural changes in the protein secondary structure of different treatments. HIGs: conventional heat-induced gels; PG300/1%: salt concentration 1% and 300 MPa/10 °C/10 min; PG300/2%: salt concentration 1% and 300 MPa/10 °C/10 min; PG400/1%: salt concentration 1% and 400 MPa/10 °C/10 min; PG400/2%: salt concentration 1% and 400 MPa/10 °C/10 min; PG500/1%: salt concentration 1% and 500 MPa/10 °C/10 min; PG500/2%: salt concen-tration 1% and 500 MPa/10 °C/10 min.

The results in Table 3 show the percentage of the secondary structure of the pangasius protein treated at different HPP conditions and HIGs. The α-helix content was significantly higher in the HPP samples compared to the HIG control ($p < 0.05$). The stabilization of α-helix under high-pressure conditions can be attributed to the ability of HPP to maintain intramolecular hydrogen bonds, preventing thermal denaturation. Moreno et al. [41] showed that high pressure also increased α-helix content in flying fish surimi that was pressurized and then cooked. Moreover, He et al. [47] suggested that high pressure helps preserve α-helix structures in proteins by restricting the unfolding process and minimizing heat-induced transitions to β-sheet or random coil structures. In contrast, Cando et al. [5] reported that the α-helix structure of isolated hake myofibrils remained similar between heated samples (90 °C for 20 min) and those subjected to pressure treatment (150–250 MPa at 10 °C for 10 min) followed by heating (90 °C for 20 min).

Unlike α-helix structures, β-sheet and β-turn contents showed no significant differences among all treatments, including HIGs. The β-sheet content remained relatively stable, ranging from 29.22% to 31.18%, while β-turn content fluctuated minimally between 19.64% and 20.14%. This suggests that these structures are stable under the thermal and high-pressure conditions. These findings align with previous reports suggesting that β-sheets are stabilized by strong inter-chain hydrogen bonds, which are resistant to both thermal and pressure-induced conformational changes [48,49].

At a 2% salt concentration, the random structure content was similar between pressure-induced gels and HIGs. However, random structure content was significantly lower in high-pressure-treated samples compared to the HIG control ($p < 0.05$) at reduced salt concentration (1%). This suggests that the less random structure of pressure-induced gels with 1% added salt was denatured compared to HIGs with 2% salt. Thus, HPP can result in molecular reorganization and influence the degree of protein denaturation in different patterns at different salt concentrations. A similar observation was also reported in flying fish surimi in which random structure content was reduced compared to unpressurized samples [50]. Chen et al. [51] also reported a decrease in random coil content in *Nemipterus virgatus* surimi when pressure increased.

### 3.9. Scanning Electron Microscope (SEM) Analysis

SEM images showed that the microstructure of pangasius gels changed with varying salt concentrations and processing conditions (Figure 7), corresponding to the mechanical properties.

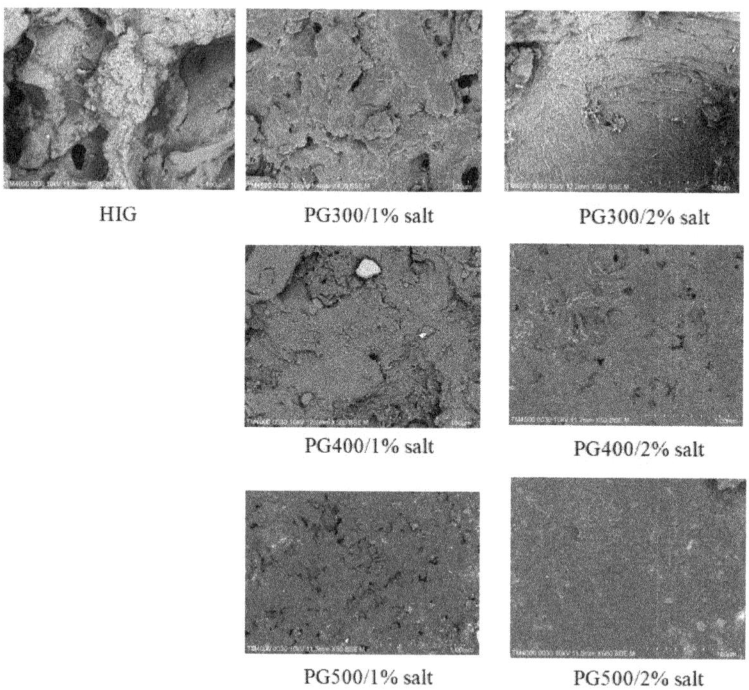


**Figure 7.** Scanning electron microscopy (SEM) images of pangasius gel at different salt concentrations and high-pressure processing conditions. HIGs: conventional heat-induced gels; PG300/1%: salt concentration 1% and 300 MPa/10 °C/10 min; PG300/2%: salt concentration 1% and 300 MPa/10 °C/10 min; PG400/1%: salt concen-tration 1% and 400 MPa/10 °C/10 min; PG400/2%: salt concentration 1% and 400 MPa/10 °C/10 min; PG500/1%: salt concentration 1% and 500 MPa/10 °C/10 min; PG500/2%: salt concen-tration 1% and 500 MPa/10 °C/10 min.

The HIG sample exhibited a relatively loose and irregular structure, with visible gaps and pores. In contrast, gels treated with high pressure (especially at 400–500 MPa) showed a denser, more homogeneous, and compact structure as compared to HIGs. This indicates that HPP treatment facilitates the formation of a well-developed gel network in pangasius gels. PG with 2% salt treated with high pressure formed a more compact microstructure with fewer voids and a smoother surface compared to PG with 1% salt, exhibiting characteristics typical of pressure-treated gels [3]. In contrast, PG with 1% salt displayed a spongier structure compared to PG with 2% salt due to the lower salt content, which limited protein solubility and resulted in a weaker gel. However, HPP still facilitated the formation of a relatively compact gel with fewer voids and a smoother surface than heat-treated gels. SEM observations also confirm the correspondence between microstructure and the mechanical properties of pangasius gels, in which HPP-treated samples exhibit improved hardness, gel strength, and WHC compared to HIGs. Similar observations in fish gels and surimi treated with HPP have been reported in previous studies, showing that pressure-induced gelation can produce finer and more stable networks compared to heat-only treatments [4,5].

## 4. Conclusions

The results of this study show that HPP prior to heat treatment is an effective method for producing high-quality pangasius gels at reduced salt conditions. HPP improved gel texture, whiteness, and water-holding capacity in pangasius gels with 2% salt while maintaining these properties of gels with 1% salt comparable to heat-induced gels (HIGs) with 2% salt. SDS-PAGE results showed that increasing pressure reduced the intensity of high molecular weight proteins in both SDS-soluble and sarcoplasmic proteins, likely due to pressure-induced aggregation. Additionally, actin from sarcoplasmic proteins was undetectable in all treatments, indicating that the effect of thermal treatment is predominant over-pressure treatment on sarcoplasmic actin protein.

HPP also facilitated the formation of a more compact gel network, as confirmed by SEM analyses. FTIR spectra showed that HPP increased α-helix content, stabilizing intramolecular hydrogen bonds and reducing protein unfolding. At 1% salt, PG had lower random structure content than HIGs, suggesting reduced protein denaturation. These structural modifications contributed to enhanced gel strength and stability, even at a lower salt concentration (1%). These results indicate that HPP can be applied to develop salt-reduced pangasius gel products with improved textural and structural properties. This approach can satisfy the growing consumer demand for healthier, lower-sodium foods and may contribute to improved product quality and greater acceptability in commercial applications. However, this study has some limitations, as it was conducted using a single fish species (pangasius) under fixed HPP conditions, including temperature and holding time. Therefore, more studies are needed to examine the changes in pangasius gels under different HPP conditions. Future research should explore the long-term storage stability and sensory attributes of high-pressure-induced pangasius gels to further validate their commercial viability. In addition, the investigation of the molecular mechanisms of protein interactions, including bond formation such as disulfide, hydrophobic, and hydrogen bonds, should be conducted for a deeper understanding of the structural changes induced by HPP and their impact on gelation behavior and functional properties.

**Author Contributions:** Conceptualization, B.Q.T. and V.C.H.; methodology, B.Q.T., R.B. and T.N.A.H.; software, B.Q.T., K.N.H.N., L.T.N. and T.N.A.H. validation, B.Q.T., R.B., K.N.H.N., L.T.N., T.N.A.H. and V.C.H.; formal analysis, B.Q.T., T.N.A.H. and V.C.H.; investigation, B.Q.T., K.N.H.N., T.N.A.H. and V.C.H.; resources, B.Q.T. and R.B.; data curation, B.Q.T., K.N.H.N. and V.C.H.; writing—original draft preparation, B.Q.T. and V.C.H.; writing—review and editing, R.B. and V.C.H.; visualization, B.Q.T. and V.C.H.; supervision, R.B. and V.C.H.; project administration, B.Q.T. and V.C.H.; funding acquisition, B.Q.T. and R.B. All authors have read and agreed to the published version of the manuscript.

**Funding:** This research was funded by the Vietnam Ministry of Education and Training under the project titled 'Research on high-pressure processing technology applied to pangasius (*Pangasianodon hypophthalmus*) muscle and its application in developing value-added products', grant No. B2023-NLS-02.

**Informed Consent Statement:** Not applicable.

**Data Availability Statement:** The original contributions presented in the study are included in the article, further inquiries can be directed to the corresponding author.

**Acknowledgments:** The corresponding author acknowledges support from Ho Chi Minh City University of Technology and Education (HCMUTE), Vietnam.

**Conflicts of Interest:** The authors declare no conflicts of interest.

## Abbreviations

The following abbreviations are used in this manuscript:

| | |
|---|---|
| HPP | High-pressure processing |
| HIGs | Heat-induced gels |
| FTIR | Fourier transform infrared spectroscopy |
| SEM | Scanning electron microscope |
| TPC | Total plate count |
| WHC | Water-holding capacity |
| MHC | Myosin heavy chain |

## References

1. Ma, X.-S.; Yi, S.-M.; Yu, Y.-M.; Li, J.-R.; Chen, J.-R. Changes in gel properties and water properties of *Nemipterus virgatus* surimi gel induced by high-pressure processing. *LWT-Food Sci. Technol.* **2015**, *61*, 377–384. [CrossRef]
2. Truong, B.Q.; Buckow, R.; Nguyen, M. Mechanical and Functional Properties of Unwashed Barramundi (*Lates calcarifer*) Gels as Affected by High-Pressure Processing at three Different Temperatures and Salt Concentrations. *J. Aquat. Food Prod. Technol.* **2020**, *29*, 373–382. [CrossRef]
3. Truong, B.Q.; Buckow, R.; Stathopoulos, C.; Nguyen, M. Advances in High-Pressure Processing of Fish Muscles. *Food Eng. Rev.* **2015**, *7*, 109–129. [CrossRef]
4. Truong, B.Q.; Buckow, R.; Nguyen, M.H.; Furst, J. Effect of high-pressure treatments prior to cooking on gelling properties of unwashed protein from barramundi (*Lates calcarifer*) minced muscle. *Int. J. Food Sci. Technol.* **2017**, *52*, 1383–1391. [CrossRef]
5. Cando, D.; Herranz, B.; Borderías, A.J.; Moreno, H.M. Effect of high pressure on reduced sodium chloride surimi gels. *Food Hydrocoll.* **2015**, *51*, 176–187. [CrossRef]
6. Sikes, A.L.; Tobin, A.B.; Tume, R.K. Use of high pressure to reduce cook loss and improve texture of low-salt beef sausage batters. *Innov. Food Sci. Emerg. Technol.* **2009**, *10*, 405–412. [CrossRef]
7. Gordon, A.; Barbut, S. Effect of chloride salts on protein extraction and interfacial protein film formation in meat batters. *J. Sci. Food Agric.* **1992**, *58*, 227–238. [CrossRef]
8. Pilar, M.; Gomez-Guillen, M.C. High-Pressure Applications on Myosystems. In *Novel Food Processing Technologies; Food Science and Technology*; CRC Press: Boca Raton, FL, USA, 2004; pp. 311–342.
9. Macfarlane, J.J.; McKenzie, I.J.; Turner, R.H.; Jones, P.N. Binding of comminuted meat: Effect of high pressure. *Meat Sci.* **1984**, *10*, 307–320. [CrossRef]
10. Xu, M.; Ni, X.; Liu, Q.; Chen, C.; Deng, X.; Wang, X.; Yu, R. Ultra-high pressure improved gelation and digestive properties of Tai Lake whitebait myofibrillar protein. *Food Chem. X* **2024**, *21*, 101061. [CrossRef]
11. Iwasaki, T.; Noshiroya, K.; Saitoh, N.; Okano, K.; Yamamoto, K. Studies of the effect of hydrostatic pressure pretreatment on thermal gelation of chicken myofibrils and pork meat patty. *Food Chem.* **2006**, *95*, 474–483. [CrossRef]
12. Huppertz, T.; Fox, P.F.; Kelly, A.L. High pressure-induced denaturation of α-lactalbumin and β-lactoglobulin in bovine milk and whey: A possible mechanism. *J. Dairy Res.* **2004**, *71*, 489–495. [CrossRef] [PubMed]
13. Pérez-Mateos, M.; Lourenço, H.; Montero, P.; Borderías, A.J. Rheological and Biochemical Characteristics of High-Pressure- and Heat-Induced Gels from Blue Whiting (*Micromesistius poutassou*) Muscle Proteins. *J. Agric. Food Chem.* **1997**, *45*, 44–49. [CrossRef]
14. Nguyen, H.T.; Bao, H.N.D.; Dang, H.T.T.; Tómasson, T.; Arason, S.; Gudjónsdóttir, M. Protein Characteristics and Bioactivity of Fish Protein Hydrolysates from Tra Catfish (*Pangasius hypophthalmus*) Side Stream Isolates. *Foods* **2022**, *11*, 4102. [CrossRef]

15. Dang, H.T.T.; Gudjónsdóttir, M.; Tómasson, T.; Nguyen, M.V.; Karlsdóttir, M.G.; Arason, S. Influence of processing additives, packaging and storage conditions on the physicochemical stability of frozen Tra catfish (*Pangasius hypophthalmus*) fillets. *J. Food Eng.* **2018**, *238*, 148–155. [CrossRef]
16. Rathod, N.B.; Pagarkar, A.U.; Pujari, K.H.; Shingare, P.; Satam, S.; Phadke, G.G.; Gaikwad, B. Status of valuable components from pangasius: A review. *Int. J. Curr. Microbiol. Appl. Sci.* **2018**, *7*, 2106–2120. [CrossRef]
17. Li, Q.; Chen, S.; You, J.; Hu, Y.; Yin, T.; Liu, R. A comprehensive unraveling of the mystery of reduced-salt surimi gels: From molecular mechanism to future prospects. *Trends Food Sci. Technol.* **2024**, *154*, 104783. [CrossRef]
18. Wu, D.; Xiong, J.; Li, P.; Zhang, Y.; Li, F.; Yin, T.; Huang, Q. Dual enhancement effects of different yeast extract on gel properties and saltiness perception of low-salt surimi gel from silver carp. *Food Hydrocoll.* **2024**, *152*, 109925. [CrossRef]
19. Zhang, C.; Chen, L.; Lu, M.; Ai, C.; Cao, H.; Xiao, J.; Teng, H. Effect of cellulose on gel properties of heat-induced low-salt surimi gels: Physicochemical characteristics, water distribution and microstructure. *Food Chem. X* **2023**, *19*, 100820. [CrossRef]
20. Zhang, X.; Guo, Q.; Shi, W. Ultrasound-assisted processing: Changes in gel properties, water-holding capacity, and protein aggregation of low-salt *Hypophthalmichthys molitrix* surimi by soy protein isolate. *Ultrason. Sonochem.* **2023**, *92*, 106258. [CrossRef]
21. Nelce Mailoa, M.; Marthina Tapotubun, A.; Matrutty, T.E.A.A. Analysis Total Plate Counte (TPC) On Fresh Steak Tuna Applications Edible Coating Caulerpa sp During Stored at Chilling Temperature. *IOP Conf. Ser. Earth Environ. Sci.* **2017**, *89*, 012014. [CrossRef]
22. Ramirez-Suarez, J.C.; Morrissey, M.T. Effect of high pressure processing (HPP) on shelf life of albacore tuna (*Thunnus alalunga*) minced muscle. *Innov. Food Sci. Emerg. Technol.* **2006**, *7*, 19–27. [CrossRef]
23. Cruz-Romero, M.; Kelly, A.L.; Kerry, J.P. Effects of high-pressure and heat treatments on physical and biochemical characteristics of oysters (*Crassostrea gigas*). *Innov. Food Sci. Emerg. Technol.* **2007**, *8*, 30–38. [CrossRef]
24. Pérez-Mateos, M.; Montero, P. High-pressure-induced gel of sardine (*Sardina pilchardus*) washed mince as affected by pressure-time-temperature. *J. Food Sci.* **1997**, *62*, 1183–1188. [CrossRef]
25. Uresti, R.M.; Velazquez, G.; Vázquez, M.; Ramírez, J.A.; Torres, J.A. Effect of sugars and polyols on the functional and mechanical properties of pressure-treated arrowtooth flounder (*Atheresthes stomias*) proteins. *Food Hydrocoll.* **2005**, *19*, 964–973. [CrossRef]
26. Lowry, O.H.; Rosebrough, N.J.; Farr, A.L.; Randall, R.J. Protein measurement with the Folin phenol reagent. *J. Biol. Chem.* **1951**, *193*, 265–275. [CrossRef]
27. Pazos, M.; Méndez, L.; Vázquez, M.; Aubourg, S.P. Proteomics analysis in frozen horse mackerel previously high-pressure processed. *Food Chem.* **2015**, *185*, 495–502. [CrossRef]
28. Iwasaki, T.; Washio, M.; Yamamoto, K.; Nakamura, K. Rheological and Morphological Comparison of Thermal and Hydrostatic Pressure-Induced Filamentous Myosin Gels. *J. Food Sci.* **2005**, *70*, e432–e436. [CrossRef]
29. Muntean, M.; Ovidiu, M.; Barbieru, V.; Catunescu, G.; Ranta, O.; Drocas, I.; Terhes, S. High Pressure Processing in Food Industry–Characteristics and Applications. *Agric. Agric. Sci. Procedia* **2016**, *10*, 377–383. [CrossRef]
30. Pinto, C.A.; Moreira, S.A.; Fidalgo, L.G.; Inácio, R.S.; Barba, F.J.; Saraiva, J.A. Effects of high-pressure processing on fungi spores: Factors affecting spore germination and inactivation and impact on ultrastructure. *Compr. Rev. Food Sci. Food Saf.* **2020**, *19*, 553–573. [CrossRef]
31. Yu, L.; Muralidharan, S.; Lee, N.A.; Lo, R.; Stokes, J.R.; Fitzgerald, M.A.; Turner, M.S. The impact of variable high pressure treatments and/or cooking of rice on bacterial populations after storage using culture-independent analysis. *Food Control* **2018**, *92*, 232–239. [CrossRef]
32. Kunnath, S.; Jaganath, B.; Panda, S.K.; Ravishankar, C.N.; Gudipati, V. Modifying textural and functional characteristics of fish (*Nemipterus japonicus*) mince using high pressure technology. *J. Food Sci. Technol.* **2022**, *59*, 4122–4133. [CrossRef] [PubMed]
33. de Oliveira, F.A.; Neto, O.C.; dos Santos, L.M.R.; Ferreira, E.H.R.; Rosenthal, A. Effect of high pressure on fish meat quality—A review. *Trends Food Sci. Technol.* **2017**, *66*, 1–19. [CrossRef]
34. Uresti, R.M.; Velazquez, G.; Ramírez, J.A.; Vázquez, M.; Torres, J.A. Effect of high-pressure treatments on mechanical and functional properties of restructured products from arrowtooth flounder (*Atheresthes stomias*). *J. Sci. Food Agric.* **2004**, *84*, 1741–1749. [CrossRef]
35. Buckow, R.; Bull, M. Advanced food preservation technologies. *Microbiol. Aust.* **2013**, *34*, 108–111. [CrossRef]
36. Cui, Y.; Zhu, L.; Shang, H.; Xuan, X.; Lin, X. Effects of Combined ε-Polylysine and High Hydrostatic Pressure Treatment on Microbial Qualities, Physicochemical Properties, Taste, and Volatile Flavor Profile of Large Yellow Croaker (*Larimichthys crocea*). *Food Bioprocess Technol.* **2024**, *18*, 3610–3627. [CrossRef]
37. Zhou, Y.; Watkins, P.; Oiseth, S.; Cochet-Broch, M.; Sikes, A.L.; Chen, C.; Buckow, R. High pressure processing improves the sensory quality of sodium-reduced chicken sausage formulated with three anion types of potassium salt. *Food Control* **2021**, *126*, 108008. [CrossRef]
38. Tabilo-Munizaga, G.; Barbosa-Cánovas, G.V. Ultra high pressure technology and its use in surimi manufacture: An overview. *Food Sci. Technol. Int.* **2004**, *10*, 207–222. [CrossRef]
39. Ramírez, J.A.; Martín-Polo, M.O.; Bandman, E. Fish myosin aggregation as affected by freezing and initial physical state. *J. Food Sci.* **2000**, *65*, 556–560. [CrossRef]

40. Sagné, C.; Isambert, M.-F.; Gasnier, B. SDS-resistant aggregation of membrane proteins: Application to the purification of the vesicular monoamine transporter. *Biochem. J.* **1996**, *316 Pt 3*, 825–831. [CrossRef]
41. Herranz, B.; Tovar, C.A.; Borderias, A.J.; Moreno, H.M. Effect of high-pressure and/or microbial transglutaminase on physicochemical, rheological and microstructural properties of flying fish surimi. *Innov. Food Sci. Emerg. Technol.* **2013**, *20*, 24–33. [CrossRef]
42. Ojagh, S.M.; Núñez-Flores, R.; López-Caballero, M.E.; Montero, M.P.; Gómez-Guillén, M.C. Lessening of high-pressure-induced changes in Atlantic salmon muscle by the combined use of a fish gelatin–lignin film. *Food Chem.* **2011**, *125*, 595–606. [CrossRef]
43. Kong, J.; Yu, S. Fourier Transform Infrared Spectroscopic Analysis of Protein Secondary Structures. *Acta Biochim. Biophys. Sin.* **2007**, *39*, 549–559. [CrossRef] [PubMed]
44. Hernández, B.; Pflüger, F.; Adenier, A.; Kruglik, S.G.; Ghomi, M. Vibrational Analysis of Amino Acids and Short Peptides in Hydrated Media. VIII. Amino Acids with Aromatic Side Chains: L-Phenylalanine, l-Tyrosine, and l-Tryptophan. *J. Phys. Chem. B* **2010**, *114*, 15319–15330. [CrossRef] [PubMed]
45. Sadat, A.; Joye, I.J. Peak Fitting Applied to Fourier Transform Infrared and Raman Spectroscopic Analysis of Proteins. *Appl. Sci.* **2020**, *10*, 5918. [CrossRef]
46. Venyaminov, S.Y.; Kalnin, N.N. Quantitative IR spectrophotometry of peptide compounds in water ($H_2O$) solutions. II. Amide absorption bands of polypeptides and fibrous proteins in α-, β-, and random coil conformations. *Biopolymers* **1990**, *30*, 1259–1271. [CrossRef]
47. He, J.-S.; Mu, T.-H.; Guo, X.; Zhu, S.; Azuma, N.; Kanno, C. Comparison of the gel-forming ability and gel properties of α-lactalbumin, lysozyme and myoglobin in the presence of β-lactoglobulin under high pressure. *Food Hydrocoll.* **2013**, *33*, 415–424. [CrossRef]
48. Bolumar, T.; Middendorf, D.; Toepfl, S.; Heinz, V. *Structural Changes in Foods Caused by High-Pressure Processing*; Springer: New York, NY, USA, 2016; pp. 509–537.
49. Knorr, D.; Heinz, V.; Buckow, R. High pressure application for food biopolymers. *Biochim. Biophys. Acta (BBA)-Proteins Proteom.* **2006**, *1764*, 619–631. [CrossRef]
50. Moreno, H.M.; Bargiela, V.; Tovar, C.A.; Cando, D.; Borderias, A.J.; Herranz, B. High pressure applied to frozen flying fish (*Parexocoetus brachyterus*) surimi: Effect on physicochemical and rheological properties of gels. *Food Hydrocoll.* **2015**, *48*, 127–134. [CrossRef]
51. Chen, H.; Zhou, A.; Benjakul, S.; Zou, Y.; Liu, X.; Xiao, S. The mechanism of low-level pressure coupled with heat treatment on water migration and gel properties of *Nemipterus virgatus* surimi. *LWT* **2021**, *150*, 112086. [CrossRef]

*applied sciences*

MDPI

*Article*

# Effect of Ozonized Water against Pathogenic Bacteria and Filamentous Fungi on Stainless Steel

**Elettra Berni [1,*], Chiara Moroni [2], Massimo Cigarini [1], Demetrio Brindani [3], Claudia Catelani Cardoso [3] and Davide Imperiale [3]**

1 Microbiology—Quality, Safety and Pre-Industrialization Area, SSICA, Stazione Sperimentale per l'Industria delle Conserve Alimentari—Fondazione di Ricerca, Viale F. Tanara, 31/A, 43121 Parma, Italy; massimo.cigarini@ssica.it
2 Dipartimento di Scienze Chimiche, Della vita e Della Sostenibilità Ambientale, Parco Area delle Scienze, Università degli Studi di Parma, 31/A, 43124 Parma, Italy
3 Environment—Sustainability and Valorization Area, SSICA, Stazione Sperimentale per l'Industria delle Conserve Alimentari—Fondazione di Ricerca, Viale F. Tanara, 31/A, 43121 Parma, Italy; demetrio.brindani@ssica.it (D.B.); claudia.cardoso@ssica.it (C.C.C.); davide.imperiale@ssica.it (D.I.)
* Correspondence: elettra.berni@ssica.it; Tel.: +39-0521795269; Fax: +39-0521795218

**Abstract:** Ozone is a molecule that has gained increasing interest in recent years by food industries for sanitization of food-grade surfaces. Compared to chemical sanitizers such as chlorine, hydrogen peroxide, or peracetic acid, ozone shows undeniable advantages, such as the absence of by-products that should affect human health or the possibility of generating it when needed. Therefore, the aim of this paper was the assessment of the resistance to ozonized water of two pathogenic bacteria (*Listeria monocytogenes, Salmonella)* and of three airborne food-spoiling fungi (*Aspergillus brasiliensis, Hyphopichia burtonii,* and *Penicillium nordicum*) inoculated on stainless steel tiles and treated in static conditions with 1 to 6 mg $L^{-1}$ (pathogens) or 8.5 mg $L^{-1}$ (filamentous fungi). Ozonized water gave different results based on the tested microorganisms: pathogenic bacteria proved markedly more sensible to ozone than filamentous fungi, even if great differences were observed at inter- and intra-specific levels for both categories of microorganisms. Nevertheless, the non-linear inactivation kinetics of the studied strains made the calculation of a punctual F-value difficult, so in industrial practice, adequate tailoring of the treatments to be applied, based on the registered extrinsic factors and the industrial bio-burden, would be appropriate.

**Keywords:** ozonized water; *Aspergillus brasiliensis* ATCC 16404; *Hyphopichia burtonii; Listeria monocytogenes; Penicillium nordicum; Salmonella enterica;* sanitizing

## 1. Introduction

The food industry is always looking for alternatives to traditional chemical sanitizers, which can guarantee the safety of foods, both maintaining their chemical–physical properties as well as their microbiological stability and reducing the water consumption rates in large-scale processes [1]. Ozone, either in its gaseous or in its liquid form, is a sanitizing agent that has gained interest in recent years for such applications since it can be generated when needed and does not form appreciable levels of by-products (mainly hydroperoxide molecules and unsaturated aldehydes, but also substances such as dibromo-acetonitrile, which is formed when bromide levels are over 1 ppm) [2]. However, it has one of the highest oxidation potentials [for instance, it proved to be 1 to 2 times stronger than chlorine and significantly stronger (>3000 times) than hypochlorous acid] [3]. As a comparison, the use of chlorine is associated with the production of carcinogenic by-products, including trihalomethanes and halo-acetic acids [4], whereas ozone eventually decomposes it into non-toxic molecular oxygen.

Aqueous ozone has been known for over a hundred years as a sanitizing agent. At the beginning, it was used for the sanitization of drinking water, whereas in the last decades its usage has been extended to a wide range of activities like swimming pool disinfection, food industry processing, clinical medical applications, and even industrial laundry operations [5,6]. Nowadays, food industries use aqueous ozone to reduce microbial loads on wastewater [7,8] and a wide range of foods such as fresh fruits [9–11], dried fruits [12], nuts [13], vegetables [14–16], and meats [17,18]. Its use concentrations are usually up to 10 ppm for application times varying from some minutes to one hour, alone or in combination with other sanitizing methods. Concerning this aspect, the UV-C light is the most common coupling [8,19], even if the use of aqueous ozone in combination with chlorine [20], organic acids [21,22], or hydrogen peroxide [7] has also been registered.

Similarly, its use as a sanitizing agent on different types of surfaces has been deeply investigated. The papers concerning this topic were referred to as food-grade steel [23–27], multi-laminated food packaging [28], polyethylene terephthalate [29], polystyrene [30], and polypropylene [31]. Naitou and Takahara [32] even tested the effectiveness of ozonized water up to 30 ppm in inactivating microbes in a food-packaging film sterilizing machine. Unfortunately, most of the works supply punctual data without assessing the inactivation kinetics of the studied microorganisms; only the paper by Shelobolina et al. [33] assessed the D-value of the pathogen *Pseudomonas aeruginosa* on glass, ceramics, and medical industry-related plastics. In addition to this, filamentous fungi were not included in the group of investigated microorganisms when aqueous ozone was tested on materials, even if they can contaminate working surfaces and machinery interiors as well as bacteria.

Therefore, the aim of this paper was the assessment of the resistance to aqueous ozone of two pathogenic bacteria, *Listeria monocytogenes* and *Salmonella* spp., and of two airborne food-spoiling fungi, *Hyphopichia burtonii* and *Penicillium nordicum*, inoculated on stainless steel (SS) and treated in static conditions. The inactivation kinetics of the mycetes were also compared with those of *Aspergillus brasiliensis* ATCC 16404, which already proved to withstand physical stresses such as UV-C irradiation [34] or chemical stresses such as the gaseous ozone itself [35] and had been adopted by the European Standards (EN) as a target microorganism for checking the effectiveness of chemical disinfectants and antiseptics against microbes [36–38].

## 2. Materials and Methods

### 2.1. Microbial Strains

The tested microorganisms were the ones used in previous tests performed with gaseous ozone on SS by the same authors [35].

#### 2.1.1. Bacteria

Among bacteria, the following strains were selected for aqueous ozone treatments:

- *Listeria monocytogenes* (Murray et al.) Pirie ATCC 7644, isolated from a source not specified by the ATCC website.
- *Listeria monocytogenes* Scott A (Murray et al.) Pirie ATCC 49594, derived from an existing strain.
- *Salmonella enterica* subsp. *enterica* serotype Senftenberg NCTC 9959 (=ATCC 43845 = Col Sal 385/57 = DSM 10062 = JT 493; 775/W), isolated from a source not specified by the ATCC website.
- *Salmonella enterica* subsp. *enterica* serotype Typhimurium ATCC 14028 (=CIP 104115 = DSM 19587 = CDC 6516-60 = NCTC 12023) isolated from the pooled heart and liver tissue of four-week-old chickens.

#### 2.1.2. Filamentous Fungi

Among filamentous fungi, the following strains were selected for aqueous ozone treatments:

- *Penicillium nordicum* SSICA 1169, isolated from dry-cured meat production environments in Italy;
- *Penicillium nordicum* SSICA B4798, isolated from a fermented meat product in Italy;
- *Hyphopichia burtonii* SSICA 175717, isolated from spoiled sandwich bread in Italy;
- *Hyphopichia burtonii* SSICA 251105, isolated from a spreadable fat used to cover dry-cured hams during seasoning in Italy;
- *Aspergillus brasiliensis* ATCC 16404, isolated from blueberries in North Carolina (USA).

### 2.2. Preparation of the Microbial Suspensions

Microbial suspensions were prepared according to the protocols previously used by the same authors [35].

#### 2.2.1. Bacteria

Bacterial suspensions were prepared from strains preserved at −20 °C in porous beads (Cryoinstant; WVR, Milan, Italy). One bead for each strain was added to 10 mL of Tryptic Soy Broth (TSB—Oxoid, Cambridge, UK) and incubated at 37 °C for 24 h. After incubation, 1 mL of bacterial suspension was centrifuged at 4100 rpm for 15 min, and the pellet was resuspended in 1 mL of a physiological salt solution (8.0 g $L^{-1}$ sodium chloride) to obtain a concentration between $10^7$ and $10^8$ CFU/mL. The suspensions were discharged just after their use and freshly prepared each time.

#### 2.2.2. Filamentous Fungi

Fungal suspensions were prepared from strains preserved at −20 °C in glycerol–water suspensions. Each strain was defrosted and separately inoculated on Malt Extract Agar (MEA; OXOID, Cambridge, UK) and incubated at 25 °C for up to seven days. Conidia and mycelium were harvested with a sterile loop in a 0.1% ($v/v$) Tween 80 aqueous solution, filtered through sterile glass wool, and counted on MEA supplemented with 0.01% chlortetracycline (Sigma-Aldrich, St. Louis, MO, USA), to obtain a concentration between $10^7$ and $10^8$ CFU/mL. The filtered conidial suspensions of *A. brasiliensis* or *H. burtonii* were stored at 2 °C up to 30 days, whereas that of *P. brevicompactum* was stored at −18 °C up to six months, in accordance with the different physiological characteristics of the selected strains.

### 2.3. Ozonized Water Treatments

The ozone generator used was a waterproof unit model with a generation capacity of 8 g/h (230 VAC, 50 Hz, 750 W) (MET Srl, Bologna, Italy). The generator was supplied with purified oxygen obtained from a Pressure Swing Adsorption (PSA) apparatus that separates oxygen from atmospheric air, providing the generator with a higher oxygen concentration in order to increase its efficiency. The ozone produced was then delivered at the bottom of a conical filter funnel Robu (Borosilicate Glass 3.3, diameter = 120 mm, capacity = 2 L) equipped with a sintered disc filter of porosity 3. The conical filter funnel was filled with sterile water that was ozonized due to the microbubbles produced by the ozone flow passing through the disc filter. In this way, stable ozonized water was generated. Its concentration was measured by means of an Ozone Vacu-vials® Kit (K-7423, CHEMetrics Inc., Midland, VA, USA). The spectrophotometric measurement and the determination of the ozone concentration in the water solution were performed using a V-2000 Photometer (CHEMetrics Inc., Midland, VA, USA). From the stock solution of ozonized water, solutions at the desired ozone concentrations were obtained by dilution. The final concentration of the diluted solution was measured again before starting the treatment.

SS tiles (15 × 20 × 1 mm) were utilized as germ carriers. They were previously sterilized at 121 °C for 15 min, inoculated with one microbial suspension at a time, and dried for one hour in sterile conditions under a laminar flow hood. Some of the inoculated tiles, marked as "positive controls," were analyzed without being treated (see the correspondence with the "time 0" bars in each figure) to consent to a comparison with those that were treated. Treatments were carried out just after the drying of the inoculum in order to avoid the loss of viable cells (Figure 1).

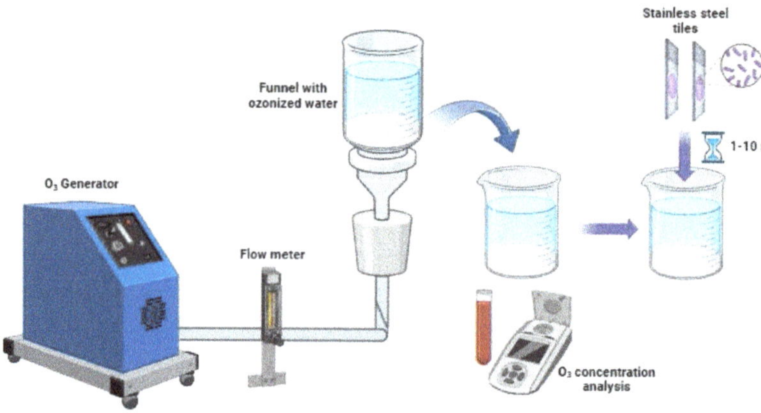


**Figure 1.** Graphical representation of the operations carried out to test the effectiveness of ozonized water against pathogenic bacteria and filamentous fungi. (Created with BioRender.com—https://biorender.com/; accessed on 12 April 2024).

#### 2.3.1. Bacteria

Tests were carried out at 25 °C in ultrapure water added with ozone (1.0, 3.0, 6.0 ± 0.5 mg $L^{-1}$) by inoculating tiles with a multi-layered (0.01 mL of the undiluted bacterial suspension) drop. The spotted tiles were positioned into screw-capped test tubes filled with 1.5 mL of ozonized water that was left to react for up to 9 min. Such a maximum treatment time was established based on preliminary tests showing that partial ozone degradation started after 10 min (unpublished data). After treatments, 1.5 mL of a 1.0% sodium thiosulfate (powder from Sigma-Aldrich, St. Louis, MO, USA) solution was added to each tube in order to neutralize the ozone, together with 0.5 g of sterile glass microbeads. Each tube was then vortexed for two minutes by an infrared apparatus (Starlab, Milan, Italy), the appropriate decimal dilutions being plated on Agar Listeria acc. to Ottaviani and Agosti (ALOA®, Biolife Italiana srl, Milan, Italy) and incubated at 37 °C for 48 h. Each combination was tested four times.

#### 2.3.2. Filamentous Fungi

Tests were carried out at 25 °C in neutral or acidified ultrapure water [1% *w*/*v* citric acid (powder from Sigma-Aldrich, St. Louis, MO, USA)] in order to assess any potential synergistic effect of citric acid with dissolved ozone (final concentration 8.5 ± 0.5 mg $L^{-1}$). The tiles were inoculated by applying either a multi-layered (0.01 mL of the undiluted suspension) or a single-layered (0.10 mL of the 1:50 diluted suspension) drop of each suspension. The experiment was carried out as it was for bacteria, even if in tests with acidified ozonized water, a 0.1% (*v*/*v*) Tween 80 aqueous solution was used as a neutralizer in place of sodium thiosulphate since it is known to react with citric acid-forming molecules with an inactivating activity against molds, such as $SO_2$ and S. After treating and vortexing, the appropriate decimal dilutions were plated on MEA supplemented with 0.01% chlortetracycline, and the colonies were counted after incubation at 25 °C for five days. Each combination was tested four times.

### *2.4. Statistical Analysis*

Microsoft® Excel 2013 (Microsoft, Redmond, WA, USA) was used to draw the inactivation curves based on the raw thermal reduction data that were changed into logarithmic mean values. The inactivation achieved, given as Logarithmic Count Reductions (LCR), was calculated for both bacteria and filamentous fungi based on the difference between the decimal logarithm of the initial average concentration of cells deposited and the decimal logarithm of the average concentration of cells surviving the treatment.

The tested strains showed different behavior in their inactivation curves, so log-linear models (followed or not by a tail) and Weibull models were used to determine the 1D-value (defined as the time required to bring the first 1-log reduction in the population of a given microorganism at a given temperature) of each strain at the corresponding aqueous ozone concentration. For such microorganisms, the calculation of the 1D values and the statistical analysis were performed using GInaFiT (Geeraerd and Van Impe Inactivation Model Fitting Tool, version 1.7), a freeware add-in tool for Microsoft® Excel (https://cit.kuleuven.be/biotec/software/GinaFit accessed on 11 March 2024). The goodness-of-fit for each elaboration was assessed using the root mean squared error (RMSE) and the regression coefficient ($R^2$). In calculating the time needed for the first decimal reduction (1D-value), $k_{max}$ was the parameter taken into consideration for log-linear models with tailing where the first logarithmic reduction occurred during the inactivation of the major, less resistant population. In this case, the 1D-value was calculated as $2.303/k_{max}$ [39]. On the contrary, delta (δ) was considered the 1D-value itself when Weibull-type models were applied [39,40].

Statistical significance between means within each treatment group was identified by applying an ANOVA multivariate analysis with the Tukey post hoc test. The software package SPSS v24 (IBM website, www.ibm.com/software/it/analytics/spss/downloads.html accessed on 5 September 2024) was used for this analysis.

## 3. Results and Discussion

The results obtained against pathogenic bacteria are reported in Table 1 and Figure 2. The ones against filamentous fungi are reported in Table 2 and Figure 3, with reference to the tests with ultrapure water, and in Table 3 and Figure 4, with reference to the tests with acidified ultrapure water. The best-fitting microbial survival models applied by GInaFiT were derived from the ones elaborated by different authors (see footnotes in each table for the corresponding bibliographical reference).

**Table 1.** Inactivation parameters of the tested strains inoculated as a multi-layer on SS tiles and treated with different concentrations of aqueous ozone (Lm = *Listeria monocytogenes* ATCC 7644; Ls = *Listeria monocytogenes* Scott A; Se = *Salmonella enterica* subsp. *enterica* serotype Senftenberg NCTC 9959; St = *Salmonella enterica* subsp. *enterica* serotype Typhimurium ATCC 14028).

| Ozone Concentration ($mg\ L^{-1}$) | Strain | Inactivation Curve | | | | Inactivation Model |
|---|---|---|---|---|---|---|
| | | Parameter | RMSE (-) | $R^2$ (-) | 1D-Value (min) | |
| 1.0 | Ls | δ | 0.1404 | 0.97 | 5.20 | [41] |
| 3.0 | | $K_{max}$ | 0.2867 | 0.96 | 0.44 | [42] |
| 6.0 | | $K_{max}$ | 0.2604 | 0.99 | 0.17 | [42] |
| 1.0 | Lm | δ | 0.1910 | 0.98 | 0.81 | [41] |
| 3.0 | | $K_{max}$ | 0.0520 | 1.00 | 0.27 | [42] |
| 6.0 | | $K_{max}$ | 0.1735 | 1.00 | 0.10 | [42] |
| 1.0 | Se | δ | 0.2390 | 0.97 | 0.82 | [41] |
| 3.0 | | δ | 0.1701 | 0.99 | 0.26 | [41] |
| 6.0 | | $K_{max}$ | 0.2217 | 0.99 | 0.22 | [42] |
| 1.0 | St | δ | 0.0300 | 1.00 | 7.27 | [41] |
| 3.0 | | δ | 0.1241 | 0.98 | 1.70 | [41] |
| 6.0 | | δ | 0.0926 | 0.99 | 1.68 | [41] |

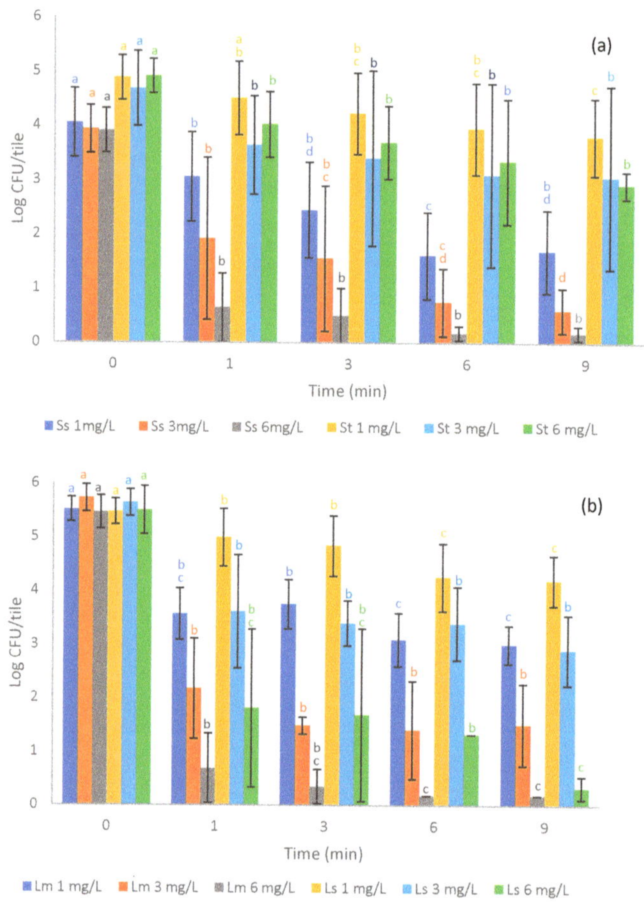


**Figure 2.** Behavior of *Salmonella* (**a**) and *Listeria monocytogenes* (**b**) strains treated with ozonized water (Lm = *Listeria monocytogenes* Pirie ATCC 7644; Ls = *Listeria monocytogenes* Scott A; Se = *Salmonella enterica* subsp. *Enterica* serotype Senftenberg NCTC 9959; St = *Salmonella enterica* subsp. *Enterica* serotype Typhimurium ATCC 14028). Vertical error bars indicate the standard deviation for mean values. Within each treatment group, different letters indicate significant differences between means at $p < 0.05$ (ANOVA and Tukey HSD post hoc test).

**Table 2.** Inactivation parameters of the fungal strains inoculated as a multi-layer or a single-layer on SS tiles and treated with 8.5 mg $L^{-1}$ ozone in neutral ultrapure water (Ab = *Aspergillus brasiliensis* ATCC 16404; Hb1 = *Hyphopichia burtonii* SSICA 175717; Hb2 = *Hyphopichia burtonii* SSICA 251105; Pn1 = *Penicillium nordicum* SSICA 1169; Pn2 = *Penicillium nordicum* SSICA B4798).

| Inoculum | Strain | Inactivation Curve | | | | Inactivation Model |
|---|---|---|---|---|---|---|
| | | Parameter | RMSE (-) | $R^2$ (-) | 1D-Value (min) | |
| **multi-layer** | Ab | nd | nd | nd | nd | - |
| | Hb1 | nd | nd | nd | nd | - |
| | Hb2 | nd | nd | nd | nd | - |
| | Pn1 | nd | nd | nd | nd | - |
| | Pn2 | nd | nd | nd | nd | - |

**Table 2.** *Cont.*

| Inoculum | Strain | Inactivation Curve | | | | Inactivation Model |
|---|---|---|---|---|---|---|
| | | Parameter | RMSE (-) | $R^2$ (-) | 1D-Value (min) | |
| | Ab | nd | nd | nd | nd | - |
| | Hb1 | $K_{max}$ | 1.1924 | 0.54 | 1.47 | [42] |
| **single-layer** | Hb2 | δ | 0.3046 | 0.98 | 0.52 | [43] |
| | Pn1 | $K_{max}$ | 0.4085 | 0.95 | 1.77 | [44] |
| | Pn2 | $K_{max}$ | 0.3567 | 0.89 | 3.66 | [44] |

Note. Nd: not determined since at least 1.0 LCR was not reached within the time considered.

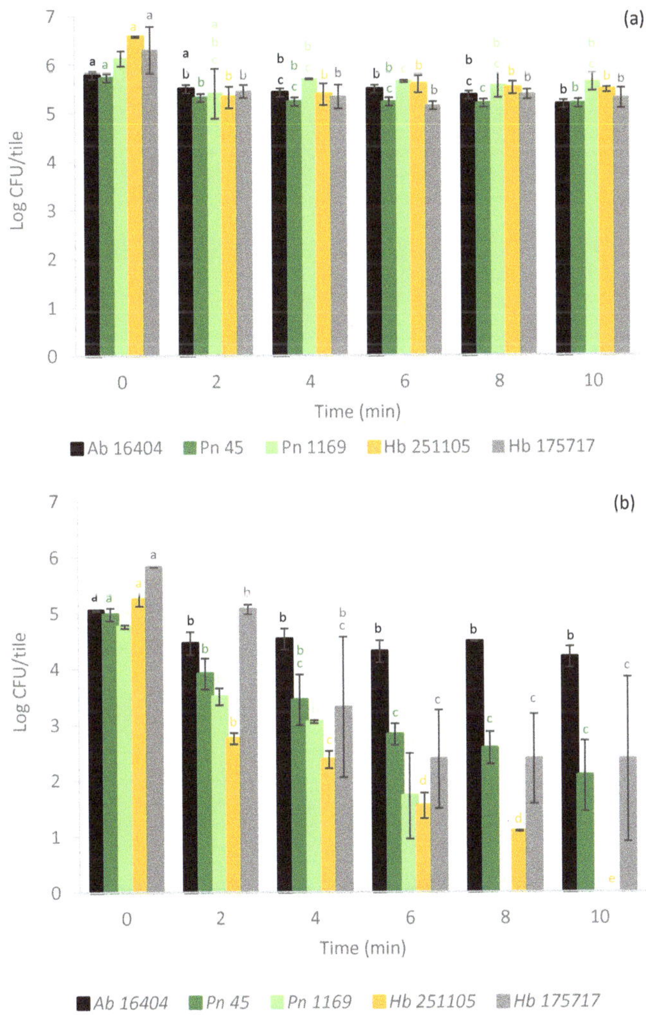


**Figure 3.** Behavior of filamentous fungi inoculated as a multi-layer (**a**) or a single-layer (**b**) and treated with 8.5 mg $L^{-1}$ ozone in neutral ultrapure water. Vertical error bars indicate the standard deviation for mean values. Within each treatment group, different letters indicate significant differences between means at $p < 0.05$ (ANOVA and Tukey HSD post hoc test).

**Table 3.** Inactivation parameters of the fungal strains inoculated as a multi-layer or a single-layer on SS tiles and treated with 8.5 mg $L^{-1}$ ozone in acidified ultrapure water (Ab = *Aspergillus brasiliensis* ATCC 16404; Hb1 = *Hyphopichia burtonii* SSICA 175717; Hb2 = *Hyphopichia burtonii* SSICA 251105; Pn1 = *Penicillium nordicum* SSICA 1169; Pn2 = *Penicillium nordicum* SSICA B4798).

| Inoculum | Strain | Inactivation Curve | | | | Inactivation Model |
|---|---|---|---|---|---|---|
| | | Parameter | RMSE (-) | $R^2$ (-) | 1D-Value (min) | |
| multi-layer | Ab | nd | nd | nd | nd | - |
| | Hb1 | δ | 0.5613 | 0.96 | 0.48 | [43] |
| | Hb2 | δ | 0.3724 | 0.97 | 3.86 | [43] |
| | Pn1 | nd | nd | nd | nd | - |
| | Pn2 | nd | nd | nd | nd | - |
| single-layer | Ab | δ | 0.1842 | 0.92 | 5.00 | [41] |
| | Hb1 | $K_{max}$ | 0.0949 | 1.00 | 0.18 | [45] |
| | Hb2 | δ | 0.4658 | 0.91 | 0.09 | [41] |
| | Pn1 | δ | 0.5224 | 0.89 | 1.62 | [41] |
| | Pn2 | δ | 0.3875 | 0.94 | 4.89 | [41] |

Note. Nd: not determined since the minimum value of 1.0 LCR was not reached within the time considered.

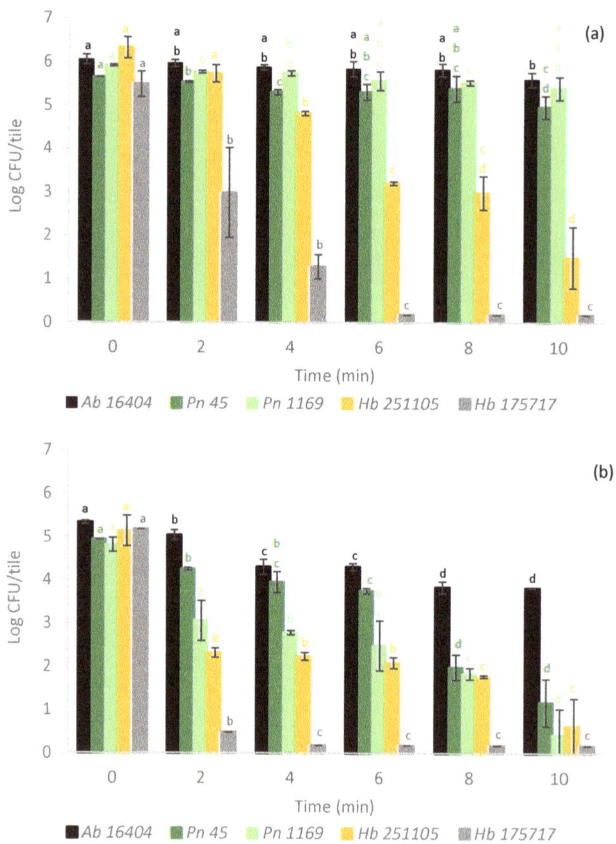


**Figure 4.** Behavior of filamentous fungi inoculated as a multi-layer (**a**) or a single-layer (**b**) and treated with 8.5 mg $L^{-1}$ ozone in acidified ultrapure water. Vertical error bars indicate the standard deviation for mean values. Within each treatment group, different letters indicate significant differences between means at $p < 0.05$ (ANOVA and Tukey HSD post hoc test).

### 3.1. Bacteria

For the tested strains, a marked tailing was observed at all the tested ozone concentrations. This phenomenon was already observed by the same authors when sanitizing agents such as UV-C radiation [46] or the ozone itself [35] were used. Similarly, it has been registered by other authors and related to heterogeneous treatments, to the aggregation of microorganisms, to some intrinsic mechanism of bacterial physiology leading to resistant sub-populations [47–49]. Together with the tailing, a huge variability in the concentrations of the surviving microorganisms at each of the considered combinations was highlighted, albeit the high number of repetitions for each tested combination. This variability had not been registered in tests carried out on bacterial and fungal suspensions (unpublished data), so it was hypothesized that the multi-layering of the inoculum could have given protection to the lower layers of cells against the oxidative processes exerted by ozone, even considering that variations within the same train of tests have been already registered by other authors who tested aqueous ozone [6,25].

For all the considered strains, a progressive reduction in the surviving cells was observed, and the reduction proved more marked as the ozone concentration increased. When ozone concentrations increased from 1.0 to 6.0 mg $L^{-1}$, the LCR moved from 2.4 to more than 3.9 for the tested *S. senftemberg* and from 1.1 to 2.0 for the tested *S. typhimurium*, the former showing a higher sensibility to aqueous ozone compared to the latter. Analogously, considering the same increase in ozone concentration, the LCR moved from 1.3 to 5.2 for the tested Listeria monocytogenes Scott A and from 2.5 to 5.3 for the tested *Listeria monocytogenes ATCC* 7644, the latter strain proving more sensible to the ozone applied in its liquid form.

A huge variability was observed considering the 1-D values of the different strains or species tested. At the lowest ozone concentration, the resistance of *L. monocytogenes Scott A* and *S. typhimurium* proved sensibly higher (from 6.5 to 9 times) than those registered for the other two tested strains, whereas at higher concentrations the calculated resistance values prove comparable.

The comparison with previous works on the same topic is not easy since literature data focused on biofilms rather than free cells inoculated on specific surfaces. The only paper concerning inoculated cells was that by Marino et al. [25], where cells of *L. monocytogenes* attached to SS were subjected to 2 to 3 LCR, yet in the first 2 min of treatment, 1 ppm of aqueous ozone in static conditions was applied. Differently, on biofilms, more papers are available. With regard to *L. monocytogenes*, ozonized water proved active at a concentration of 4 ppm for 3 min on cells attached to SS, whereas a four-fold increase in sanitizer was needed to destroy all biofilm cells [50]. On polystyrene, a reduction of about 0.9, 3.4, and 4.1 Log was reached when ozonized water was applied at 1.0, 2.0, 1 md, and 4.0 ppm, respectively [30]. With regard to *Salmonella* spp., the effectiveness of ozonized water seemed less marked, a reduction less than 0.8 CFU/$cm^2$ being obtained on cells adhered to SS after 20 min of exposure [26].

### 3.2. Filamentous Fungi

On filamentous fungi, a concentration of ozone equal to 8.5 mg $L^{-1}$ was applied since preliminary tests at the concentrations used against bacteria did not prove effective at all in inactivating fungal conidia.

The tested strains were not affected by aqueous ozone if inoculated as a multi-layer of cells, with the exception of *H. burtonii* strains that proved subjected to a partial inactivation when acidified ozonized water was used. Dissimilarly, on single layers of cells, a gradual decrease was observed for all strains belonging to *P. nordicum* or *H. burtonii*. A statistically significant difference within the same fungal species was observed comparing data obtained testing either not-acidified or acidified ozonated water. On multi-layered cells, differences were observed only for *Hyphopichia* strains at any treatment time, with a marked decrease in survivors registered when citric acid was added to ozonized water. Differently, on single-layered cells, the observed statistical differences seemed randomized: they were registered at 8 min for *A. brasiliensis*, at 6 min for *P. nordicum* 45, or from 4 to 8 min for *P. nordicum* 1169, and from 2 to 8 min for *Hyphopichia* strains.

When the multi-layered cells were tested in both liquid matrices, the *A. brasiliensis* and *P. nordicum* strains did not reach even 1.0 LCR after the longest treatment time. Conversely, for *H. burtonii* strains, the results differed when not-acidified and acidified ozonated water were tested: from 1.0 to 1.1 LCR after 10 min in the former case or more than 5.0 LCR after 6–10 min in the latter case were reached, depending on the isolate considered.

When the single-layered cells were tested, *A. brasiliensis* inactivation kinetics did not change compared to the multi-layered spatial arrangement. On the contrary, strain- and species-specificity was observed for the tested *Hyphopichia* and *Penicillium* strains. *P. nordicum* isolates underwent from 2.9 to 4.7+ LCR in not-acidified water and from 3.8 to 4.40 LCR in acidified water. Similarly, *H. burtonii* strains were subjected to a minimum of 3.4 LCR just after 6 min of treatment up to 5.0+ LCR after 10 min in not-acidified water and to 3.8–4.4 LCR in acidified water.

When compared with the other tested strains, *Aspergillus brasiliensis* showed the greatest resistance to the sanitizing agent applied whenever arranged in a multi-layer or in a single layer. It yet resulted in a point of reference when not-ionizing radiation such as the UVC light [34,46] or chemicals such as the ozone itself [35] have to be tested. This was probably due to the great amounts of melanin and melanin-like pigments in its outer cellular wall [51,52]. Furthermore, it must be taken into account that, like other filamentous fungi in the genus *Aspergillus*, it possesses good amounts of ROS-scavenging enzymes such as superoxide dismutases (SOD) and catalases (CAT) [53]. SOD proved critical to cell survival following ozone exposure by eliminating the superoxide radicals and thereby preventing the formation of hydroxyl radicals, whereas CAT proved necessary for protection against ozone exposure [54]. Therefore, the comprehension of melanin and of major ROS-scavenging enzymes could be responsible for its higher robustness against ozonized water.

With regard to the literature, few papers have dealt with the effect of ozonized water on filamentous fungi either in static or dynamic conditions, and none of them reported tests on metallic or plastic supports.

In static conditions, Hageskal et al. [55] tested filamentous fungi inoculated in ozonized water and registered a great variability in fungal response. All the tested strains (*Fusarium solani*, *Trichoderma viride*, and *Aspergillus calidoustus*) were reduced from one to three Logs after 10 min of treatment at 8 ppm, with the exception of *P. spinulosum* that did not prove affected by aqueous ozone. In agreement with our observations, [56] found that *Aspergillus brasiliensis* ATCC 16404 spores were still alive in freshly ozonized water at 1.5–3.0 ppm after 30 min, requiring longer times than those used for totally inactivating both [57] bacterial and yeast cells.

If dynamic fluxes were considered, the point of reference could be the paper by Beuchat et al. [58], where the D-values of two aflatoxigenic aspergilli (*A. flavus* and *A. parasiticus*) exposed to 1.74 ppm ozone were determined. D-values of *A. flavus* conidia were 1.72 and 1.54 min at pH 5.5 and 7.0, respectively, whereas D-values of *A. parasiticus* were 2.08 and 1.71 min, respectively. One of the most interesting ones is that by Restaino et al. [57], where the effect of the aqueous zone was tested on mycetes, and their strain of *Aspergillus niger* behaved similarly to the one used in this paper, its concentration being affected by less than one logarithmic reduction after 5 min, even if the ozone concentration tested by Restaino's group was sensibly lower (less than 0.2 ppm).

## 4. Conclusions

Aqueous ozone has a Generally Regarded as Safe (GRAS) designation from the U.S. Food and Drug Administration (FDA, 2001), and it is allowed in many countries, including Japan, Australia, France, and Canada [59]. On the contrary, in Europe, no regulation exists about the use of ozonized water for agro-industrial purposes, with a recent EFSA report (European Union, 2021) indicating that "no conclusion could be drawn concerning the risk assessment to non-target species from potentially occurring by-products of the chemical interaction between ozone and substances occurring in water." Nevertheless, the use of ozone in its liquid form for sanitization purposes is gaining an increasing interest in the food and feed industry due to its undeniable advantages compared with chemicals, such as the absence of by-products that should affect human health or the possibility to generate it when needed.

In our paper, ozonized water gave different results based on the tested microorganism, as observed yet for gaseous ozone [35]. Pathogenic bacteria proved markedly more sensible to ozone than filamentous fungi, even if great differences were observed at the inter- and intra-specific level for both categories of microorganisms, as in the case of *L. monocytogenes* and *P. nordicum*. For a better interpretation of the obtained results, it could be taken into consideration the European Standard EN 13697:2015+A1 [36], which establishes that any product with bactericidal and/or fungicidal activity used in food, industrial, domestic, and institutional areas should demonstrate at least a four-decimal log reduction for bacteria and at least a three-decimal log reduction for fungi. In this perspective, an appropriate number of log reductions were achieved only for three out of four pathogenic strains when higher concentrations of ozonized water were used or for some *P. nordicum* and *H. burtonii* at different treatment times, depending on their spatial arrangement of the conidia and the use of neutral or acidified pure water. Nevertheless, the non-linear inactivation kinetics of the studied strains made the calculation of a punctual F-value difficult, so in industrial practice, adequate tailoring of the treatments to be applied, based on the registered extrinsic factors and the industrial bio-burden, would be appropriate.

**Author Contributions:** E.B.: conceptualization (filamentous fungi), microbiological software, and writing; C.M.: execution of microbiological analyses and acquisition of data; M.C.: conceptualization (pathogens); D.B.: execution of chemical analyses and acquisition of data (ozone treatments); C.C.C.: conceptualization and writing (ozone treatments); D.I.: project supervision, conceptualization, statistical analysis, and review and editing. All authors have read and agreed to the published version of the manuscript.

**Funding:** This research received no external funding.

**Institutional Review Board Statement:** Not applicable.

**Informed Consent Statement:** Not applicable.

**Data Availability Statement:** The raw data supporting the conclusions of this article will be made available by the authors on request.

**Conflicts of Interest:** The authors declare no conflicts of interest.

## References

1. Ölmez, H.; Kretzschmar, U. Potential Alternative Disinfection Methods for Organic Fresh-Cut Industry for Minimizing Water Consumption and Environmental Impact. *LWT Food Sci. Technol.* **2009**, *42*, 686–693. [CrossRef]
2. Richardson, S.D.; Thruston, A.D., Jr.; Caughran, T.V.; Chen, P.H.; Collette, T.W.; Schenck, K.M.; Lykins, B.W., Jr.; Rav-Acha, C.; Glezer, V. Identification of New Drinking Water Disinfection by—Products from Ozone, Chlorine Dioxide, Chloramine, and Chlorine. *Water Air Soil. Pollut.* **2000**, *123*, 95–102. [CrossRef]
3. Suslow, T. *Ozone Applications for Postharvest Disinfection of Edible Horticultural Crops*; UCANR Publications: Irvine, CA, USA, 2004. [CrossRef]
4. Martinelli, M.; Giovannangeli, F.; Rotunno, S.; Trombetta, C.M.; Montomoli, E. Water and Air Ozone Treatment as an Alternative Technology. *J. Prev. Med. Hyg.* **2017**, *58*, E48. [CrossRef]
5. Bryce, E.A.; Dorovoni-Zis, K.; Trudeau, D.; Sinclair, M.; Roberts, F.J. Disinfection of Hospital Laundry Using Ozone Microbiological Evaluation. *Infect. Control Hosp. Epidemiol.* **2000**, *21*, 248. [CrossRef]
6. Epelle, E.I.; Emmerson, A.; Nekrasova, M.; Macfarlane, A.; Cusack, M.; Burns, A.; Mackay, W.; Yaseen, M. Microbial Inactivation: Gaseous or Aqueous Ozonation? *Ind. Eng. Chem. Res.* **2022**, *61*, 9600–9610. [CrossRef]
7. Nahim-Granados, S.; Rivas-Ibáñez, G.; Antonio Sánchez Pérez, J.; Oller, I.; Malato, S.; Polo-López, M.I. Synthetic Fresh-Cut Wastewater Disinfection and Decontamination by Ozonation at Pilot Scale. *Water Res.* **2020**, *170*, 115304. [CrossRef]

8. Sousa, J.M.; Macedo, G.; Pedrosa, M.; Becerra-Castro, C.; Castro-Silva, S.; Pereira, M.F.R.; Silva, A.M.T.; Nunes, O.C.; Manaia, C.M. Ozonation and UV254 Nm Radiation for the Removal of Microorganisms and Antibiotic Resistance Genes from Urban Wastewater. *J. Hazard. Mater.* **2017**, *323*, 434–441. [CrossRef]
9. Bialka, K.L.; Demirci, A. Decontamination of Escherichia Coli O157:H7 and Salmonella Enterica on Blueberries Using Ozone and Pulsed UV-Light. *J. Food Sci.* **2007**, *72*, M391–M396. [CrossRef]
10. Smilanick, J.L.; Margosan, D.M.; Mlikota Gabler, F. Impact of Ozonated Water on the Quality and Shelf-Life of Fresh Citrus Fruit, Stone Fruit, and Table Grapes. *Ozone Sci. Eng.* **2002**, *24*, 343–356. [CrossRef]
11. Alexandre, E.M.C.; Santos-Pedro, D.M.; Brandão, T.R.S.; Silva, C.L.M. Influence of Aqueous Ozone, Blanching and Combined Treatments on Microbial Load of Red Bell Peppers, Strawberries and Watercress. *J. Food Eng.* **2011**, *105*, 277–282. [CrossRef]
12. Zorlugenç, B.; Kiroğlu Zorlugenç, F.; Öztekin, S.; Evliya, I.B. The Influence of Gaseous Ozone and Ozonated Water on Microbial Flora and Degradation of Aflatoxin B1 in Dried Figs. *Food Chem. Toxicol.* **2008**, *46*, 3593–3597. [CrossRef] [PubMed]
13. Freitas-Silva, O.; Morales-Valle, H.; Venâncio, A. Potential of Aqueous Ozone to Control Aflatoxigenic Fungi in Brazil Nuts. *ISRN Biotechnol.* **2013**, *2013*, 859830. [CrossRef] [PubMed]
14. Alexopoulos, A.; Plessas, S.; Ceciu, S.; Lazar, V.; Mantzourani, I.; Voidarou, C.; Stavropoulou, E.; Bezirtzoglou, E. Evaluation of Ozone Efficacy on the Reduction of Microbial Population of Fresh Cut Lettuce (*Lactuca Sativa*) and Green Bell Pepper (*Capsicum Annuum*). *Food Control* **2013**, *30*, 491–496. [CrossRef]
15. Bhilwadikar, T.; Pounraj, S.; Manivannan, S.; Rastogi, N.K.; Negi, P.S. Decontamination of Microorganisms and Pesticides from Fresh Fruits and Vegetables: A Comprehensive Review from Common Household Processes to Modern Techniques. *Compr. Rev. Food Sci. Food Saf.* **2019**, *18*, 1003–1038. [CrossRef] [PubMed]
16. Neal, J.A.; Marquez-Gonzalez, M.; Cabrera-Diaz, E.; Lucia, L.M.; O'Bryan, C.A.; Crandall, P.G.; Ricke, S.C.; Castillo, A. Comparison of Multiple Chemical Sanitizers for Reducing Salmonella and *Escherichia Coli* O157:H7 on Spinach (*Spinacia Oleracea*) Leaves. *Food Res. Int.* **2012**, *45*, 1123–1128. [CrossRef]
17. Cano, C.; Meneses, Y.; Chaves, B.D. Ozone-Based Interventions To Improve the Microbiological Safety and Quality of Poultry Carcasses and Parts: A Review. *J. Food Prot.* **2019**, *82*, 940–947. [CrossRef]
18. Gorman, B.M.; Sofos, J.N.; Morgan, J.B.; Schmidt, G.R.; Smith, G.C. Evaluation of Hand-Trimming, Various Sanitizing Agents, and Hot Water Spray-Washing as Decontamination Interventions for Beef Brisket Adipose Tissue. *J. Food Prot.* **1995**, *58*, 899–907. [CrossRef]
19. Watanabe, M.; Masaki, H.; Mori, T.; Tsuchiya, T.; Konuma, H.; Hara-Kudo, Y.; Takatori, K. Inactivation Effects of UV Irradiation and Ozone Treatment on the Yeast and the Mold in Mineral Water. *J. Food Prot.* **2010**, *73*, 1537–1542. [CrossRef]
20. Liang, Z.; Xu, X.; Cao, R.; Wan, Q.; Xu, H.; Wang, J.; Lin, Y.; Huang, T.; Wen, G. Synergistic Effect of Ozone and Chlorine on Inactivating Fungal Spores: Influencing Factors and Mechanisms. *J. Hazard. Mater.* **2021**, *420*, 126610. [CrossRef]
21. Tadepalli, S.; Bridges, D.F.; Anderson, R.; Zhang, R.; Wu, V.C.H. Synergistic Effect of Sequential Wash Treatment with Two Different Low-Dosage Antimicrobial Washes in Combination with Frozen Storage Increases Salmonella Typhimurium and Listeria Monocytogenes Reduction on Wild Blueberries. *Food Control* **2019**, *102*, 87–93. [CrossRef]
22. Britton, H.C.; Draper, M.; Talmadge, J.E. Antimicrobial Efficacy of Aqueous Ozone in Combination with Short Chain Fatty Acid Buffers. *Infect. Prev. Pract.* **2020**, *2*, 100032. [CrossRef] [PubMed]
23. Crapo, C.; Himelbloom, B.; Vitt, S.; Pedersen, L. Ozone Efficacy as a Bactericide in Seafood Processing. *J. Aquat. Food Prod. Technol.* **2004**, *13*, 111–123. [CrossRef]
24. Greene, A.K.; Few, B.K.; Serafini, J.C. A Comparison of Ozonation and Chlorination for the Disinfection of Stainless Steel Surfaces. *J. Dairy Sci.* **1993**, *76*, 3617–3620. [CrossRef] [PubMed]
25. Marino, M.; Maifreni, M.; Baggio, A.; Innocente, N. Inactivation of Foodborne Bacteria Biofilms by Aqueous and Gaseous Ozone. *Front. Microbiol.* **2018**, *9*, 406181. [CrossRef] [PubMed]
26. Shao, L.; Dong, Y.; Chen, X.; Xu, X.; Wang, H. Modeling the Elimination of Mature Biofilms Formed by Staphylococcus Aureus and Salmonella Spp. Using Combined Ultrasound and Disinfectants. *Ultrason. Sonochem.* **2020**, *69*, 105269. [CrossRef]
27. Wirtanen, G.; Langsrud, S.; Salo, S.; Olofson, U.; Alnås, H.; Neuman, M.; Homleid, J.P.; Mattila-Sandholm, T. *Evaluation of Sanitation Procedures for Use in Dairies*; VTT Publications 481; Julkaisija Utgivare Publisher: Helsinki, Finland, 2002.
28. Khadre, M.A.; Yousef, A.E. Decontamination of a multilaminated aseptic food packaging material and stainless steel by ozone. *J. Food Saf.* **2001**, *21*, 1–13. [CrossRef]
29. Botti, L.C.M.; Bócoli, P.F.J.; Leber, A.S.M.L.; Parada; Tribst, A.A.L.; Cristianini, M. Sanitation of Plastic Bottles Using Ozonated Water. *Int. Food Res. J.* **2014**, *21*, 1375–1379.
30. Korany, A.M.; Hua, Z.; Green, T.; Hanrahan, I.; El-Shinawy, S.H.; El-Kholy, A.; Hassan, G.; Zhu, M.J. Efficacy of Ozonated Water, Chlorine, Chlorine Dioxide, Quaternary Ammonium Compounds and Peroxyacetic Acid against Listeria Monocytogenesbiofilm on Polystyrene Surfaces. *Front. Microbiol.* **2018**, *9*, 411379. [CrossRef]
31. Cabo, M.L.; Herrera, J.J.; Crespo, M.D.; Pastoriza, L. Comparison among the Effectiveness of Ozone, Nisin and Benzalkonium Chloride for the Elimination of Planktonic Cells and Biofilms of Staphylococcus Aureus CECT4459 on Polypropylene. *Food Control* **2009**, *20*, 521–525. [CrossRef]
32. Naitou, S.; Takahara, H. Recent Developments in Food and Agricultural Uses of Ozone as an Antimicrobial Agent-Food Packaging Film Sterilizing Machine Using Ozone. *Ozone Sci. Eng.* **2008**, *30*, 81–87. [CrossRef]

33. Shelobolina, E.S.; Walker, D.K.; Parker, A.E.; Lust, D.V.; Schultz, J.M.; Dickerman, G.E. Inactivation of Pseudomonas Aeruginosa Biofilms Formed under High Shear Stress on Various Hydrophilic and Hydrophobic Surfaces by a Continuous Flow of Ozonated Water. *Biofouling* **2018**, *34*, 826–834. [CrossRef] [PubMed]
34. Belloli, M.; Cigarini, M.; Milesi, G.; Mutti, P.; Berni, E. Effectiveness of Two UV-C Light-Emitting Diodes (LED) Systems in Inactivating Fungal Conidia on Polyethylene Terephthalate. *Innov. Food Sci. Emerg. Technol.* **2022**, *79*, 103050. [CrossRef]
35. Berni, E.; Belloli, M.; Cigarini, M.; Brindani, D.; Cardoso, C.C.; Mutti, P.; Imperiale, D. Sanitizing of Stainless Steel Surfaces in the Food Industry: Effect of Gaseous Ozone against Pathogens and Filamentous Fungi. *J. Food Saf.* **2024**, *44*, e13106. [CrossRef]
36. *EN 13697:2015+A1*; Chemical Disinfectants and Antiseptics—Quantitative Nonporous Surface Test for the Evaluation of Bactericidal and/or Fungicidal Activity of Chemical Disinfectants Used in Food, Industrial, Domestic and Institutional Areas—Test Method and Requirements without Mechanical Action (Phase 2, Step 2). UNI—Ente Nazionale di Normazione: Milan, Italy, 2019.
37. *EN 1650*; Chemical Disinfectants and Antiseptics—Quantitative Suspension Test for the Evaluation of Fungicidal or Yeasticidal Activity of Chemical Disinfectants and Antiseptics Used in Food, Industrial, Domestic and Institutional Areas—Test Method and Requirements (Phase 2, Step 1). UNI—Ente Nazionale di Normazione: Milan, Italy, 2019.
38. *EN 17272*; Chemical Disinfectants and Antiseptics—Methods of Airborne Room Disinfection by Automated Process—Determination of Bactericidal, Mycobactericidal, Sporicidal, Fungicidal, Yeasticidal, Virucidal and Phagocidal Activities. UNI—Ente Nazionale di Normazione: Milan, Italy, 2020.
39. Geeraerd, A.H.; Valdramidis, V.P.; Van Impe, J.F. GInaFiT, a Freeware Tool to Assess Non-Log-Linear Microbial Survivor Curves. *Int. J. Food Microbiol.* **2005**, *102*, 95–105. [CrossRef]
40. John, D.; Ramaswamy, H.S. Comparison of Pulsed Light Inactivation Kinetics and Modeling of Escherichia Coli (ATCC-29055), Clostridium Sporogenes (ATCC-7955) and Geobacillus Stearothermophilus (ATCC-10149). *Curr. Res. Food Sci.* **2020**, *3*, 82–91. [CrossRef]
41. Mafart, P.; Couvert, O.; Gaillard, S.; Leguerinel, I. On Calculating Sterility in Thermal Preservation Methods: Application of the Weibull Frequency Distribution Model. *Int. J. Food Microbiol.* **2002**, *72*, 107–113. [CrossRef]
42. Geeraerd, A.H.; Herremans, C.H.; Van Impe, J.F. Structural Model Requirements to Describe Microbial Inactivation during a Mild Heat Treatment. *Int. J. Food Microbiol.* **2000**, *59*, 185–209. [CrossRef]
43. Coroller, L.; Leguerinel, I.; Mettler, E.; Savy, N.; Mafart, P. General Model, Based on Two Mixed Weibull Distributions of Bacterial Resistance, for Describing Various Shapes of Inactivation Curves. *Appl. Environ. Microbiol.* **2006**, *72*, 6493–6502. [CrossRef]
44. Bigelow, W.D.; Esty, J.R. The Thermal Death Point in Relation to Time of Typical Thermophilic Organisms. *J. Infect. Dis.* **1920**, *27*, 602–617. [CrossRef]
45. Cerf, O.; METRO, F. Tailing of Survival Curves of Bacillus Licheniformis Spores Treated with Hydrogen Peroxide. *J. Appl. Bacteriol.* **1977**, *42*, 405–415. [CrossRef]
46. Racchi, I.; Scaramuzza, N.; Hidalgo, A.; Cigarini, M.; Berni, E. Sterilization of Food Packaging by UV-C Irradiation: Is Aspergillus Brasiliensis ATCC 16404 the Best Target Microorganism for Industrial Bio-Validations? *Int. J. Food Microbiol.* **2021**, *357*, 109383. [CrossRef] [PubMed]
47. Baysal, A.H.; Molva, C.; Unluturk, S. UV-C Light Inactivation and Modeling Kinetics of Alicyclobacillus Acidoterrestris Spores in White Grape and Apple Juices. *Int. J. Food Microbiol.* **2013**, *166*, 494–498. [CrossRef] [PubMed]
48. Hijnen, W.A.M.; Beerendonk, E.F.; Medema, G.J. Inactivation Credit of UV Radiation for Viruses, Bacteria and Protozoan (Oo)Cysts in Water: A Review. *Water Res.* **2006**, *40*, 3–22. [CrossRef] [PubMed]
49. Oppezzo, O.J.; Abrevaya, X.C.; Giacobone, A.F.F. An Alternative Interpretation for Tailing in Survival Curves for Bacteria Exposed to Germicidal Radiation. *Photochem. Photobiol.* **2024**, *100*, 129–136. [CrossRef] [PubMed]
50. Robbins, J.B.; Fisher, C.W.; Moltz, A.G.; Martin, S.E. Elimination of Listeria Monocytogenes Biofilms by Ozone, Chlorine, and Hydrogen Peroxide. *J. Food Prot.* **2005**, *68*, 494–498. [CrossRef]
51. Krijgsheld, P.; Bleichrodt, R.; van Veluw, G.J.; Wang, F.; Müller, W.H.; Dijksterhuis, J.; Wösten, H.A.B. Development in Aspergillus. *Stud. Mycol.* **2013**, *74*, 1–29. [CrossRef]
52. Pombeiro-Sponchiado, S.R.; Sousa, G.S.; Andrade, J.C.; Lisboa, H.F.; Gonçalves, R.C. Production of Melanin Pigment by Fungi and Its Biotechnological Applications. *Melanin* **2017**, *4*, 47–75. [CrossRef]
53. De Obeso Fernandez Del Valle, A.; Scheckhuber, C.Q. Superoxide Dismutases in Eukaryotic Microorganisms: Four Case Studies. *Antioxidants* **2022**, *11*, 188. [CrossRef]
54. Fisher, C.W.; Lee, D.; Dodge, B.A.; Hamman, K.M.; Robbins, J.B.; Martin, S.E. Influence of Catalase and Superoxide Dismutase on Ozone Inactivation of *Listeria Monocytogenes*. *Appl. Environ. Microbiol.* **2000**, *66*, 1405–1409. [CrossRef]
55. Hageskal, G.; Tryland, I.; Liltved, H.; Skaar, I. No Simple Solution to Waterborne Fungi: Various Responses to Water Disinfection Methods. *Water Supply* **2012**, *12*, 220–226. [CrossRef]
56. Platforma Medyczna, P.; Dariusz Jerzy, B.; Ewa, B.; Bożena, B.; Magdalena, C.; Beata, S.-L.; Stefan, T.; Białoszewski, D.; Bocian, E.; Bukowska, B.; et al. Antimicrobial Activity of Ozonated Water. *Med. Sci. Monit* **2010**, *16*, 71–75.
57. Restaino, L.; Frampton, E.W.; Hemphill, J.B.; Palnikar, P. Efficacy of Ozonated Water against Various Food-Related Microorganisms. *Appl. Environ. Microbiol.* **1995**, *61*, 3471–3475. [CrossRef] [PubMed]

58. Beuchat, L.R.; Chmielewski, R.; Keswani, J.; Law, S.E.; Frank, J.F. Inactivation of Aflatoxigenic Aspergilli by Treatment with Ozone. *Lett. Appl. Microbiol.* **1999**, *29*, 202–205. [CrossRef] [PubMed]
59. O'Donnell, C.; Tiwari, B.K.; Cullen, P.J.; Rice, R.G. *Ozone in Food Processing*; Wiley: Hoboken, NJ, USA, 2012. [CrossRef]

applied sciences

MDPI

Article

# The Application of Osmodehydrated Tomato and Spinach in Ready-to-Eat Mixed Salad Products: Design, Development, and Shelf Life Study

**Alexandros Katsimichas, George Dimopoulos, Efimia Dermesonlouoglou * and Petros Taoukis**

Laboratory of Food Chemistry and Technology, School of Chemical Engineering, National Technical University of Athens (NTUA), 15780 Athens, Greece; alkats@chemeng.ntua.gr (A.K.); gdimop@chemeng.ntua.gr (G.D.); taoukis@chemeng.ntua.gr (P.T.)
* Correspondence: efider@chemeng.ntua.gr

**Abstract:** Osmotically dehydrated cherry tomatoes and spinach leaves were incorporated into Greek salad-type (including OD-treated and air-dried feta cheese trimmings and air-dried olive rings) and green salad-type (including OD-treated and air-dried feta cheese trimmings and roasted ground peanuts) ready-to-eat (RTE) product prototypes, respectively. The osmotic dehydration of cherry tomatoes and spinach leaves was conducted in a pilot scale setting (100 L) in a 60% glycerol-based solution at 35 °C and 25 °C for 180 min and 60 min, respectively. To quantify the moisture transfer between the three ingredients of different moisture content (and water activity), the moisture equilibrium curves for each ingredient of the RTE product were determined. The equilibrium water activity of RTE products was 0.86 and 0.76, respectively. The quality of the RTE products (more specifically, tomato and spinach color and texture, instrumentally measured and sensorially perceived, sensory characteristics) was evaluated. The shelf life of the prototypes (from 4 °C to 20 °C) was kinetically modeled based on sensory deterioration and microbial growth, using the zero-order kinetic model and the Gompertz model, respectively. In the case of the tomato-based product, a shelf life of 54 days (based on sensory deterioration) was achieved at 4 °C, a shelf-life extension of 40 days compared to untreated, fresh-cut tomato. The shelf life of the spinach-based product (based on sensory deterioration) was 36 days at 4 °C, 30 days longer when compared to untreated spinach. Our results indicate that osmotic dehydration was successful in significantly extending the shelf life of such products, contributing to the increased temperature resilience of their keeping quality and allowing for their distribution and storage in a variable cold chain.

**Keywords:** *Lycopersicon esculentum*; *Spinacia oleracea* L.; osmotic dehydration; ready-to-eat salad; shelf-life determination

## 1. Introduction

Osmotic dehydration (OD) is a well-established food processing technique involving the immersion of food items in a hypertonic solution [1–3]. The selection of appropriate conditions (such as the type and concentration of the osmotic solute, solution/product mass ratio, temperature, time, pressure, shape and size of the vegetables, agitation level, etc.) depends on achieving effective dehydration within a reasonably short time while preserving the quality characteristics of the food product [3]. In addition, the combination of mild OD conditions and the selective enrichment of the osmosed tissue with solutes of the osmotic solution can produce modified food products with desired quality attributes [4,5]. The osmotic dehydration process also ensures moisture removal with lower energy consumption compared to the conventional dehydration processes [6]. Its nonthermal nature offers the ability to prolong the shelf life of fruits and vegetables without sacrificing the fresh-like sensory attributes that are commonly associated with the consumption of fresh-cut salads. In most cases, the effect of osmotic dehydration on mass transfer kinetics (during OD) and/or quality characteristics of the dehydrated tissue (after OD and during storage) was studied. The potential use of osmotically dehydrated fruits and vegetables in ready-to-eat (RTE) mixed salad products has not been investigated. The marketing of osmo-dehydrated products (intermediate moisture, ready-to-eat) is limited to fruit products/ingredients used in dairy, candy and bakery industries, as well as utilized in the production of fruit and vegetable concentrates and jams [1].

Ready-to-eat (RTE) mixed salads are fresh cut leafy vegetables subjected to processing such as washing, cutting, mixing, and packaging [7,8]. Due to rapid urbanization and an increasing demand for healthy, convenient, and diverse plant-based diets, the RTE salads market is growing [9]. In 2022, fresh-cut vegetables represented a global market of USD 346.05 billion [8]. According to a new research report by Global Market Insights, the market for processed fruits and vegetables, including RTE salads, will exceed USD 392 billion by 2025. This advance is attributed to several factors; among them are technological innovations in production that have enabled the development of products with greater nutritional value and longer shelf life (Global Market Insights Inc., Selbyville, DE, USA, 2019) [10]. In terms of composition and nutritional importance, RTE salads are low in calories, rich in fiber, vitamins, minerals, antioxidants, and other phytochemicals [11–13]. However, they are highly perishable with a storage life of about 7–10 days under refrigeration at temperatures $\leq 5$ °C [14,15]. The shelf life of RTE salads is determined using microbial and chemical changes. The complex indigenous spoilage flora of RTE salads comprises *Pseudomonas*, lactic acid bacteria, *Enterobacteriaceae*, yeasts and molds. Other causes of quality degradation include browning and enzymatic softening, which may also partially be attributed to enzymes released by microorganisms [15]. A limited number of models that describe the effect of processing and storage conditions on the quality and the microbial risk in RTE fresh-cut vegetables has been reported [16–20]. A challenge for food scientists is to explore the potential of preserving the quality and extending the shelf life of RTE salads [8,21,22].

Spinach (*Spinacia oleracea* L.) is one of the most common leafy vegetables present in RTE salads, well recognized as a functional food owing to its diverse nutritional composition, including lipophilic (such as carotenoids and chlorophylls) and hydrophilic bioactive compounds (such as flavonoids and phenolic acids) [23,24]. It is consumed raw, minimally processed (in RTE salads), boiled, canned, frozen or in other food formulations. The high perishability of fresh spinach makes handling, transportation, and storage difficult and costly, and this limits utilization and marketability [24]. The application of osmotic dehydration of leafy vegetables (such as spinach) has not been reported [25]. Tomatoes (*Solanum lycopersicum* L.), a globally significant agricultural commodity, are prized for their high nutritional value and appealing sensory attributes. More specifically, cherry tomatoes are characterized by a unique texture and palatability, qualities susceptible to damage during processing [26]. Among the different types of commercial products, dehydrated tomatoes (via sun drying, air drying, osmotic dehydration) are among the most appreciated [6,27–31]. The application of osmotic dehydration (as a pre-processing step to cooling, freezing, air-drying) of tomatoes has been widely reported [25,28–30]. OD of tomatoes has been proved to have several advantages such as the conservation of bioactive compounds, the maintenance of color and texture properties, and lower energy consumption [3,32].

Successfully dehydrated tomatoes as well as spinach leaves could potentially find application in pre-packed RTE mixed salads with an extended shelf life. Due to the perishability of each individual component, such a product would not be feasible without a nonthermal dehydration step.

The aim of this study is to design and develop two innovative ready-to-eat (RTE) mixed salad product prototypes of extended shelf life based on osmotically dehydrated (OD) tomatoes and spinach. The quality and stability (both sensory and physicochemical, and microbial stability) of a developed RTE salad product prototype, with the use of untreated and OD-treated cherry tomatoes and spinach, was monitored during storage in a wide range of temperatures (from 4 °C to 20 °C), and the shelf life was calculated. Osmotically dehydrated (OD) cherry tomatoes were used for the development of a Greek salad-inspired composite product, and the OD spinach leaves were used for the development of a green salad-inspired composite product.

## 2. Materials and Methods

### 2.1. Raw Materials: Tomatoes and Spinach Leaves

Cherry tomatoes were obtained from retail, washed with tap water, and manually halved. Commercially available RTE spinach leaves (washed, fresh-cut and packed in a modified atmosphere) were obtained from retail.

### 2.2. Other Materials—Ingredients of RTE Salad Products

Other ingredients of the RTE salad products (feta cheese, olive rings in brine, roasted peanuts) were supplied by the local supermarket. OD-treated and air-dried feta cheese (trimmings) and air-dried olive rings were prepared according to the OD experimental procedure presented in Table 1 [33].

**Table 1.** OD-treated tomato and spinach-based ready-to-eat (RTE) salad product formulation and characteristics (the serving size was 100 g of net product weight).

| Ingredient | Processing Characteristics | Mass Percentage in the Product |
|---|---|---|
| | **OD-treated tomato-based ready-to-eat salad product** | |
| OD-treated tomatoes | Halved cherry tomatoes were osmotically dehydrated for 180 min at 35 °C | 70% |
| OD-treated and air-dried feta cheese trimmings | Feta cheese trimmings were osmotically dehydrated for 60 min at room temperature in an osmotic medium containing 40% glycerol, and further air-dried for 24 h at 50 °C | 25% |
| Air-dried olives | Commercially available olive rings in brine, were air-dried for 24 h at 50 °C | 5% |
| | **OD-treated spinach-based ready-to-eat salad product** | |
| OD-treated spinach | Spinach leaves were osmotically dehydrated for 60 min at room temperature | 50% |
| OD-treated and air-dried feta cheese trimmings | Feta cheese trimmings were osmotically dehydrated for 60 min at room temperature in an osmotic medium containing 40% glycerol, and further air-dried for 24 h at 50 °C. | 25% |
| Roasted ground peanuts | Commercially available roasted peanuts, ground to fine particles. | 25% |

### 2.3. Osmotic Dehydration of Tomatoes and Spinach Leaves

Osmotic dehydration of cherry tomatoes and spinach leaves was conducted in a pilot scale setting. A cylindrical (Ø28 × 60 cm) stainless steel tank of 100 L was used. The tank was fitted with spigots that allowed draining of the solution. Stirring was provided using an electrical motor stirrer fitted with a paddle and submerged in the solution. To avoid damaging the delicate plant tissues, stirring was kept to a minimum and the motor was set at a rotational speed of 120 rpm. Where required, heating capabilities were provided to the tank using a submerged stainless steel helical tube, and heated via a continuous circulation of water from a 3 kW water bath. The temperature of the solution was monitored with a thermocouple.

The solution required for each osmotic dehydration was prepared in situ within the stainless steel tank by dosing the ingredients using a high-capacity electronic balance.

Gentle heating was employed (temperature set at 35 °C and room temperature for tomatoes and spinach, respectively) to ensure the homogeneous dissolution of the osmotic medium constituents. The osmotic dehydration medium for the OD of cherry tomato contained 60% *w*/*w* glycerol, 10% *w*/*w* vinegar (8% acetic acid), 10% *w*/*w* maltodextrin (DE 47), 3.5% sodium chloride, and 1.5% calcium chloride [31]. The osmotic dehydration medium for the OD of spinach contained 60% *w*/*w* glycerol, 10% *w*/*w* maltodextrin, 1.5% calcium chloride, 1% sodium chloride, 2.5% vinegar and 0.05% *w*/*w* sodium metabisulfite [25].

The vegetables were manually loaded and unloaded in the tank. After OD treatment, the vegetables were ringed with a centrifugal household vegetable wringer and placed on a perforated rack for 30 min to drain off excess osmotic solution.

### *2.4. RTE Salad Product Formulation*

#### 2.4.1. OD-Treated Tomato-Based RTE Salad Product

The OD-treated tomato-based RTE salad combined 70% osmotically dehydrated halved cherry tomatoes, 25% OD-dehydrated feta cheese trimmings, and 5% dehydrated olive rings (Table 1).

#### 2.4.2. OD-Treated Spinach-Based RTE Salad Product

The OD-treated spinach-based RTE salad combined 50% osmotically dehydrated spinach, 25% OD-dehydrated feta trimmings, and 25% ground peanuts (Table 1, Figure 1).

(a)

(b)

**Figure 1.** OD-treated (**a**) tomato- and (**b**) spinach-based ready-to-eat (RTE) salad product prototypes.

#### 2.4.3. Determination of Moisture Sorption Characteristics of OD-Treated Vegetables and Other Ingredients

To quantify the moisture transfer between ingredients of different moisture content (and water activity), it was necessary to determine the moisture equilibrium curves for each ingredient of the RTE salad products. The methodology proposed by Barbosa-Cánovas et al. (2020) was followed for the determination of the ingredients' sorption isotherms [34]. One-two grams of each ingredient was placed in pre-weighed 25 mL glass beakers and left to equilibrate in a sealed container with a controlled water activity (relative humidity) at 20 °C. Constant relative humidity of (%) 43, 53, 68, 75, 81, 85 and 93 in the containers was achieved using saturated salt solutions of potassium carbonate ($K_2CO_3$), magnesium nitrate ($MgNO_3$), potassium iodide (KI), sodium chloride (NaCl), ammonium sulfate ($(NH_4)_2SO_4$), potassium chloride (KCl), and potassium nitrate ($KNO_3$), respectively. After introducing the samples in the containers, they were periodically weighed until a constant weight was achieved, whereupon equilibrium was achieved. Thus, diagrams of water activity versus moisture content were plotted. The initial characteristics of the samples were also determined (water activity, initial moisture content). The dependence of moisture content from water activity was mathematically modeled with the Guggenheim–Anderson–de Boer (GAB) model via nonlinear regression to the experimental data (Equation (1)) [35]:

$$X_e = \frac{X_m \cdot C \cdot K \cdot a_w}{(1 - K \cdot a_w) \cdot (1 - K \cdot a_w + C \cdot K \cdot a_w)} \quad (1)$$

where $X_e$ the equilibrium moisture content on a dry basis (g water/g dry weight), $X_m$ the water content corresponding to an adsorbed monolayer, $a_w$ the water activity and $K$ and $C$ are constants related to the effect of temperature.

### 2.5. *Quality Evaluation of the Developed OD-Treated Tomato and Spinach-Based RTEsalad Products*

#### 2.5.1. Determination of Physicochemical Parameters

Moisture was determined gravimetrically after drying at 105 °C for 24 h. Water activity was measured using a water activity meter (Rotronic AM3 Hygrometer, Bassersdorf, Switzerland).

#### 2.5.2. Determination of Objective Color

Objective color was measured on the surface of both tomato and spinach using an Xrite-i1 portable digital colorimeter (Gretag-Macbeth, Grand Rapids, MI, USA) and expressed in the CIE-Lab scale. Color was measured at three equidistant equatorial points on at least five replicates. Color change in samples during storage was expressed using the total color difference $\Delta E$, given by the following equation (Equation (2)) [36]:

$$\Delta E = \sqrt{(a - a_0)^2 + (b - b_0)^2 + (L - L_0)^2} \quad (2)$$

where $L$, $a$, $b$ are the measured CIE-Lab color parameters and $L_0$, $a_0$, $b_0$ are the measured color parameters of the sample at time zero.

#### 2.5.3. Determination of Objective Texture

The texture of both tomato and spinach was determined using a texture analyzer (TA XT2i Stable Micro Systems, Godalming, Surrey, UK). Halved tomatoes were compressed to a deformation of 20% using a cylindrical probe at a speed of 0.1 mm/s (diameter of 1 cm). The maximum force recorded during the analysis was taken as the punching force. For spinach leaf texture determination, the Film Support Rig (HDP/FSR), which consists of two hole-bearing plates and a punching rod, was used to measure the burst strength. A spinach leaf was fixed between the perforated plates and pierced with the descending probe of the analyzer. The maximum force recorded during the analysis was taken as the punching force. At least 5 replicates for both tomato and spinach were performed per measurement.

#### 2.5.4. Determination of Sensory Characteristics

The sensory characteristics of the developed OD-treated tomato and spinach-based RTE salad products were carried out using a panel of 8 members. The scored characteristics for RTE salad products are described in Supplementary Files (Tables S1 and S2). The panelists had been previously trained in sensory assessment and were asked to independently evaluate the characteristics of the samples. Only adults participated in the recruitment to the sensory team. Participation in the tests and assessments was voluntary. Informed written consent was obtained from the participants in the sensory evaluation study. Each of them could withdraw their consent without providing any justification. All participants obtained a detailed description of the test and were informed about the food samples that would be assessed. Each of the participants was obliged to report any indispositions and allergies and if such was the case, the subject did not participate in the tests. The samples were analyzed following ISO 6579:2002 [37] for the detection of *Salmonella* spp. and ISO 11290-2 [38] for the enumeration of *Listeria monocytogenes* to ensure that the samples were safe for consumption. No pathogens were detected.

2.5.5. Determination of Microbial Load

The microbial growth of the developed OD-treated tomato and spinach-based RTE salad products was determined using the method of agar plate colony enumeration. For each sample, 10 g was homogenized with 90 mL sterile Ringer's solution in a bag homogenizer. For the determination of each type of microbial growth, a volume (100 μL or 1000 μL) of at least two appropriate serial decimal dilutions (prepared in Ringer's solution) were plated on the appropriate agar plate substrate: total viable counts (TVC) were enumerated using spread plate methodology on tryptic glucose yeast agar (Biolife, Milan, Italy) after aerobic incubation at 25 °C for 72 h. Yeasts and molds were enumerated using spread plate methodology on Rose Bengal Chloramphenicol agar (Biolife, Milan, Italy) after aerobic incubation at 25 °C for 5 days. Filamentous fungi were determined using spread plate methodology on Potato Dextrose Agar (PDA) after aerobic incubation at 25 °C for 5 days. *Enterobacteriaceae* spp. were enumerated using pour plate methodology in Violet Red Bile Glucose agar (VRBG, Merck, Darmstadt, Germany) after incubation under a facultatively anaerobic condition at 37 °C for 24 h. *Pseudomonas* spp. were enumerated using spread plate methodology on Cetrimide agar (CFC, Merck, Darmstadt, Germany) after aerobic incubation at 25 °C for 48 h.

2.6. *Shelf-Life Determination*

OD-treated tomato and spinach-based RTE salad products were subjected to shelf-life determination. Approximately 100 g of each product was packed into polyethylene–polypropylene sachets and stored in incubators at constant temperatures 4, 8, 12 and 20 °C. At regular time intervals, samples were withdrawn and analyzed in terms of microbial growth and sensory quality loss described in Section 2.5. Microbial growth versus storage time for each storage temperature was mathematically modeled using the Gompertz model [39]. The evolution of the sensory total acceptance versus storage time for each storage temperature was mathematically modeled using a simple zero order equation of the form $S = S_0 - k_s\,t$, where $S$ is the sensory attribute at time $t$, $S_0$ is the sensory attribute at time zero and $k_s$ the deterioration rate constant. The dependence of the rate constant and lag phase parameters of the Gompertz equation, as well as the dependence of the sensory deterioration rate constant from storage temperature, was mathematically modeled using the Arrhenius equation. Combining the mathematical models used, the following expressions for the shelf life were derived [25]:

$$t(SL_{MG}) = \frac{logN_l - logN_0}{\mu_m(T)} + \lambda(T) \quad (3)$$

$$t(SL_s) = \frac{S_0 - S_L}{k_{s_{Tref}} \cdot \exp\left(-\frac{E_{a_S}}{R} \cdot \left(\frac{1}{T} - \frac{1}{T_{ref}}\right)\right)} \quad (4)$$

where $t_{SL_S}$ is the shelf life determined in terms of sensory deterioration, $t_{SL_{MG}}$ is the shelf life determined in terms of microbial growth, $S_0$ is the initial sensory score, $S_L$ is the acceptance limit of the sensory score, $logN_l$ the acceptable maximum microbial load limit of the product, $logN_0$ the initial microbial load of the product and $\mu_m(T)$, $\lambda(T)$ the dependence of the rate constant and the lag phase parameter from storage temperature through the Arrhenius equation, $R$ is the universal gas constant, and $E_a$ and $k_{ref}$ the activation energy and rate constants at the reference temperature, respectively, for each measured parameter. Since different processing conditions were expected to lead to different shelf-life estimations based on the two parameters, both values were calculated. The parameter leading to the lowest shelf life estimate at each storage temperature was considered as the limiting parameter [25].

### 2.7. Statistical Analysis

Results were expressed as means ± standard deviation of three experimental replicates. For the estimation of the main interaction effects of the investigated factors, factorial analysis of variance (Factorial ANOVA) was used. As a post hoc analysis for the separation of means with significant differences ($p < 0.05$), Duncan's multiple range test was used. For all statistical analyses, the Statistica 7 software (StatSoft, Hamburg, Germany) package was used. For all the mathematical regressions, the IBM SPSS Statistics Version 19 software package (IBM Corporation, Armonk, NY, USA) was used, and $R^2$ and standard errors of model parameters were calculated. The Pearson's test was applied to the data, specifically for the discussion of the correlation between the objective measurements and the sensorial test results (color and texture) for both RTE products.

## 3. Results

### 3.1. Moisture Sorption Characteristics of OD-Treated Cherry Tomato and Spinach-Based RTE Products

#### 3.1.1. OD-Treated Tomato-Based RTE Product

OD-treated tomatoes serve as the foundational constituent, providing a vibrant and nutrient-rich component to the salad. Complementing the tomatoes, OD-treated feta cheese trimmings and dehydrated olive rings bring distinct tangy and savory flavors, infusing the salad with a Mediterranean-inspired twist. Another aspect of the product's processing is that OD enables the introduction of components into the product (e.g., sodium chloride, vinegar), which act as both flavoring agents and as functional ingredients that contribute to shelf-life extension.

The three ingredients of the RTE salad products were subjected to a determination of their water sorption characteristics. The resulting equilibrium curves at 25 °C, obtained at different water activities for the three ingredients (OD-treated tomatoes, dehydrated olives, OD-treated and air-dried cheese) are presented in Figure 2a–c. The dependence of moisture content from water activity for the tomato-based RTE product was mathematically modeled using the GAB equation (Equation (1)) and the model parameters are presented in Table 2. The GAB model was reported to describe the water vapor adsorption isotherms of the dehydrated tomatoes at 25 °C (for the entire $a_w$ range) [40,41].

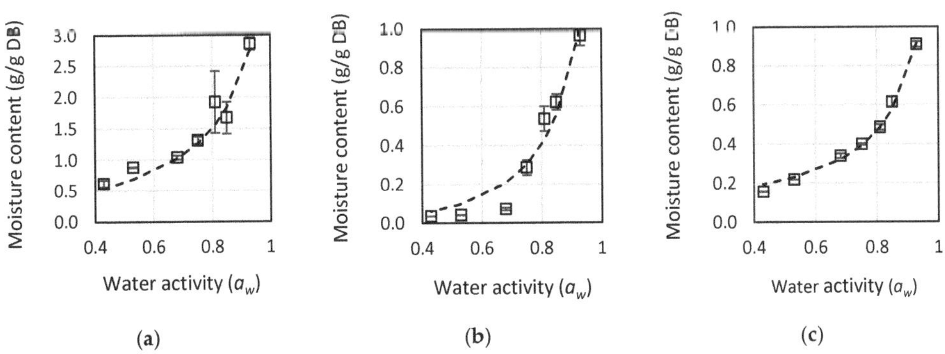


**Figure 2.** Apparent moisture equilibrium curves at 25 °C for (**a**) OD-treated tomatoes, (**b**) dehydrated olives and (**c**) OD-treated and air-dried cheese. The dashed lines represent the fitting of the GAB model (Equation (1)) to the experimental data.

**Table 2.** Water activity characteristics of the ingredients for the OD-treated tomato-based RTE product before mixing and at equilibrium. The predicted equilibrium value results from the water mass balances combined with the fitting of the GAB equation to the experimental data.

| Ingredient | Initial $a_w$ | Equilibrium $a_w$ | GAB Model Parameters | | | Predicted Equilibrium $a_w$ (GAB) |
|---|---|---|---|---|---|---|
| | | | $W_m$ (g/g) | $C$ (-) | $K$ (-) | |
| OD-Treated Tomato | $0.920 \pm 0.002$ [a] | | $0.446 \pm 0.005$ [a] | $4.30 \pm 0.11$ [a] | $0.908 \pm 0.012$ [a] | |
| OD-Treated and Air-Dried Cheese | $0.533 \pm 0.002$ [b] | $0.852 \pm 0.002$ | $0.116 \pm 0.002$ [b] | $4.04 \pm 0.09$ [a] | $0.939 \pm 0.010$ [a] | $0.859 \pm 0.001$ |
| Air-Dried Olives | $0.267 \pm 0.001$ [c] | | $2.25 \pm 0.04$ [c] | $0.03 \pm 0.01$ [b] | $0.835 \pm 0.009$ [b] | |

Mean value ± standard deviation. Different superscript letters in the same column indicate significant differences between means as calculated using Duncan's multiple range test for a significance level of $p = 0.05$.

The results revealed that the three ingredients exhibited different characteristics in terms of moisture transfer. Upon mixing the three ingredients of the final product, moisture is transferred between the ingredients. Specifically, it was expected that OD-tomato would lose water while olives and cheese would gain moisture, to finally reach an equilibrium water activity. Given the product's formulation, the initial water content of each ingredient, the initial water activity values and the apparent moisture equilibrium curves, it was possible to predict the final equilibrium water activity of the product (Table 2). This value was also confirmed via experimental measurements after the equilibrium had been established. Periodic water activity measurements revealed that at 20 °C the equilibrium was achieved in 5 days, while at 4 °C the equilibrium was achieved in 13 days.

As observed by the final values presented in Table 2, the moisture transfer between the ingredients of the developed product reaches an equilibrium at a water activity of 0.852. Compared to the value for OD-treated tomato (0.920), this value significantly surpasses the lower 0.9 limit commonly set for the inhibition of bacterial proliferation [42].

3.1.2. OD-Treated Spinach-Based RTE Product

OD-treated spinach serves as the foundational green base, contributing a vibrant and nutrient-rich element to the salad. Its osmotic dehydration process ensures moisture removal while preserving essential nutrients, enhancing the spinach's shelf stability without compromising its nutritional value. Complementing the spinach, osmotically dehydrated and air-dried feta trimmings bring a distinctive tangy and creamy flavor profile, infusing the salad with a hint of Mediterranean-inspired richness. The combination of these feta trimmings with ground peanuts adds a satisfying textural dimension, with the peanuts providing a crunchy contrast to the softer spinach and feta components.

The resulting equilibrium curves at 25 °C obtained at different water activities for the three ingredients of OD-treated spinach-based RTE product are presented in Figure 3. The dependence of moisture content from water activity was mathematically modeled using the GAB equation (Equation (1)) and the model parameters are presented in Table 3. The GAB model was reported to describe the water vapor adsorption isotherms of dehydrated/dried fruits/vegetables (including dried spinach) [41]. The results revealed that the three ingredients exhibited different characteristics in terms of moisture transfer. The moisture content of the OD-spinach material remained relatively high within the whole range of $a_w$ values studied, namely between 2 and 5 g/g DW. On the other hand, both peanuts and dehydrated cheese had moisture contents ranging between 0.1 and 0.9 g/g DW. This shows that peanuts are much more hydrophobic compared to the other two materials, which is expected due to their high oil content (up to 50%).

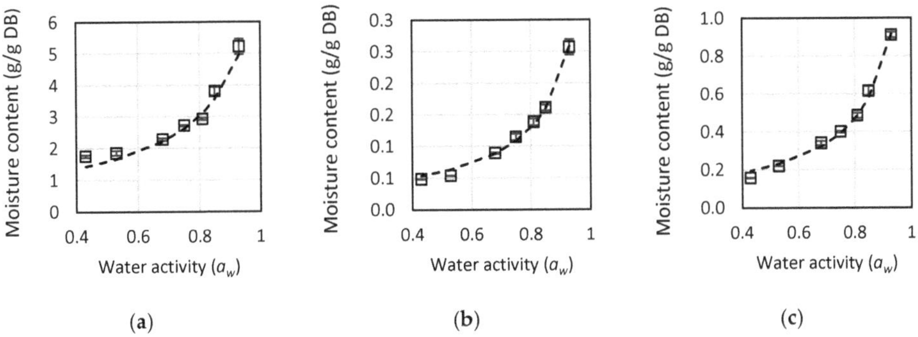


**Figure 3.** Apparent moisture equilibrium curves at 25 °C for (**a**) OD-treated spinach, (**b**) peanuts and (**c**) OD-treated and air-dried cheese. The dashed lines represent the fitting of the GAB model to the experimental data.

**Table 3.** Water activity characteristics of the raw materials for the OD-treated spinach-based RTE product before mixing and at equilibrium. The predicted equilibrium value results from the water mass balances combined with the fitting of the GAB equation to the experimental data.

| Ingredient | Initial $a_w$ | Equilibrium $a_w$ | GAB Model Parameters | | | Predicted Equilibrium $a_w$ (GAB) |
|---|---|---|---|---|---|---|
| | | | $W_m$ (g/g) | $C$ (-) | $K$ (-) | |
| **OD-treated spinach** | 0.930 ± 0.003 [a] | 0.754 ± 0.002 | 0.883 ± 0.011 [a] | 12.4 ± 0.9 [a] | 0.884 ± 0.003 [a] | 0.758 ± 0.003 |
| **OD-treated and air-dried cheese** | 0.533 ± 0.002 [b] | | 0.116 ± 0.008 [b] | 4.04 ± 0.3 [b] | 0.939 ± 0.002 [b] | |
| **Peanuts** | 0.431 ± 0.002 [c] | | 0.031 ± 0.001 [c] | 27.5 ± 1.3 [c] | 0.944 ± 0.003 [b] | |

Mean value ± standard deviation. Different superscript letters in the same column indicate significant differences between means as calculated using Duncan's multiple range test for a significance level of $p = 0.05$.

Upon mixing the three ingredients of the final product, moisture is transferred between the ingredients. Specifically, it was expected that OD-spinach would lose water while peanuts and feta would gain moisture, to finally reach an equilibrium water activity. Given the product's formulation, the initial water content of each ingredient, the initial water activity values, and the apparent moisture equilibrium curves, it was possible to estimate the final equilibrium water activity of the product (Table 3). This value was also confirmed by experimental measurements after the equilibrium had been established. As observed by the final values presented in Table 3, the water activity of the product was significantly lower (0.754) than the respective water activity of the OD-treated spinach (0.930).

### 3.2. *Evolution of Microbial Growth of OD-Treated Cherry Tomato and Spinach-Based RTE Products*

#### 3.2.1. OD-Treated Tomato-Based RTE Product

The evolution of microbial growth for the OD-treated tomato-based RTE product with a storage time at all storage temperatures studied is presented in Figure 4a–c. The similar growth profile observed between the three substrates tested reveal that yeasts and molds are the dominant spoilage microflora for the tomato-based product. The dominant microorganisms in fruits (including tomatoes) are yeasts and molds. According to microbiological criteria recommendations, the maximum limits are $1 \times 10^3$ CFU/g for yeasts and molds and $1 \times 10^5$ (or $1 \times 10^6$) CFU/g for aerobic mesophiles [43,44]. The yeast and molds are undesirable in food, because they are able to produce a wide variety of enzymes, which cause the deterioration of food, impairing the sensory characteristics of the product. Moreover, many yeasts and molds can produce toxic metabolites in food, when the total count of these microorganisms is more than $10^6$ CFU/g.

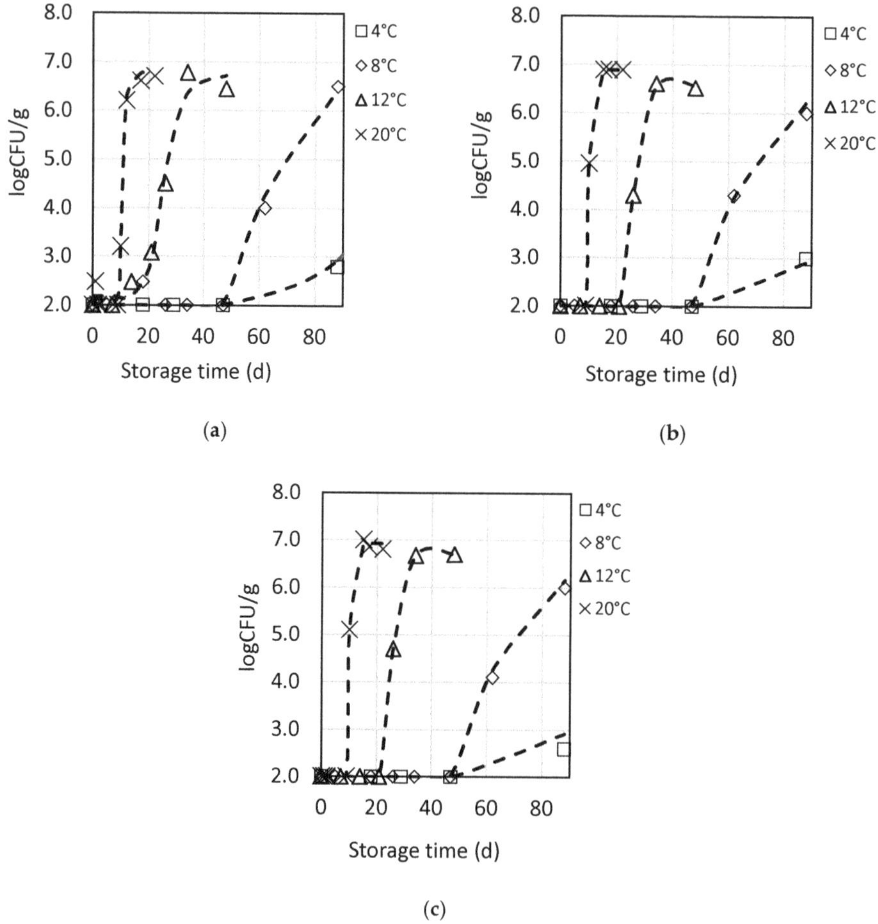


**Figure 4.** Evolution of microbial load of the tomato-based composite product with storage time for all storage temperatures studied (4, 8, 12, 20 °C): (**a**) total viable counts, (**b**) total molds (**c**) yeasts and molds. Dashed lines represent the fitting of the Gompertz model to the experimental data.

In this study, microbial growth remains under a measurable level from approximately 10 to 40 days depending on the storage temperature, after which it rapidly increases. This can be attributed to the unfavorable growth conditions present within the OD-treated tomato-based product.

The evolution of microbial growth (total viable counts) was mathematically modeled using the Gompertz model and the exponential rate constant ($\mu_m$) and lag phase ($\lambda$) were calculated (Table 4). The dependence of the Gompertz model parameters on storage temperature was modeled using the Arrhenius equation and the activation energy ($E_a$) was determined. The reduction in storage temperature from 20 to 4 °C led to a significant reduction in the exponential growth rate up to 95%. Additionally, the lag phase was increased up to approximately 70 d leading to a significant delay of microbial growth.

**Table 4.** Gompertz model parameters resulting from fitting Gompert equation to the experimental data for total viable counts of the tomato-based composite product. The dependence of the parameters $\mu_m$ and $\lambda$ from storage temperature was modeled with the Arrhenius equation, with activation energy values given in the last row of the table.

| Storage Temperature (°C) | $\mu_m$ $\left(d^{-1}\right)$ | $\lambda$ (d) | $R^2$ |
|---|---|---|---|
| 4 | $0.11 \pm 0.01$ [a] | $79.8 \pm 5.1$ [a] | 0.928 |
| 8 | $0.21 \pm 0.02$ [b] | $51.2 \pm 3.2$ [b] | 0.934 |
| 12 | $0.39 \pm 0.02$ [c] | $19.32 \pm 1.1$ [c] | 0.982 |
| 20 | $2.08 \pm 0.10$ [d] | $9.44 \pm 0.7$ [d] | 0.993 |
| | $E_{a,\mu m} = -124.4 \pm 9.3 \frac{kJ}{mol}$ | $E_{a,\lambda} = 62.2 \pm 4.8 \frac{kJ}{mol}$ | |

Mean value ± standard deviation. Different superscript letters in the same column indicate significant differences between means as calculated using Duncan's multiple range test for a significance level of $p = 0.05$.

#### 3.2.2. OD-Treated Spinach-Based RTE Product

The major issue of osmo-dehydrated food products (fruit or vegetable cuts) is the microbial load. It was found that the suitable combination of water activity and (low) pH significantly reduces the microbial load [45]. The increase in shelf life of osmotically dehydrated fruits and vegetables could be attributed to lower water activity and pH reduction, as well as solids uptake [25,46]. The approach proposed in our study resulted in the production of products with low water activity values ranging from 0.75 to 0.86 and pH values ranging from 4.4 to 4.6.

For all conditions studied, the microbial load of the spinach-based product remained stagnant during a 40-day storage period. This was attributed to the equilibration of moisture between the product's components after mixing and storage. However, it must be noted that the microbial load is not eliminated during the whole storage period but rather remains constant at approximately 5.5 logCFU/g.

### 3.3. *Evolution of Sensory Characteristics of OD-Treated Cherry Tomato and Spinach-Based RTE Products*

#### 3.3.1. OD-Treated Tomato-Based RTE Product

The evolution of the sensory characteristics of the tomato-based product with storage time at all storage temperatures studied is presented in Figure S1a–m. All measured attributes exhibited a deterioration with storage time. Examining the figures, it becomes apparent that all constituents of the product are affected. Specifically, for tomatoes, a deterioration of color, flavor and texture is observed, with a gradual loss of red color and an intense softening. The development of off-flavors was also reported, which is attributed to microbial growth and enzymatic activity. The quality of the cheese trimmings also deteriorates, as the cheese absorbs moisture becoming softer and its fats develop rancidity. This is also reflected in the development of a yellow color (Figure S1e), which might also be a result of the different ingredients being in contact with each other in the product packaging. Panelists reported the presence of brown spots on the feta, possibly due to the absorption of dark pigments (e.g., polyphenols) from the olive slices. The deterioration of the sensory quality of each ingredient leads to the deterioration of the quality of the product as a whole, as seen in Figure S1i–m. The appearance of the product, its aroma, its flavor and its aftertaste are all affected by storage time and temperature. Although it was hard to discern individual off-flavors in the product, panelists reported a deterioration in the product's fresh aroma (imparted mostly by the tomato) and the development of a slight fermented odor. The development of bitterness was also reported, which was attributed to the concentration of calcium chloride from the osmotic medium due to the moisture loss of the tomato during equilibration. Finally, panelists reported the occurrence of irregularities within the packaging of the product, namely the presence of liquid and the appearance of smudges and fragments of the individual components. The deterioration of the overall acceptance of the product is driven by all the aforementioned factors.

### 3.3.2. OD-Treated Spinach-Based RTE Product

The evolution of the sensory characteristics of the spinach-based composite product with storage time at all storage temperatures studied is presented in Figure S2a–n. The complexity of the product necessitated the determination of a wide array of parameters pertaining to the individual components (spinach, cheese, peanuts), as well as the product as a whole. The deterioration of the sensory quality of the product is a combination of several physicochemical alterations such as color deterioration (Figure S2a,d), development of gumminess (Figure S2c), loss of crunchiness (Figure S2n) and the development of rancidity (Figure S2j). Loss of textural integrity can be attributed to the transfer of moisture between the product constituents, as the constituents with the lower moisture content (feta cheese and peanut) absorb water while the spinach loses water. These alterations lead to the degradation of the product's texture, flavor and aroma and culminate in the storage temperature-dependent deterioration of the overall product acceptance (Figure S2n). Panelists also reported the development of product defects within the packaging (Figure S2m), namely the appearance of a liquid phase resulting from losses of osmotic solution and moisture. This effect was prominent at higher storage temperatures. Flavor degradation also occurred in a temperature-dependent manner (Figure S2i) and was considered a combination of all factors discussed. The development of a bitter-metallic flavor was also reported, most likely attributed to the concentration of salts taken up by the spinach during OD (especially calcium chloride).

The evolution of the overall sensory liking of both RTE products was mathematically modeled using the zero-order kinetic model. The model parameter (sensory deterioration rate $k_s$) is presented in Table 5. The dependence of the sensory deterioration rate was mathematically modeled using the Arrhenius equation.

**Table 5.** Sensory deterioration rates (overall sensory liking) for the OD-treated tomato and spinach RTE product at storage temperatures 4, 8, 12, 20 °C. The dependence of the rate constants from storage temperature was modeled using the Arrhenius equation and the fitting parameters are given in the last row of the table.

| | OD-Tomato-Based RTE Product | | OD-Spinach-Based RTE Product | |
|---|---|---|---|---|
| **Storage Temperature (°C)** | **Sensory Deterioration Rate $k_s$ $\left(d^{-1}\right)$** | $R^2$ | **Sensory Deterioration Rate $k_s$ $\left(d^{-1}\right)$** | $R^2$ |
| 4 | $0.074 \pm 0.003^{a}$ | 0.955 | $0.103 \pm 0.009^{d}$ | 0.897 |
| 8 | $0.093 \pm 0.004^{b}$ | 0.975 | $0.170 \pm 0.011^{b}$ | 0.938 |
| 12 | $0.226 \pm 0.012^{c}$ | 0.982 | $0.409 \pm 0.031^{c}$ | 0.948 |
| 20 | $0.551 \pm 0.031^{d}$ | 0.993 | $0.807 \pm 0.062^{d}$ | 0.962 |
| | $E_\alpha = 89.3 \pm 11.2 \frac{kJ}{mol}$ | 0.968 | $E_\alpha = 88.9 \pm 11.6 \frac{kJ}{mol}$ | 0.975 |

Mean value ± standard deviation. Different superscript letters in the same column indicate significant differences between means as calculated using Duncan's multiple range test for a significance level of $p = 0.05$.

The deterioration of the overall sensory liking of the product was presented in Figure 5. As observed in Figure 5, the reduction in the storage temperature of OD-treated tomato (Figure 5a) and spinach (Figure 5b) at 4 °C led to a higher score of sensory acceptance compared to other temperatures during storage. The overall sensory score of OD-treated spinach fell below 5 after 30 d at 4 °C, while the score of OD-treated tomato did not fall below 5 for up to 47 d at 4 °C. To quantify the effect of storage temperature on the sensory deterioration, the data were mathematically modeled, and the sensory deterioration rate ($k_s$) was calculated. The reduction in storage temperature from 20 to 4 °C significantly reduced the sensory deterioration rate of both RTE products by 87%, indicating a strong temperature dependence. This strong dependance is reflected in activation energy which is calculated at approximately 89 kJ/mol for both RTE products. This kinetic approach was very important for the estimation of the shelf life based on sensorial characteristics, since the shelf life of the RTE products was found to be limited by sensory deterioration.

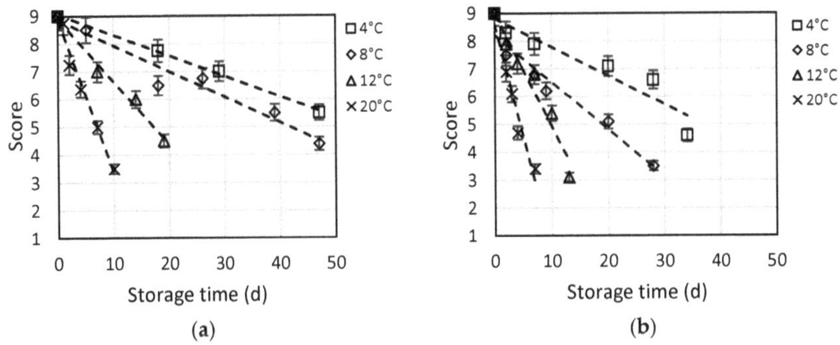


**Figure 5.** Evolution of sensory liking of the (**a**) OD-treated tomato and (**b**) OD-treated spinach-based RTE product with storage time for storage temperatures 4, 8, 12, 20 °C. Dashed lines represent fitting of the zero-order kinetic model to the experimental data, where applicable.

### 3.4. Evolution of Selected Quality Indices of OD-Treated Tomato and Spinach-Based RTE Products

#### 3.4.1. OD-Treated Tomato-Based RTE Product: Tomato Color and Texture

Figure 6a presents the evolution of the total color difference $\Delta E$ for tomato in the tomato-based RTE product with the storage time at all storage temperatures studied. A significant temperature-dependent color change was observed for the OD-treated tomatoes, attributed to the complex interactions of the raw ingredients upon mixing during product formulation, as well as to the reduction in the lightness of the tomatoes. This total color difference was found to be significantly correlated (negative correlation) with the intensity of the red color score, as validated using the Pearson's test (Table 6). Similarly, cheese trimmings exhibited a shift in their color, caused by the uptake of moisture and the development of yellowness, possibly due to the oxidation of fats. These observations are consistent with the results obtained from the sensory evaluation of the product. The loss of firmness of the tomatoes during storage reported during the sensory evaluation was not correlated with objective firmness measurements in Figure 6b (Pearson coefficient close to zero). This result indicates that the panelists did not manage to detect the change in texture, possibly due to the solid uptake of OD-tomatoes that offered a crunchy character during chewing, which is an advantage for this product.

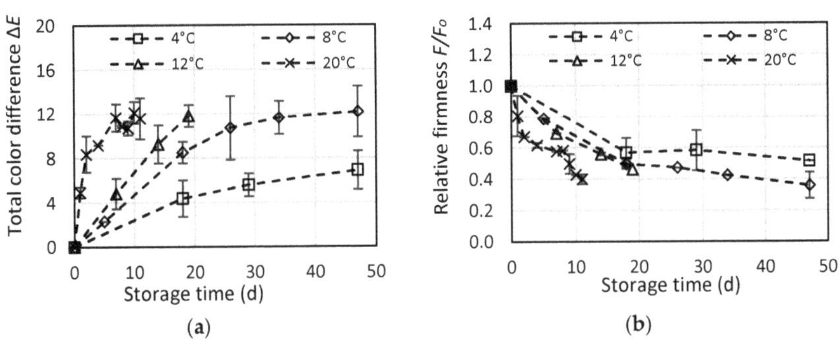


**Figure 6.** Evolution of (**a**) the total color difference $\Delta E$ and (**b**) the relative firmness ($F/F_0$) of tomatoes in the tomato-based RTE product with storage time at storage temperatures 4, 8, 12, 20 °C.

**Table 6.** Correlation matrix of Pearson coefficients of total color difference ($\Delta E$), intensity of red or green color, relative firmness ($F/F_0$), and intensity of perceived firmness or gumminess.

| | | | | |
|---|---|---|---|---|
| | **OD-tomatoes of the tomato-based RTE product** | | | |
| | **Total color difference ($\Delta E$)** | **Intensity of red color** | **Relative firmness ($F/F_0$)** | **Intensity of perceived firmness** |
| Total color difference ($\Delta E$) | | −0.854 ** | −0.803 ** | −0.289 |
| Intensity of red color | −0.854 ** | | 0.877 ** | 0.220 |
| Relative firmness ($F/F_0$) | −0.803 ** | 0.877 ** | | −0.031 |
| Intensity of perceived firmness | −0.289 | 0.220 | −0.031 | |
| | **OD-spinach of the spinach-based RTE product** | | | |
| | **Total color difference ($\Delta E$)** | **Intensity of green color** | **Relative firmness ($F/F_0$)** | **Intensity of perceived gumminess** |
| Total color difference ($\Delta E$) | | −0.697 ** | 0.720 ** | 0.630 ** |
| Intensity of green color | −0.697 | | −0.740 ** | −0.888 ** |
| Relative firmness ($F/F_0$) | 0.720 ** | −0.740 ** | | 0.786 ** |
| Intensity of perceived gumminess | 0.630 ** | −0.888 ** | 0.786 ** | |

** Significant correlation at level of $p = 0.01$.

### 3.4.2. OD-Treated Spinach-Based RTE Product: Spinach Color and Texture

A temperature-dependent color deterioration was observed in the OD-treated spinach leaves, which was more pronounced for OD-treated samples. This color change was attributed to both a reduction in sample lightness due to enzymatic browning, and to a degradation of chlorophylls to pheophytins. In fact, this deterioration is favored under acidic pH which is the case with osmotically dehydrated spinach. The evolution of the total color difference ($\Delta E$) in spinach samples at all treatments with storage time is presented in Figure 7a and it was found to be significantly correlated (negative correlation) with the intensity of the green color score using the Pearsons test (Table 6). Figure 7b presents the evolution of the relative burst strength of OD-treated spinach samples with storage time at all storage temperatures studied. For all treatments, it was observed that the burst strength of the spinach leaves increased with storage time. This effect was more pronounced when an OD treatment was performed with the relative burst strength of the treated leaves, almost doubling after 25 days of storage. Macroscopically, this can manifest as an increased elasticity and gumminess of the leaves, which was confirmed via the sensory evaluation of the samples. Due to the loss of water during storage (untreated samples) or dehydration (OD-treated samples), the fibers in the plant tissue become compacted, leading to an increased mechanical strength of the tissue [47]. This behavior of the relative burst strength was detected by the panelists and it was positively correlated with the intensity of the perceived gumminess score.

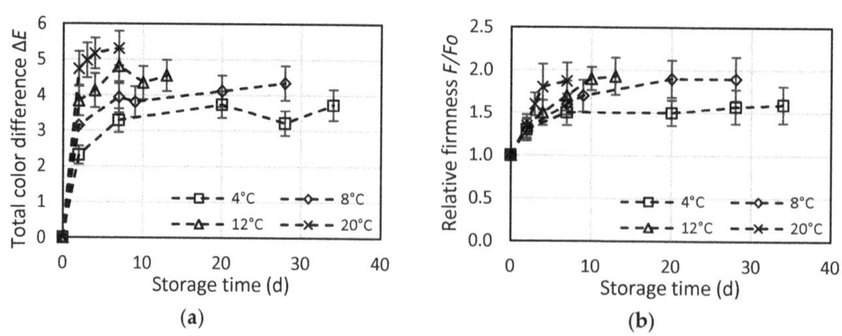


**Figure 7.** Evolution of (**a**) the total color difference ($\Delta E$) and (**b**) the relative burst strength of OD-treated spinach samples with storage time at storage temperatures 4, 8, 12, and 20 °C.

### *3.5. Shelf-Life Determination and Modeling of OD-Treated Cherry Tomato and Spinach-Based RTE Products*

#### 3.5.1. OD-Treated Tomato-Based RTE Product

The two shelf-life-defining characteristics for the tomato-based product were considered to be microbial growth and sensory deterioration. As discussed, microbial growth was severely retarded due to the various hurdles present in the product. This becomes evident in the shelf-life curves presented in Figure 8a, where the curve corresponding to the shelf life of the product based on sensory deterioration appears to be shifted towards lower storage temperatures. Therefore, the shelf life of the product was limited by sensory deterioration. This is not uncommon, since other physicochemical alterations often precede microbial growth, especially in dehydrated products [47].

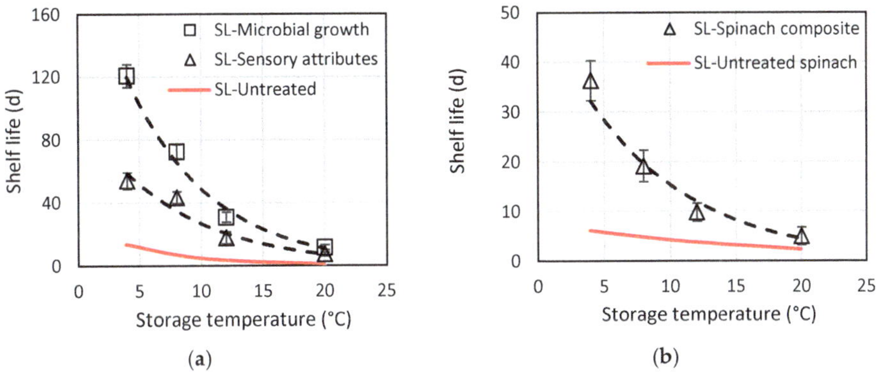


**Figure 8.** Shelf-life (SL) curves of the developed (**a**) tomato- (based on microbial growth and sensory criteria) and (**b**) spinach- (based on sensory criteria) based RTE products at storage temperatures 4–20 °C. Lines represent the prediction of the shelf life based on the overall mathematical models developed. The continuous red line represents the expected shelf life of the product if untreated tomatoes had been used in the formulation.

In Table 7, the shelf life of the RTE tomato-based product is presented and compared with the shelf life of untreated tomatoes, as calculated from the results of the experiments on the untreated tomatoes. It is observed that at 4 °C, the shelf life of the OD-tomato-based RTE product was calculated to be 54.0 d compared to the 13.8 d of the shelf life of the respective untreated tomato-based RTE product. This shelf life extension achieved through applying OD to RTE product materials is due to the significant reduction in the water activity that led to the significant delay of microbial growth.

**Table 7.** Calculated shelf life of the tomato- and spinach-based RTE product for each storage temperature studied. Values are compared with the estimated shelf-life values of the shelf life of the untreated tomatoes.

| | OD-Tomato-Based RTE Product | | OD-Spinach-Based RTE Product | |
|---|---|---|---|---|
| **Storage Temperature (°C)** | **Shelf Life (d)** | **Estimated Shelf Life with Untreated Tomatoes (d)** | **Shelf Life (d)** | **Estimated Shelf Life with Untreated Spinach (d)** |
| 4 | $54.0 \pm 5.3^{a}$ | $13.8 \pm 0.9^{c}$ | $36.3 \pm 4.0^{a}$ | $4.7 \pm 0.3^{d}$ |
| 8 | $43.2 \pm 3.8^{b}$ | $7.2 \pm 0.4^{d}$ | $19.1 \pm 3.2^{b}$ | $4.0 \pm 0.3^{d}$ |
| 12 | $17.7 \pm 3.7^{c}$ | $3.8 \pm 0.2^{de}$ | $9.8 \pm 1.8^{c}$ | $3.2 \pm 0.2^{d}$ |
| 20 | $7.3 \pm 1.2^{d}$ | $1.0 \pm 0.1^{e}$ | $5.0 \pm 1.2^{d}$ | $1.4 \pm 0.1^{e}$ |

Different superscript letters indicate significant differences between means of shelf life ± standard deviation as calculated using Duncan's multiple range test for a significance level of $p < 0.05$.

### 3.5.2. OD-Treated Spinach-Based RTE Product

In the case of the OD-treated spinach-based RTE product, since microbial growth did not yield results that could define the shelf life of the product, this was determined based on the sensory deterioration rate. For the determination of shelf life, a sensory score limit equal to 5 was used, which is commonly accepted as the limit for a product's end-of-life [25]. In Table 7, the shelf life of the product is presented and compared with the shelf life of untreated spinach, as calculated from the results of the experiments on the untreated spinach. It is observed that at 4 °C, the shelf life of the OD-spinach-based RTE product was calculated to be 36.3 d compared to the 4.7 days of the shelf life of the respective untreated spinach-based RTE product.

## 4. Conclusions

This research presents opportunities to enhance fresh vegetable or salad processing techniques and diversify their culinary uses, aligning with consumer preferences for high-quality, nutritious options with reduced energy consumption. Two ready-to-eat salad product prototypes based on osmotically dehydrated (OD) cherry tomatoes and spinach leaves with an extended shelf life were developed. At a common storage temperature of 4 °C, a shelf life extension of 40 and 30 days was calculated for the OD-tomato and OD-spinach-based RTE product compared to the product with untreated tomatoes and spinach, respectively. This extension also significantly expands the temperature tolerance of the samples, with a shelf life of approximately 5–7 days at 20 °C for OD-treated samples, comparable to the values obtained at 4 °C for untreated samples. Such fresh cut salads in the form of a pre-packed RTE meal would not be feasible without the application of the osmotic dehydration.

**Supplementary Materials:** The following supporting information can be downloaded at: https://www.mdpi.com/article/10.3390/coatings14135863/s1, Table S1. Tested sensory characteristics for the tomato-based ready-to-eat (RTE) salad product. Table S2. Tested sensory characteristics for the spinach-based ready-to-eat (RTE) salad product. Figure S1. Evolution of sensory characteristics of the tomato-based composite product with storage time for all studied storage temperatures. The subfigure is divided into sections marked with the characteristics of the individual product components as well as the characteristics of the whole product. Dashed lines represent fitting of the zero-order kinetic model to the experimental data, where applicable. Figure S2. Evolution of sensory characteristics of the spinach-based composite product over storage time for each storage temperature studied. The figure is subdivided into sections describing the characteristics of the individual product constituents as well as the characteristics evaluated for the product as a whole. Dashed lines represent the fitting of a zero-order kinetic model where applicable.

**Author Contributions:** Conceptualization, P.T.; methodology, A.K., G.D. and E.D.; validation, A.K., G.D. and E.D.; investigation, A.K. and G.D.; data curation, A.K. and G.D.; writing—original draft preparation, A.K. and G.D.; writing—review and editing, E.D.; visualization, A.K.; supervision, E.D. and P.T.; project administration, P.T. All authors have read and agreed to the published version of the manuscript.

**Funding:** This research has received funding from the General Secretariat for Research and Innovation of the Ministry of Development and Investments under the PRIMA Programme (Project FRUALGAE Code: PRIMA2019-03). PRIMA is an Art.185 initiative supported and co-funded under Horizon 2020, the European Union's Programme for Research and Innovation.

**Institutional Review Board Statement:** Not applicable.

**Informed Consent Statement:** Not applicable.

**Data Availability Statement:** The data presented in this study are available on request from the corresponding author.

**Conflicts of Interest:** The authors declare no conflicts of interest.

## References


1. Ahmed, I.; Qazi, I.M.; Jamal, S. Developments in osmotic dehydration technique for the preservation of fruits and vegetables. *Innov. Food Sci. Emerg. Technol.* **2016**, *34*, 29–43. [CrossRef]
2. Ciurzyńska, A.; Kowalska, H.; Czajkowska, K.; Lenart, A. Osmotic dehydration in production of sustainable and healthy food. *Trends Food Sci. Technol.* **2016**, *50*, 186–192. [CrossRef]
3. Asghari, A.; Assana Zongo, P.; Osse, E.F.; Aghajanzadeh, S.; Raghavan, V.; Khalloufi, S. Review of osmotic dehydration: Promising technologies for enhancing products' attributes, opportunities, and challenges for the food industries. *Compr. Rev. Food Sci. Food Saf.* **2024**, *23*, e13346. [CrossRef] [PubMed]
4. Artés, F.; Allende, A. Processing lines and alternative preservation techniques to prolong the shelf-life of minimally fresh processed leafy vegetables. *Eur. J. Hortic. Sci.* **2005**, *70*, 231–245.
5. Lončar, B.; Nićetin, M.; Filipović, V.; Knežević, V.; Pezo, L.; Šuput, D.; Kuljanin, T. Osmotic dehydration in sugarbeet molasses-Food safety and quality benefits. *J. Hyg. Eng. Des.* **2021**, *34*, 16–20.
6. Nabnean, S.; Janjai, S.; Thepa, S.; Sudaprasert, K.; Songprakorp, R.; Bala, B.K. Experimental performance of a new design of solar dryer for drying osmotically dehydrated cherry tomatoes. *Renew. Energy* **2016**, *94*, 147–156. [CrossRef]
7. Nicola, S.; Fontana, E. Chapter 9—Fresh-cut produce quality: Implications for a systems approach. In *Postharvest Handling: A Systems Approach*, 3rd ed.; Florkowski, W., Shewfelt, R., Brueckner, B., Prussia, S., Eds.; Academic Press: Cambridge, MA, USA, 2014.
8. Tang, T.; Zhang, M.; Lim Law, C.; Mujumdar, A.S. Novel strategies for controlling nitrite content in prepared dishes: Current status, potential benefits, limitations and future challenges. *Food Res. Int.* **2023**, *170*, 112984. [CrossRef]
9. Lorente-Mento, J.M.; Serrano, V.M.; Pretel, M.T. Fresh-cut salads: Consumer acceptance and quality parameter evolution during storage in domestic refrigerators. *Sustainability* **2022**, *14*, 3473. [CrossRef]
10. Global Market Insights. Processed Fruits and Vegetables Market to Hit $392 Billion by 2025: Global Market Insights, Inc. Selbyville, Delaware, April 15, 2019 (GLOBE NEWSWIRE). 2019. Available online: https://www.globenewswire.com/news-release/2019/04/15/1803744/0/en/Processed-Fruits-and-Vegetables-Market-to-hit-392-billion-by-2025-Global-Market-Insights-Inc.html (accessed on 24 June 2024).
11. Romano, R.; Pizzolongo, F.; De Luca, L.; Cozzolino, E.; Rippa, M.; Ottaiano, L.; Mormile, P.; Mori, M.; Di Mola, I. Bioactive compounds and antioxidant properties of wild rocket (*Diplotaxis tenuifolia* L.) grown under different plastic films and with different UV-B radiation postharvest treatments. *Foods* **2022**, *11*, 4093. [CrossRef]
12. Santos, J.; Oliveira, M.B.P.P.; Ibanez, E.; Herrero, M. Phenolic profile evolution of different ready-to-eat baby-leaf vegetables during storage. *J. Chromatogr. A* **2014**, *1327*, 118–131. [CrossRef]
13. Saini, R.K.; Ko, E.S.; Keum, Y. Minimally processed ready-to-eat baby-leaf vegetables: Production, processing, storage, microbial safety and nutritional potential. *Food Res. Int.* **2017**, *33*, 644–663. [CrossRef]
14. Krasackoopt, W.; Bhandari, B. Fresh Cut Vegetables. In *Handbook of Vegetables and Vegetable Processing*; Sinha, N.K., Ed.; Blackwell Publishing Ltd.: Hoboken, NJ, USA, 2011; pp. 221–242.
15. Tsironi, T.; Dermesonlouoglou, E.; Giannoglou, M.; Gogou, E.; Katsaros, G.; Taoukis, P. Shelf-life prediction models for ready-to-eat fresh cut salads: Testing in real cold chain. *Int. J. Food Microbiol.* **2017**, *240*, 131–140. [CrossRef] [PubMed]
16. Barry-Ryan, C.; O'Beirne, D. Quality and shelf-life of fresh cut carrot slices as affected by slicing method. *J. Food Sci.* **1998**, *63*, 51–56. [CrossRef]
17. Puerta-Gomez, A.F.; Moreira, R.G.; Kim, J.; Castell-Perez, E. Modeling the growth rates of *Escherichia coli* spp. and *Salmonella* Typhimurium LT2 in baby spinach leaves under slow cooling. *Food Control* **2013**, *29*, 11–17. [CrossRef]
18. Kebede, B.T.; Grauwet, T.; Magpusao, J.; Palmers, S.; Michiels, C.; Hendrickx, M.; Van Loey, A. Chemical changes of thermally sterilized broccoli puree during shelf-life: Investigation of the volatile fraction by fingerprinting-kinetics. *Food Res. Int.* **2015**, *67*, 264–271. [CrossRef]
19. Zhan, L.; Li, Y.; Hu, J.; Pang, L.; Fan, H. Browning inhibition and quality preservation of fresh-cut romaine lettuce exposed to high intensity light. *Innov. Food Sci. Emerg. Technol.* **2012**, *14*, 70–76. [CrossRef]
20. Zilelidou, E.A.; Tsourou, V.; Poimenidou, S.; Loukou, A.; Skandamis, P.N. Modeling transfer of *Escherichia coli* O157:H7 and *Listeria monocytogenes* during preparation of fresh-cut salads: Impact of cutting and shredding practices. *Food Microbiol.* **2015**, *45*, 254–265. [CrossRef]
21. Jacxsens, L.; Castro Ibañez, I.; Gómez-López, V.M.; Araujo Fernandes, J.; Allende, A.; Uyttendaele, M.; Huybrechts, I. Belgian and Spanish consumption data and consumer handling practices for fresh fruits and vegetables useful for further microbiological and chemical exposure assessment. *J. Food Prot.* **2015**, *78*, 784–795. [CrossRef]
22. Tudela, J.A.; Marín, A.; Martínez-Sánchez, A.; Luna, M.C.; Gil, M.I. Preharvest and postharvest factors related to off-odours of fresh-cut iceberg lettuce. *Postharvest Biol. Technol.* **2013**, *86*, 463–471. [CrossRef]
23. Chu, Y.-F.; Sun, J.; Wu, X.; Liu, R.H. Antioxidant and antiproliferative activities of common vegetables. *J. Agric. Food Chem.* **2002**, *50*, 6910–6916. [CrossRef]
24. Grace, M.H.; Hoskin, R.T.; Hayes, M.; Iorizzo, M.; Kay, C.; Ferruzzi, M.G.; Lila, M.A. Spray-dried and freeze-dried protein-spinach particles; effect of drying technique and protein type on the bioaccessibility of carotenoids, chlorophylls, and phenolics. *Food Chem.* **2022**, *388*, 133017. [CrossRef] [PubMed]

25. Dimopoulos, G.; Katsimichas, A.; Balachtsis, K.; Dermesonlouoglou, E.; Taoukis, P. Effect of pulsed electric fields on the shelf stability and sensory acceptability of osmotically dehydrated spinach: A mathematical modeling approach. *Foods* **2024**, *13*, 1410. [CrossRef] [PubMed]
26. Chang, C.-K.; Tsai, S.-Y.; Gavahian, M.; Cheng, K.-C.; Hou, C.-Y.; Yudhistira, B.; Lin, S.-H.; Santoso, S.P.; Hsieh, C.-W. Direct and alternating current electric fields affect pectin esterase and cellulase in tomato (*Solanum lycopersicum* L.) fruit during storage. *Postharvest Biol. Technol.* **2023**, *205*, 112495. [CrossRef]
27. Azumah, L.C.; Abu, M.; Kaburi, S.A.; Lamptey, F.D. Effect of pre-drying treatments on the quality of solar-dehydrated tomato (*Lycopersicon esculentum* L.) fruits. *Appl. Food Res.* **2024**, *4*, 100422. [CrossRef]
28. Dermesonlouoglou, E.K.; Giannakourou, M.C.; Taoukis, P. Stability of dehydrofrozen tomatoes pretreated with alternative osmotic solutes. *J. Food Eng.* **2007**, *78*, 272–280. [CrossRef]
29. Derossi, A.; Severini, C.; De Pilli, T. Study and optimization of osmotic dehydration of cherry tomatoes in complex solution by response surface methodology and desirability approach. *LWT* **2015**, *60*, 641–648. [CrossRef]
30. Heredia, A.; Peinado, I.; Escriche, I. Volatile profile of dehydrated cherry tomato: Influences of osmotic pre-treatment and microwave power. *Food Chem.* **2012**, *130*, 889–895. [CrossRef]
31. Katsimichas, A.; Dimopoulos, G.; Dermesonlouoglou, E.; Taoukis, P. Modelling and evaluation of the effect of pulsed electric fields and high pressure processing conditions on the quality parameters of osmotically dehydrated tomatoes. *Appl. Sci.* **2023**, *13*, 11397. [CrossRef]
32. Abrahão, F.R.; Corrêa, J.L.G. Osmotic dehydration: More than water loss and solid gain, Review. *Crit. Rev. Food Sci. Nutr.* **2023**, *63*, 2970–2989. [CrossRef]
33. Giannoglou, M.; Koumandraki, H.; Andreou, V.; Dermesonlouoglou, E.; Katsaros, G.; Taoukis, P. Combined osmotic and air dehydration for the production of shelf-stable white cheese. *Food Bioprocess. Technol.* **2020**, *13*, 1435–1446. [CrossRef]
34. Barbosa-Cánovas, G.V.; Fontana, A.J., Jr.; Schmidt, S.J.; Labuza, T.P. (Eds.) *Water Activity in Foods: Fundamentals and Applications*; John Wiley & Sons: Hoboken, NJ, USA, 2008.
35. Maroulis, Z.B.; Tsami, E.; Marinos-Kouris, D.; Saravacos, G.D. Application of the GAB model to moisture sorption isotherms for dried fruits. *J. Food Eng.* **1988**, *7*, 63–78. [CrossRef]
36. Vega-Gálvez, A.; Miranda, M.; Clavería, R.; Quispe, I.; Vergara, J.; Uribe, E.; Paez, H.; Di Scala, K. Effect of air temperature on drying kinetics and quality characteristics of osmo-treated jumbo squid (*Dosidicus gigas*). *Food Sci. Technol.* **2011**, *44*, 16–23. [CrossRef]
37. Mooijman, K.A.; Pielaat, A.; Kuijpers, A.F.A. Validation of EN ISO 6579-1—Microbiology of the food chain—Horizontal method for the detection, enumeration and serotyping of *Salmonella*—Part 1 detection of *Salmonella* spp. *Int. J. Food Microbiol.* **2019**, *288*, 3–12. [CrossRef] [PubMed]
38. Rollier, P.; Lombard, B.; Guillier, L.; François, D.; Romero, K.; Pierru, S.; Bouhier, L.; Gnanou Besse, N. Validation of standard method EN ISO 11290—Part 2 for the enumeration of Listeria monocytogenes in food. *Int. J. Food Microbiol.* **2019**, *288*, 22–31. [CrossRef]
39. Gil, M.M.; Brandão, T.R.S.; Silva, C.L.M. A modified Gompertz model to predict microbial inactivation under time-varying temperature conditions. *J. Food Eng.* **2006**, *76*, 89–94. [CrossRef]
40. Akanbi, C.T.; Adeyemi, R.S.; Ojo, A. Drying characteristics and sorption isotherm of tomato slices. *J. Food Eng.* **2006**, *73*, 157–163. [CrossRef]
41. Krokida, M.T.; Karathanos, V.T.; Maroulis, Z.B.; Marinos-Kouris, D. Drying kinetics of some vegetables. *J. Food Eng.* **2003**, *59*, 391–403. [CrossRef]
42. Jasper, J.; Elmore, J.S.; Wagstaff, C. Determining the quality of leafy salads: Past, present and future. *Postharvest Biol. Technol.* **2021**, *180*, 111630. [CrossRef]
43. Bierhals, V.S.; Chiumarelli, M.; Hubinger, M.D. Effect of cassava starch coating onquality andshelf life offresh-cut pineapple (*Ananas comosus* L. merril cv Pérola). *J. Food Sci.* **2011**, *76*, E62–E72. [CrossRef]
44. Moreno, J.; Simpson, R.; Pizarro, N.; Parada, K.; Pinilla, N.; Reyes, J.E.; Almonacid, S. Effect of ohmic heating and vacuum im pregnation on the quality and microbial stability of osmotically dehydrated strawberries (cv. *Camarosa*). *J. Food Eng.* **2012**, *110*, 310–316. [CrossRef]
45. Castelló, M.L.; Igual, M.; Fito, P.J.; Chiralt, A. Influence of osmotic dehydration on texture, respiration and microbial stability of apple slices (var. Granny Smith). *J. Food Eng.* **2009**, *91*, 1–9. [CrossRef]
46. Tiganitas, A.; Zeaki, N.; Gounadaki, A.S.; Drosinos, E.H.; Skandamis, P.N. Study of the effect of lethal and sublethal pH and aw stresses on the inactivation or growth of *Listeria monocytogenes* and *Salmonella* typhimurium. *Int. J. Food Microbiol.* **2009**, *134*, 104–112. [CrossRef] [PubMed]
47. Kim, S.; Jeong, S.H.; Choi, H.S.; Yeo, H.; Lee, D.U. Accelerated brining kinetics and NaCl distribution of Chinese cabbage (*Brassica rapa* ssp. *pekinensis*) using pulsed electric field. *LWT* **2023**, *179*, 114634. [CrossRef]

*applied sciences*

MDPI

*Article*

# Polyphenol-Rich Extracts and Essential Oil from Egyptian Grapefruit Peel as Potential Antioxidant, Antimicrobial, and Anti-Inflammatory Food Additives

**Faten Mohamed Ibrahim [1,*], Eman Abdelsalam [2], Reda Sayed Mohammed [3], Wedian El Sayed Ashour [3], Ana A. Vilas-Boas [4], Manuela Pintado [4,*] and El Sayed El Habbasha [5]**

1 Medicinal and Aromatic Plants Research Department, Pharmaceutical and Drug Industries Research Institute, National Research Centre, Cairo P.O. Box 12622, Egypt
2 Chemistry of Natural and Microbial Products Department, Pharmaceutical and Drug Industries Research Institute, National Research Centre, Cairo P.O. Box 12622, Egypt; dr.emanabdelsalamali@gmail.com
3 Pharmacognosy Department, Pharmaceutical and Drug Industries Research Institute, National Research Centre, Cairo P.O. Box 12622, Egypt; redamohammed2015@gmail.com (R.S.M.); dr.wedian.ashour@gmail.com (W.E.S.A.)
4 CBQF—Centro de Biotecnologia e Química Fina—Laboratório Associado, Escola Superior de Biotecnologia, Universidade Católica Portuguesa, Rua Diogo Botelho 1327, 4169-005 Porto, Portugal; avboas@ucp.pt
5 Field Crops Research Department, National Research Centre, Cairo P.O. Box 12622, Egypt; elsayedelhabasha@yahoo.com
* Correspondence: fatenmibrahim@gmail.com (F.M.I.); mpintado@ucp.pt (M.P.)

**Featured Application: The bio-based food additives obtained from upcycling grapefruit peels can be used mainly in the food industry as natural antioxidant and antimicrobial alternatives. Additionally, these ingredients also show anti-inflammatory potential, underscoring their potential as a nutraceutical.**

**Abstract:** Grapefruit (GF) processing generates significant nutrient and economic losses due to the production of 50% by-products, primarily peels. GF peels are a rich and sustainable source of bioactive compounds (BCs), such as essential oils (EOs) and phenolic compounds. Thus, finding value-added solutions based on a circular economy is paramount. This research aims to assess the antioxidant, anti-inflammatory, and antimicrobial properties of a hydroethanolic polyphenol-rich extract from crude GF peels (GF-CE), essential oil (GF-EO), and polyphenol-rich extract from GF peels after essential oil extraction (GF-PE). The GF-CE and GF-PE showed high concentrations of naringenin (7.71 and 48.60 mg/g dry extract (DE)), narirutin (15.03 and 28.73 mg/g DE), and hesperidin (0.67 and 0.29 mg/mL), respectively. Extracting firstly EOs from GF improved the release of phenolic acids (p-coumaric, ferulic, and chlorogenic acid). The GF-CE exhibited stronger free radical scavenging activity mainly in DPPH ($IC50 = 75.69 \pm 0.81$ μg/mL) than GF-EO ($1271 \pm 0.85$ μg/mL) and GF-PE ($113.45 \pm 0.85$ μg/mL). The GF-EO demonstrated moderate antimicrobial activity against Gram-positive bacteria compared to the reference standard (amoxicillin) and strong activity against the yeast *Candida albicans* (inhibition zone of 16 mm). The major compounds in the GF-EO included D-limonene (25%), nootkatone (24%), and β-pinene (8%). Both polyphenol-rich extracts showed promising activities as COX1 and COX2 inhibitors with IC50 values of $25 \pm 0.1$ and $0.28 \pm 0.00$ μg/mL (compared to celecoxib ($97.5 \pm 0.1$ and $0.31 \pm 0.01$ μg/mL) and indomethacin ($6.25 \pm 0.00$ and $0.52 \pm 0.01$ μg/mL) as the standards), respectively. The study concludes that GF peels are a valuable source of BCs with significant bioactivities, offering a sustainable multi-cascade approach to recovering value-added compounds from GF peels in alignment with circular economy principles and open opportunities as functional ingredients for food applications.

**Keywords:** grapefruit peel; polyphenols; essential oil; circular economy; bioactivities

## 1. Introduction

Citrus fruits are widely consumed worldwide and have received considerable attention recently due to high levels of bioactive compounds (BCs), such as phenolic compounds and essential oils, which showed several human health-promoting benefits [1]. According to recent FAO statistical data, in 2022, the genus Citrus had an annual production of about 166.30 million tons, whereas the worldwide production of grapefruits (*Citrus paradisi*) was 9.76 million tons during the marketing year 2021/2022 [2,3]. The Mediterranean region is one of the main GF producers (30% of total production), where Turkey, Tunisia, Spain, and Egypt are among the top countries in the ranking [3].

Consumers highly appreciate grapefruit (GF) due to its abundance of nutrients and phytochemicals, which are valuable additions to a nutritious diet. They are consumed either fresh or processed [4]. Nevertheless, the processing results in juice and by-products consisting mainly of peels, pulps, and seeds, accounting for half of the fruit weight [5]. Thereby, GF processing generates a vast amount of waste generally used in composting and animal feed. In the worst cases, it is burned or dumped in landfills, which results in serious environmental issues and financial losses for the companies [6]. To overcome these unsustainable actions, a great industrial and scientific interest has risen to upcycle these side streams into high-value-added products [7].

Peels remain the primary by-product of GF processing and contain several phytochemicals, such as flavonoids, phenolic acids, vitamin C, carotenoids, and terpenes [8]. Generally, these molecules are present in higher amounts in the peels than other fruit parts [9]. Polyphenols have attracted much attention due to their use in food industries since recent studies have highlighted citrus peel extracts' superior antioxidant capacity compared to synthetic antioxidants, along with their potent inhibitory effects on lipid oxidation and rancidity [10]. Moreover, these BCs exhibited promising antimicrobial properties against several potent foodborne pathogens, although their exact mechanism of action is yet to be fully elucidated [11]. Recently, these phytochemicals have been highly sought after by the nutraceutical industry due to their potential benefits against various oxidative stress-related disorders [12]. Over the last decade, several review studies have shown the positive effect of a high dietary intake of GF and other citrus polyphenols against inflammation processes, obesity, diabetes, neurodegenerative diseases, and cancer [8,12–15].

Additionally, citrus polyphenols have been strongly associated with cardioprotective effects [16,17]. Naringin, its aglycone, and naringenin are GF's most abundant phenolic compounds [18]. For instance, naringenin has shown more effectiveness than vitamin C in inhibiting lipopolysaccharide (LPS)-induced inflammatory responses in macrophages, including a reduced nitrite production and the suppression of cyclooxygenase-2 (COX-2) and inducible nitric oxide synthase (iNOS) expression [19]. In addition to phenolic compounds, for a very long time, EOs have been used in the food processing, preservation, and flavoring industries for multiple purposes. The *citrus* genus has garnered great attention due to its abundance of EOs with high antibacterial, antifungal, and insecticidal properties [20]. However, the EOs extracted from GF peels also showed a great antioxidant capacity [21], and recently, a study from Nikolic et al. [22] suggested that EO extracted from GF peels showed a protective action against LPS-induced inflammation by attenuating the gene expression and concentrations of pro-inflammatory interleukin-6 (IL-6) and tumor necrosis factor-alpha (TNF-$\alpha$) cytokines. The main advantage of EOs compared to polyphenol-rich extracts is their recognized interest in food applications as it has been used since ancient times. Nevertheless, EOs possess certain limitations primarily associated with the dosage and administration methods, which lack uniformity [23].

In accordance with emerging trends, there is a noticeable shift in consumer attitudes towards a heightened interest in natural food products free from synthetic additives. Concurrently, consumers are increasingly seeking food options that align with the principles of health, sustainability, and social responsibility [24]. As previously reported, the EOs and polyphenol-rich extracts from GF have shown impressive positive health benefits and interesting functional properties, making them desirable as a natural alternative for the food industry [25]. These options enhance the shelf-life of food. For instance, an edible coating of alginate/chitosan enriched with a polyphenol-rich extract from GF seeds reduced the off-flavor of shrimp and the bacterial count by 2 log CFU during the storage time (15 days under refrigeration (4 °C)), prolonging the shelf-life of shrimp [26]. Another study from Durmus et al. [27] showed that nanoemulsions based on GF-EO increased, in 6 days (compared to the control), the shelf-life of rainbow trout filets. In addition, these bioactive extracts obtained from GF by-products can also reduce the risk of microbial contamination and eliminate the need for artificial additives [28]. On the other hand, the BCs present in GF extracts/EOs prevent and/or improve human health and enhance the food's nutritional value [29]. Over the past decade, there has been a notable increase in the accessibility of natural food additives, functional foods, nutraceuticals, and supplements in the market. Projections anticipate that this industry will expand to approximately USD 210 billion by 2026 [30]. Therefore, strategies to reduce food waste by upcycling them to produce new food ingredients/extracts are paramount [31]. This aligns with multiple Sustainable Development Goals (SDGs), contributing to the global agenda for sustainable development and addressing interconnected challenges, such as hunger, environmental degradation, and economic inequality. For instance, by upcycling GF peels into new extracts/ingredients, we contribute to reducing hunger and ensuring access to nutritious food for all (SDG 2), while enhancing human well-being with bioactive ingredients in the upcycled food (SDG 3) and promoting sustainable consumption and production patterns (SDG 12). In addition, food waste generates significant greenhouse gas emissions when it decomposes in landfills. Therefore, using GF by-products to produce new value-added ingredients will mitigate climate change (SDG 13), while contributing to the preservation of terrestrial ecosystems (SDG 15) [32,33]. Promoting the integration of upcycling practices within food systems holds significant promise for enhancing circularity within the food industry. Exploring this potential has emerged as a prominent topic of interest throughout the food value chain.

This research work aims to evaluate the antioxidant, antimicrobial, and anti-inflammatory activity of an EO and polyphenol-rich extracts from GF peels cultivated in Egypt. In addition, a multi-compound cascade approach is employed to collect bioactive extracts from the same by-product, increasing the circular economy value. The antioxidant capacity was evaluated using two standard methods: DPPH and nitric oxide; the antimicrobial activity was tested against both Gram-positive (Gram+) and Gram-negative (Gm ) bacteria as well as fungi; and the anti-inflammatory activities were analyzed by the cyclooxygenase enzyme expression (COX1 and COX2). The chemical composition of the EO and polyphenol-rich extracts was analyzed by GC/MS and LC/MS.

## 2. Materials and Methods

### *2.1. Chemicals*

1,1-Diphenyl-2-picrylhydrazyl (DPPH), ascorbic acid (vitamin C), celecoxib, and the Griess reagent were purchased from Sigma-Aldrich, St. Louis, MO, USA. Sodium nitroprusside (s.d. Fine-Chem. Ltd., Mumbai, India), indomethacin, and acetylsalicylic acid were purchased from (The Arab Drug Company (ADCO), Cairo, Egypt). The in vitro anti-inflammatory activity was assessed using an ELISA kit provided by Cayman Chemical Company, Ann Arbor, MI, USA. Gallic acid, Folin–Ciocalteu reagent, aluminum chloride reagent, quercetin, and anhydrous sodium sulphate were purchased from Oxford Lab Chem, Navghar, India.

### *2.2. Plant Material and Extract Preparation*

#### 2.2.1. Plant Material

The fresh fruit of *Citrus paradisi* (GF) was collected from the National Research Centre Farm at the Agricultural Production and Research Station, National Research Centre, El Nubaria Province, El Behira Governorate, in 2021. The farm applies agro-eco-friendly environment procedures, and we collected healthy fruits only. A commercial juice squeezer was used to extract the juice from the fresh GF, which leaves the peels (flavedo + albedo) and some pulp residue. The GF peels are used for this study.

#### 2.2.2. Grapefruit Crude Polyphenol-Rich (GF-CE) Extract

Three kg of fresh GF peels was mixed with ethanol 80% *(v/v)* at a 1:3 *(m/v)* ratio and left for 24 h under agitation to extract the polyphenols. The extraction process was repeated till exhaustion. After that, the extracts were combined and filtered and then dried at 45 °C under vacuum in a rotary evaporator. The final polyphenol-rich extract (GF-CE) was stored at 4 °C in a dark-colored bottle during the analysis time (Figure 1).

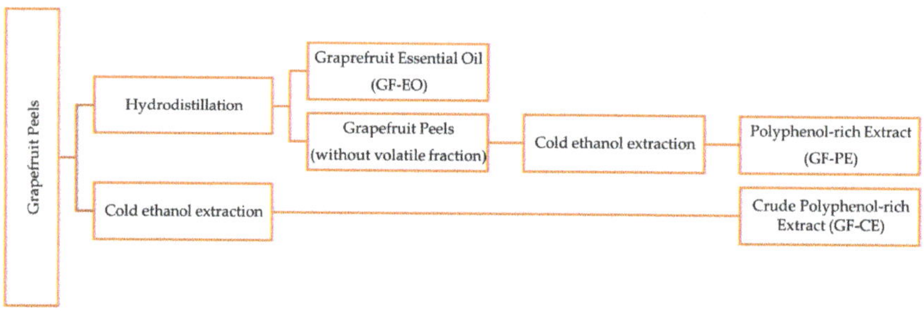


**Figure 1.** Diagram of the essential oil (GF-EO) and polyphenol-rich extract (GF-CE and GF-PE) extraction.

#### 2.2.3. Grapefruit Essential Oil (GF-EO) and Grapefruit Polyphenol-Rich Extract (GF-PE) Extraction

To perform an integrated extraction approach to increase the economic value of GF peel valorization, two different BC extracts were obtained sequentially from the same by-product (Figure 1). Firstly, the EO was extracted from two kg of fresh peels through hydrodistillation using a Clevenger-type apparatus for 3 h, as mentioned in Egyptian pharmacopeia. The obtained GF-EO was desiccated using anhydrous sodium sulfate and stored in a freezer at −20 °C for subsequent analysis. The GF peels remaining from hydrodistillation were used to extract the polyphenols (GF-PE). The extraction process was the same one as that used for GF-CE, as described in Section 2.2.2.

### 2.3. *Phytochemical Analysis*

#### 2.3.1. Essential Oil (GF-EO) Composition

The analysis of the GF-EO compounds was conducted through GC-MS using a Thermo Scientific Trace GC Ultra/ISQ Single Quadrupole MS system, equipped with a TG-5MS fused silica capillary column (30 m, 0.251 mm, and 0.1 mm film thickness) from Shimadzu Corporation, Kyoto, Japan. The GC-MS detection employed an electron ionization system with an ionization energy of 70 eV, with helium gas serving as the carrier gas at a constant flow rate of 1 mL/min. The injector and MS transfer line temperatures were maintained at 280 °C. The oven temperature program was initiated at 40 °C (held for 3 min) and increased gradually to 280 °C at a rate of 5 °C/min (held for 5 min). The tentative identification of compounds was accomplished by comparing their relative retention times and mass spectra with data from the NIST and WILLY libraries as well as the published literature (Adam, 2009 [34]). The quantification of all identified compounds was determined using relative percentage peak area calculations.

#### 2.3.2. Total Phenolic Content

The total phenolic content (TPC) for the EO and polyphenol-rich extracts was determined using the Folin–Ciocalteu reagent following the method of Farid et al. [35]. Briefly, 1 mL of GF extracts or standard solution was added to 10 mL of deionized water and 1.0 mL of Folin–Ciocaulteu phenol reagent. After 5 min of reaction, 2.0 mL of sodium carbonate

(20% ($m/v$)) was added to the mixture. Following a full hour in complete darkness, the absorbance was measured at 750 nm using a spectrophotometer. Gallic acid was used as a standard for the calibration curve. The results were calculated according to Equation (1) and are expressed as milligrams equivalent of gallic acid per gram of dry weight extract (mg GAE/g DW).

$$\text{TPC (mg GAE/g DE)} = \frac{C_{gallic\ acid} \times V \times m}{M} \quad (1)$$

$C_{gallic\ acid}$ is the standard (gallic acid) concentration established from the calibration curve; $V$ is the dilution factor; $m$ is the total extract weight; and $M$ is the DW extract concentration.

#### 2.3.3. Total Flavonoid Content

The total flavonoid content (TFC) for the polyphenol-rich extracts (GF-CE and GF-PE) was measured using the aluminum chloride reagent, as described by Farid et al. [35]. The extracts (1.0 mL), previously diluted, were mixed with 0.7 mL of $NaNO_2$ 5% ($m/v$) and 10.0 mL of ethanol 30% ($v/v$) for 5 min. Then, 0.7 mL of $AlCl_3$ 10% ($m/v$) was added and mixed altogether. After 6 min of reaction, 5.0 mL of NaOH (1 M) was added. Finally, the solution was diluted to 25 mL using ethanol 30% ($v/v$). After standing for 10 min, the absorbance of the reaction mixture was measured at 430 nm with a spectrophotometer. Quercetin (dissolved in ethanol) was used as a standard for the calibration curve. The results were calculated according to Equation (2) and are expressed as milligrams equivalent of quercetin per gram dry weight extract (mg QE/g DW).

$$\text{TFC (mg QE/g DE)} = \frac{C_{quercetin} \times V}{m} \quad (2)$$

$C_{quercetin}$ is the standard (quercetin) concentration established from the calibration curve; $V$ is the volume of extract (mL); and $m$ is the total extract weight (g).

#### Phenolic Compound Identification and Quantification

The analysis of the phenolic compounds in the polyphenol-rich extracts (GF-CE and GF-PE) was conducted using liquid chromatography–electrospray ionization–tandem mass spectrometry (LC-ESI-MS/MS) with an ExionLC ™ AC system (Sciex, Framingham, MA, USA) for separation and SCIEX Triple Quad 5500+ (SCIEX, USA) equipped with electrospray ionization (ESI). Separation was achieved using a ZORBAX Eclipse Plus $C_{18}$ Column (4.6 mm × 100 mm, 1.8 μm) (Agilent, Santa Clara, CA, USA). The mobile phases consisted of (A) 0.1% aqueous formic acid and (B) 100% acetonitrile, with the following gradient elution conditions: 2% B from 0–1 min, 2–60% B from 1–21 min, 60% B from 21–25 min, and returning to 2% B from 25.01–28 min. The flow rate was set at 0.8 mL/min, with an injection volume of 3 μL. The multiple reaction monitoring (MRM) analysis of the selected phenolic compounds was conducted in the negative ionization mode, with the following parameters: curtain gas at 25 psi, ion spray voltage at 4500 V, source temperature at 400 °C, ion source gases 1 and 2 at 55 psi with a declustering potential at 50, collision energy at 25, and collision energy spread at 10.

### 2.4. *Antioxidant Activity*

#### 2.4.1. DPPH Radical Scavenging Assay

The DPPH assay was conducted following the procedure outlined by Ibrahim et al. [36]. Firstly, the GF extracts and vitamin C were prepared in methanol at different concentrations (31.25–2000 μg/mL). After that, 1.0 mL of DPPH methanolic solution (0.1 mM) was added to the sample (3.0 mL). Subsequently, the mixture was vigorously shaken and left to incubate in darkness at room temperature for 30 min. A control sample was created following the same procedure, with methanol replacing the sample. Absorbance was measured at 517 nm using a spectrophotometer, and the DPPH radical scavenging activity was determined using Equation (3). A linear modeling of the data obtained were performed and the results expressed as the concentration required to scavenge 50% of the initial DPPH radicals ($IC_{50}$). The lower the $IC_{50}$ value, the more powerful is the extract at scavenging DPPH and this implies a higher antioxidant activity.

2.4.2. Nitric Oxide Radical Scavenging Assay

The assay operates on the principle of inducing the release of nitric oxide (NO) free radicals from sodium nitroprusside (SNP) within an aqueous solution. At physiological pH, SNP undergoes a transformation, yielding nitrite ions that are detectable using the Griess reagent, composed of 1% sulfanilamide in 5% ortho-phosphoric acid ($H_3PO_4$) and 0.1% naphthylethylene diamine dihydrochloride [36]. The method was carried out as described by Ibrahim et al. (2021). Briefly, 2 mL of sample and 2 mL of SNP (10 mM) in phosphate-buffered saline at pH 7.4 were left to react at 25 °C for 150 min. After the incubation, 1 mL of the reaction mixtures were removed and diluted with 1 mL of Griess reagent. The absorbance of these solutions was measured at 540 nm against the corresponding blank solution. The NO radical scavenging activity was calculated following Equation (3). The linear modeling of the data obtained was performed, and the results are expressed as the concentration required to scavenge 50% of the initial DPPH radicals ($IC_{50}$).

$$\text{Inhibition } (\%) = \frac{Abs_{A0} - Abs_{sample}}{Abs_{A0}} \times 100 \quad (3)$$

$Abs_{A0}$ is the absorbance of the control and $Abs_{sample}$ is the absorbance of the treated sample with the extract at different concentrations.

## 2.5. *Antimicrobial Activity*

2.5.1. GF-EO Evaluation

Qualitative evaluations were conducted on nutrient agar plates, following the methodology of Mostafa et al. [37]. The pathogenic microorganisms used in this study included Gram+ bacteria (*Bacillus cereus* (ATCC 6629), *Micrococcus leutus* (ATCC 10240), *Staphylococcus aureus* (ATCC 6538), and *Staphylococus epidermidis* (ATCC 12228)), Gram− bacteria (*Escherichia coli* (ATCC 25922), *Salmonella enterica* (ATCC 255566), and *Pseudomonas aeruginosa* (ATCC 27853)), and the pathogenic yeast *Candida albicans* (ATCC 10231). These microorganisms were obtained from fresh overnight broth cultures grown in a nutrient broth medium and incubated at 37 °C [38]. The inoculum size of this pathogenic strain was prepared and adjusted to approximately 0.5 McFarland standard ($1.5 \times 10^8$ CFU /mL) [39]. Subsequently, 25.0 μL of the microorganism inoculum was added to each plate containing 20.0 mL of sterile nutrient agar medium. Then, in the cooled and solidified agar medium, a sample of GF-EO (10 μg/mL) was added to a 0.6 cm well previously created in the agar plate using a 6.0 cm corn borer, following the well diffusion method. These plates were then refrigerated for one hour to enhance the sample diffusion, followed by incubation at 37 °C for 24 h. The zones of inhibition were measured in millimeters (mm).

2.5.2. GF Polyphenol-Rich Extracts

The antimicrobial activities of the GF polyphenol-rich extracts were assessed using Kirby–Bauer disc diffusion method [40]. Six pathogenic microorganisms were used: *Escherichia coli* (ATCC 25922), *Pseudomonas aeruginosa* (ATCC 10145), and *Salmonella enterica typhimurium* (ATCC 14028) as Gram− bacteria, and *Listeria monocytogenes* (ATCC 35152) and *Staphylococcus aureus* (ATCC 43300) as Gram+ bacteria; and *Candida albicans* (ATCC 10231) as the fungal model was used. The pathogenic microorganisms stored at −20 °C were inoculated in brain–heart infusion broth tubes (BBL, Viersen, Germany). These tubes were then placed in an incubator at 37 °C for 24 to 48 h. Subsequently, the refreshed pathogenic microorganism cultures were evenly spread on the surface of Müller–Hinton agar (MHA) plates (BBL, Germany) using sterile swabs. A sterile disc with a diameter of 6 mm was immersed in the GF polyphenol-rich extracts (GF-CE and GF-PE) at a concentration of 100 mg/mL. Subsequently, the discs were placed on the surface of the MHA plates. The incubation of the plates occurred for 24 h at 37 °C. The diameters of the inhibition zones were measured in millimeters (mm).

### 2.6. *Anti-Inflammatory Activity of GF-EO, GF-CE, and GF-PE*

The evaluation of the cyclooxygenase (COX1 and COX2) inhibition efficacy in the GF polyphenol-rich extracts and GF-EO was conducted following the procedures outlined by Blobaum and Marnett et al. [41]. Indomethacin and celecoxib served as standards for the anti-inflammatory activity against COX1 and COX2, respectively.

### 2.7. *Statistical Analysis*

The results were presented as the mean ± standard deviation. The normality of data distribution was confirmed using the Shapiro-Wilk test ($p < 0.05$). Differences between the mean values were assessed through a one-way analysis of variance (ANOVA), in which the rejection of the null hypothesis (H0) (indicating the equality of means) occurred when $p < 0.05$. Upon the rejection of H0, multiple comparisons were conducted using Duncan's post hoc test. Statistical analysis was performed using the Statistical Package for the Social Sciences (IBM SPSS version 16, Armonk, NY, USA).

## 3. Results and Discussion

### 3.1. *Grapefruit Essential Oil Composition*

The GC-MS analysis of the volatile compounds of the GF-EO is presented in Figure 2. It was possible to identify 35 compounds, accounting for 97.11% of the total peak area (Table 1). The major volatile compounds were identified as D-limonene (24.90%), nootkatone (24.33%), β-pinene (7.71%), γ-Terpinene (7.52%), trans-caryophyllene (4.40%), and α-pinene (3.98%). Nootkatone was the main oxygenated sesquiterpene present in the GF-EO, while limonene, β-pinene, γ-Terpinene, and α-pinene were the main non-oxygenated monoterpenes present. D-limonene was the main prominent compound, which is also reported by other authors [21,42].

D-limonene is the principal component of citrus EOs (orange, lemon, mandarin, lime, and GF) [43]. However, these studies reported a higher quantity of D-limonene, at about 70%. However, a recent study from Ahmed et al. [44] reported that EOs extracted from GF peels cultivated in India, depending on the variety, had 1 15% of limonene. These differences observed between studies in the compound % may be explained by the geographical area of GF cultivation, mainly by the edaphoclimatic factors. Daily sun exposure, air humidity, and soil microbiota are the parameters mostly responsible for the changes observed in secondary metabolites [45]. Additionally, the maturity of the fruit, the harvest season, and the condition of the extraction method can impact the concentrations of the GF-EO. For instance, in Turkish GF-EO, the main compounds present were shown to be the same as those reported in our study [46,47], especially limonene, β-pinene, α-pinene, and nootkatone. However, opposite to our results, in the Turkish GF-EO study, a high concentration of myrcene and sabinene was found.

D-Limonene showed several biological effects, including antioxidant, anti-inflammatory, anticancer activity, and immune modulatory effects [48]. For instance, the study in [49] showed the antioxidant and anti-lipid peroxidation activities of D-limonene (100 mg/kg body weight) for Streptozotocin-induced alterations in male Albino Wistar rats after 45 days of administration. Overall, D-limonene significantly decreased ($p < 0.05$) the thiobarbituric acid reactive substances, lipid hydroperoxides, and conjugate dienes in the plasma, liver, and kidney and improved ($p < 0.05$) the tissue antioxidants, namely superoxide dismutase, catalase, and vitamin C. In addition, at a concentration of 0.04%, D-limonene significantly ($p < 0.01$) decreased nitrite levels and reduced LPS-induced PGE2 production in the murine macrophage cell line RAW 264.7 [50]. Moreover, the D-limonene treatment led to an enhanced protein expression of TNF-α (increased by 30%), IL-1β (increased by 50%), and IL-6 (increased by 20%) compared to the LPS-induced cells left untreated.

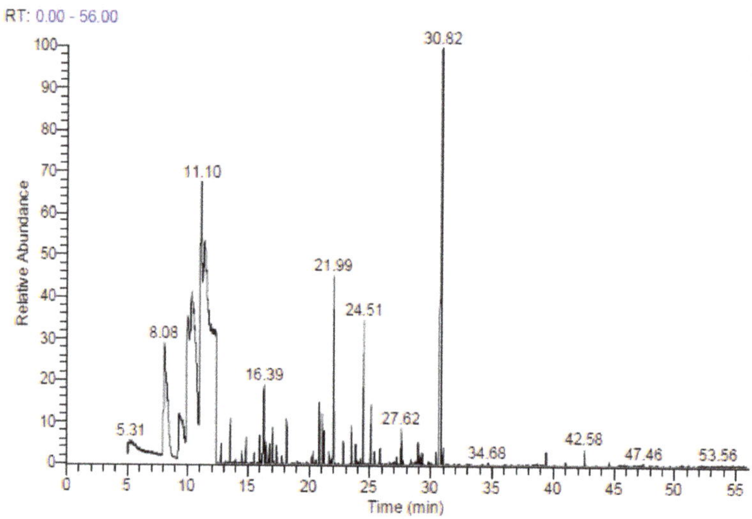


**Figure 2.** GC-MS chromatogram of the essential oil extracted from grapefruit (GF-EO).

**Table 1.** Composition of essential oils from grapefruit peels.

| No. | Retention Time | Molecular Weight ($M^+$) | Base Peak | Molecular Formula | Compound | Area (%) |
|---|---|---|---|---|---|---|
| 1 | 8.07 | 136 | 93 | $C_{10}H_{16}$ | α-Pinene | 3.98 |
| 2 | 8.33 | 136 | 93 | $C_{10}H_{16}$ | Camphene | 0.44 |
| 3 | 9.33 | 136 | 93 | $C_{10}H_{16}$ | β-Pinene | 7.71 |
| 4 | 10.37 | 128 | 83 | $C_8H_{16}O$ | 1-Octanal | 3.40 |
| 5 | 10.42 | 136 | 93 | $C_{10}H_{16}$ | D-Limonene | 24.90 |
| 6 | 12.32 | 136 | 93 | $C_{10}H_{16}$ | γ-Terpinene | 7.52 |
| 7 | 13.58 | 154 | 71 | $C_{10}H_{18}O$ | Linalool | 1.10 |
| 8 | 14.49 | 152 | 67 | $C_{10}H_{16}O$ | *Trans*-Limonene oxide | 0.29 |
| 9 | 14.91 | 154 | 69 | $C_10H_{18}O$ | Citronellal | 0.61 |
| 10 | 15.59 | 154 | 71 | $C_{10}H_{18}O$ | Terpinen-4-ol | 0.26 |
| 11 | 16.02 | 154 | 71 | $C_{10}H_{18}O$ | α-Terpineol | 0.81 |
| 12 | 16.38 | 158 | 67 | $C_{10}H_{22}O$ | Dihydro-citronellol | 2.65 |
| 13 | 16.85 | 152 | 109 | $C_{10}H_{16}O$ | Caveol | 0.41 |
| 14 | 17.12 | 184 | 69 | $C_{11}H_{20}O_2$ | Citronellyl formate | 0.93 |
| 15 | 17.38 | 152 | 69 | $C_{10}H_{16}O$ | Neral | 0.36 |
| 16 | 17.47 | 150 | 82 | $C_{10}H_{14}O$ | Carvone | 0.15 |
| 17 | 17.82 | 154 | 69 | $C_{10}H_{18}O$ | Geraniol | 0.15 |
| 18 | 18.19 | 152 | 69 | $C_{10}H_{16}O$ | Geranial (E-citral) | 1.69 |
| 19 | 20.40 | 194 | 109 | $C_{12}H_{18}O_2$ | Carvyl acetate | 0.24 |
| 20 | 20.86 | 204 | 161 | $C_{15}H_{24}$ | α-Copaene | 1.12 |
| 21 | 21.10 | 196 | 69 | $C_{12}H_{20}O_2$ | Geranyl acetate | 0.88 |
| 22 | 21.23 | 204 | 161 | $C_{15}H_{24}$ | β-Copaene | 0.86 |
| 23 | 21.99 | 204 | 93 | $C_{15}H_{24}$ | *Trans*-carophyllene | 4.40 |
| 24 | 22.80 | 204 | 93 | $C_{15}H_{24}$ | α-Humulene | 0.49 |
| 25 | 23.47 | 204 | 161 | $C_{15}H_{24}$ | Germacrene-D | 0.69 |
| 26 | 23.86 | 204 | 121 | $C_{15}H_{24}$ | Bicyclogermacrene | 0.36 |
| 27 | 24.51 | 204 | 161 | $C_{15}H_{24}$ | α-Amorphene | 2.78 |
| 28 | 25.13 | 204 | 161 | $C_{15}H_{24}$ | Elemol | 1.06 |
| 29 | 25.88 | 220 | 93 | $C_{15}H_{24}O$ | Caryophllene oxide | 0.30 |
| 30 | 27.52 | 222 | 121 | $C_{15}H_{26}O$ | α-Cadinol | 0.32 |
| 31 | 27.63 | 222 | 186 | $C_{15}H_{26}O$ | Eudesm-7(11)en-4-ol | 0.63 |
| 32 | 28.97 | 222 | 69 | $C_{15}H_{26}O$ | Farnesol | 0.65 |

**Table 1.** *Cont.*

| No. | Retention Time | Molecular Weight ($M^{+}$) | Base Peak | Molecular Formula | Compound | Area (%) |
|---|---|---|---|---|---|---|
| 33 | 29.32 | 206 | 135 | $C_{15}H_{26}$ | Nootkatol | 0.29 |
| 34 | 30.41 | 218 | 135 | $C_{15}H_{22}O$ | Curcuphenol | 0.35 |
| 35 | 30.38 | 218 | 146 | $C_{15}H_{22}O$ | Nootkatone | 24.33 |
| | | Total Area of the Identified Compounds | | | | 97.11 |

Other present compounds, such as nootkatone, are noteworthy, with nootkatone being the most important aromatic organic compound in EO extracted from GF peels [51]. This compound showed, in a mice model, an anti-inflammatory effect, mainly associated with the inhibition of IL1-β and TNF-α production, possibly due to the inhibition of COX-2 activity and antagonism of the histamine receptor type 1 [52]. In addition, this compound showed other beneficial activities, including antimicrobial, antioxidant, cardioprotective, and neuroprotective [53]. Furthermore, the GF-EO and its bioactive properties have scented flavors [54], making them an excellent flavoring agent for different industries. For instance, the cosmetics industry uses EOs for manufacturing gels, shampoos, and creams; in the food industry, they are used as natural flavoring agents to replace the use of synthetic additives [55].

### *3.2. Grapefruit Polyphenol-Rich Extract Composition*

#### 3.2.1. TFC and TPC

Phenolic compounds, found abundantly in plants and fruits like those of the Citrus genus, are recognized for their antioxidant and antimicrobial activities. Recent research has also associated them with additional health benefits, including lower blood sugar and cholesterol levels, as well as reduced inflammation [56]. The TPCs of the GF-EO and hydroethanolic extracts (GF-PE and GF-CE) are presented in Table 2. For the EO, the value was 13 mg GAE/g DE, while for the polyphenol-rich extracts GF-PE and GF-CE, the TPCs were 51 and 208 mg GAE/g DE, respectively. The crude polyphenol-rich extract (GF-CE) showed 4-fold higher values than those of the polyphenol-rich extract obtained after oil extraction (GF-PE) ($p < 0.05$). The reduction in the TPC value for the GF-PE may be due to the previously high temperature applied during the hydrodistillation process since high temperatures are the most common explanation for the degradation of polyphenols [57]. However, for the TFC, the opposite results are observed. The GF-PE had 29 ± 0.01 mg QE/g DE, while the GF-CE had only 13 ± 0.06 mg QE/g DE ($p < 0.05$). Flavonoids are sensitive to heat [58]. However, certain extraction methods and parameter combinations may enhance the release or extraction of flavonoids from the plant material because of the breakdown from the cell wall [59].

**Table 2.** Total phenolic content (TPC) and total flavonoid content (TFC) of the grapefruit essential oil (GF-EO) and polyphenol-rich extracts (GF-CE and GF-PE). Values represent the mean ± standard deviation. Different letters mean significant differences between the bioactive extracts ($p < 0.05$).

| | TPC (mg GAE/g DE) | TFC (mg QE/g DE) |
|---|---|---|
| GF-CE | $20.18 \pm 0.02^{a}$ | $13.09 \pm 0.06^{b}$ |
| GF-EO | $13.21 \pm 0.00^{c}$ | -- |
| GF-PE | $51.27 \pm 0.01^{b}$ | $29.31 \pm 0.01^{a}$ |

The TPC and TFC in extracts from GF peels can vary depending on several factors, including the GF variety, ripeness, growing conditions, and the extraction method used. To the best of our knowledge, these TPC and TFC values for the extracts obtained from the fresh GF peels cultivated in Egypt have not been reported in the literature to date. However, for other geographical locations, the value of the TPC of the grapefruit peel extracts can range from approximately 50 to 200 mg GAE/extract [60–63]. For instance, with 80% ethanolic extraction, the TPC was about of 20 mg GAE/g DE using GF peels cultivated in Pakistan [10]. On the other hand, studies from Bagdatli et al. [64] using ultrasound extraction with ethanol 70% (*v*/*v*) demonstrated a TPC of 667 mg GAE/g DE. Another study from Garcia-Castello et al. [62] with GF solid waste from Valencia, Spain reported an optimal TPC extraction of approximately 80 mg/g DE.

Overall, the TPC constituted approximately 20%, 5%, and 1% of the entire extracts of GF-CE, GF-PE, and GF-EO, respectively. Consequently, other compounds, such as sugars, minerals, organic acids, soluble fibers, and proteins, were also extracted, while terpenes were predominant in the GF-EO.

3.2.2. Phenolic Compounds' Identification and Quantification

The qualitative and quantitative analyses of the phenolic compounds in the GF peels' polyphenol-rich extracts are crucial for upcycling food ingredients. Understanding the specific BCs present can offer insights into the extract's bioactivities and elucidate the underlying biological mechanisms involved. The LC/MS chromatograms of each polyphenol-rich extract are presented in Figure 3.

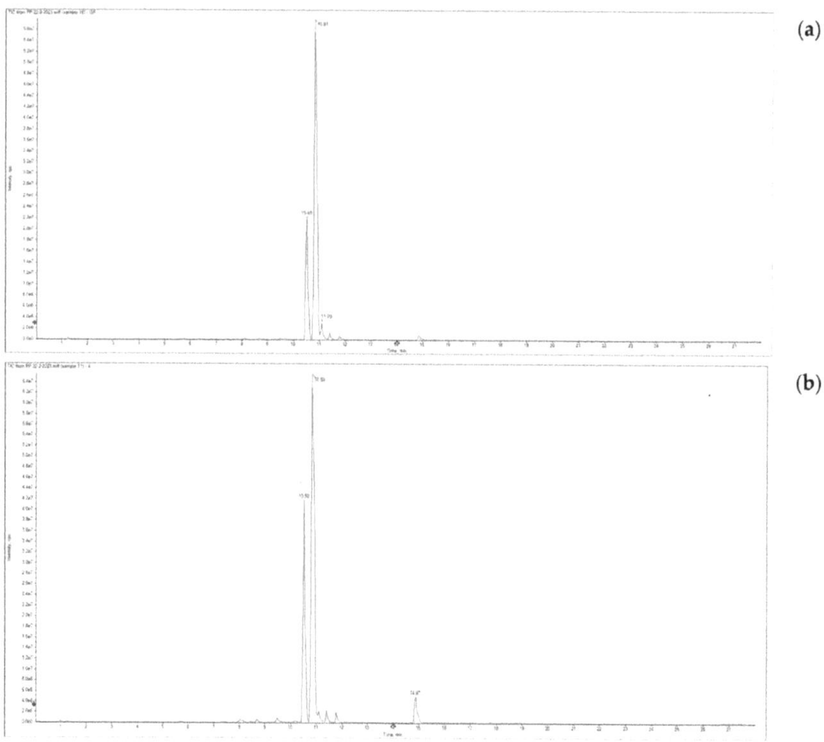


**Figure 3.** LC/MS chromatograms for (**a**) GF-CE and (**b**) GF-PE.

Compared to the commercial standards, sixteen phenolic compounds were identified and quantified in the GF-CE and fourteen in the GF-PE (Table 3). The identification of these compounds led to their distribution into three main structurally related classes, i.e., hydroxycinnamic acids (four compounds), hydroxybenzoic acids (five compounds), and flavonoids (seven compounds).

**Table 3.** Quantification of the phenolic compounds (μg/g DE) in the GF peels' polyphenol-rich extracts (GF-CE and GF-PE). Values represent the mean ± standard deviation. Different letters mean significant differences between the extracts for the same phenolic compound ($p < 0.05$).

| Phenolic Compound | GF-CE | GF-PE |
|---|---|---|
| | Hydroxycinnamic acids | |
| Chlorogenic acid | 3.88 ± 0.21 [a] | 11.22 ± 0.09 [b] |
| Caffeic acid | 7.48 ± 0.40 [b] | 43.83 ± 0.61 [a] |
| p-Coumaric acid | 12.92 ± 0.52 [b] | 52.50 ± 0.45 [a] |
| Ferulic acid | 50.99 ± 1.03 [b] | 227.89 ± 2.15 [a] |
| | Hydroxybenzoic acids | |
| Gallic acid | 2.86 ± 0.01 | n.d. |
| 3.4-Dihydroxybenzoic acid | 11.63 ± 0.18 [b] | 24.52 ± 0.08 [a] |
| Methyl gallate | 0.43 ± 0.00 [a] | 0.12 ± 0.01 [b] |
| Ellagic acid | 2.32 ± 0.00 [b] | 3.15 ± 0.03 [a] |
| Saponarin | 1.29 ± 0.00 | n.d |
| | Flavonoids | |
| Rutin | 6.43 ± 0.24 [a] | 3.71 ± 0.23 [b] |
| Narirutin | 15,026.94 ± 6.05 [b] | 28,729.79 ± 7.41 [a] |
| Hesperidin | 667.73 ± 1.53 [a] | 285.58 ± 1.98 [b] |
| Diosmin | 62.34 ± 0.81 [b] | 80.80 ± 0.78 [a] |
| Quercetin | 0.62 ± 0.01 [b] | 1.14 ± 0.11 [a] |
| Naringenin | 7716.13 ± 2.36 [b] | 48,609.40 ± 9.16 [a] |
| Hesperitin | 3.00 ± 0.02 [b] | 4.08 ± 0.31 [a] |

n.d., not detected.

The presence of such BCs is mainly associated with the antioxidant and antimicrobial activities of GF peels and their derived extracts [65–67]. The most abundant compounds found in GF peels' polyphenol-rich extracts were naringenin, narirutin, and hesperidin, which agrees with the literature available for GF [62,67–69]. The values for naringenin, narirutin, and hesperidin ranged from 7.72 to 48, from 15 to 28.7, and from 0.28 to 0.67 mg/g DE, respectively.

An effective extraction technique is essential to achieve a greater yield of BCs, significantly impacting the composition, yield, and bioactivities of the extract. In the food industry, the preference lies in utilizing non-toxic and easily manageable solvents for the extraction of by-products [70]. Water stands out as the most safe and cost-effective environmentally friendly solvent option [71]; it demonstrates effectiveness in extracting polar molecules, whereas for less polar compounds, organic solvents or combination solvent systems are utilized. However, most of the polyphenols are poorly soluble in water. Flavonoid glycosides are usually more water-soluble than aglycones, which have a better solubility in organic solvents [72]. Therefore, to enhance the extraction, mixtures of water–alcohol are used [73]. Typically, insoluble phenolic compounds, such as flavonoids, are in the cell wall, while soluble phenolic compounds are found inside the cell vacuoles [74]. At high temperatures, the integrity of the cell wall is compromised, and this may promote the delivery of cell wall polysaccharides into the solvent and allow the phenolic compounds to diffuse [75,76]; so, their extractability can also increase. Comparing our results (Table 3) to the reported data, it was observed that the concentration of phenolic acids, mainly hydroxycinnamic acids, was higher in the GF-PE obtained after the extraction of the EO by hydrodistillation during 3 h. This process may weaken the cell wall integrity, resulting in an increase in the contents of low-molecular-weight insoluble phenolics and the flavonoid aglycons when extracted by ethanol (80%) [57]. However, certain flavonoid compounds may undergo degradation when exposed to elevated temperatures. For instance, rutin initiates degradation at 75 °C, with its concentration declining progressively as the temperature rises until it becomes non-detectable at 90 °C [77]. On the other hand, gallic acid exhibits instability at higher temperatures. The degradation rates increase as the temperature increases from 60 °C to 100 °C for 4 h [78]. In our results, it was noticed that the complete decomposition of gallic acid was observed in the GF-PE, which may be due to the previous 3 h in hydrodistillation. This explains the decline in the antioxidant, antimicrobial, and anti-inflammatory activities of the GF-PE. The GF-CE was prepared and evaporated at a temperature not exceeding 45 °C, enabling the stability of heat-sensitive compounds, which allows for the synergistic activity with the flavonoid glycosides and phenolic acids present even at a lower concentration than that of the GF-PE (obtained after EO extraction) to perform higher antioxidant, antimicrobial, and anti-inflammatory activities [79–81]. Furthermore, it is crucial to account for the total activity of polyphenol-rich extracts, influenced by the synergistic or antagonistic interactions among their constituents. Typically, phytochemicals are recognized for conferring human health benefits, including anti-inflammatory, antimicrobial, antihypertensive, and antidiabetic effects [82,83]. In our results, the GF extracts contained various important flavonoids, such as naringin, quercetin, and hesperidin. Naringenin, the main phenolic compound found in the GF peel extract, is a flavanone aglycone renowned for its multifaceted bioactive effects on human health. These effects include antioxidative, anti-inflammatory, antidiabetic, and anti-neurodegenerative properties, as reported in numerous studies [84–86].

### 3.3. *Antioxidant Activity*

Certain EOs have been shown to scavenge free radicals that harm the organism and lower the risk of certain diseases brought on by oxidative stress. The peels of GF have been documented to have many phytochemicals, including flavonoids, carotenoids, vitamin C, and other BCs, which have been reported as antioxidant compounds [87]. These BCs promote human health and delay/prevent the incidence of many chronic diseases [13]. To evaluate the antioxidant capacities of GF bioactive extracts, two different methods were used: DPPH and nitric oxide (NO) assays. The antioxidants react with DPPH, transforming it into 1,1-diphenyl-2-picryl hydrazine by swiftly accepting hydrogen, thereby halting the propagation of free radical oxidation chains. This process leads to the creation of stable end products, preventing subsequent lipid oxidation [36]. In the nitric oxide assay, the capacity of GF peel polyphenol-rich extracts to scavenge nitrogen free radicals was evaluated [88]. The antioxidant activities of the GF-EO and polyphenol-rich extracts (GF-CE and GF-PE) were concentration-dependent, which means that the activity increases as the concentration increases (Tables 4 and 5). Vitamin C, used as a positive control, exhibited the same behavior as the extracts. The $IC_{50}$ for the GF-EO was 1271 ± 0.85 μg/mL for the DPPH assay and 1656 ± 0.71 μg/mL for the NO assay. Compared to vitamin C, the natural GF-EO showed a lower activity ($p < 0.05$); however, compared to the results obtained by Deng et al. [21], the Egyptian GF-EO showed 100 times more activity against DPPH radicals than the GF-EO extracted by cold-pressing. Regarding the polyphenol-rich extracts, the $IC_{50}$ values for the GF-CE were 75.69 ± 0.81 and 113.45 ± 0.71 μg/mL for the DPPH and NO assays, respectively. The $IC_{50}$ values for the GF-PE were much higher for the DPPH (1069 ± 0.56 μg/mL) and for the NO (791 ± 0.52 μg/mL) assays. Meanwhile, the GF-CE showed a higher activity than the EO at the same concentration (1000 μg/mL) for both assays, while the GF-PE showed a higher activity than the GF-EO only for the NO assay. The variations in antioxidant effectiveness between the polyphenol-rich extracts and the EO could also be related to the levels of ascorbic acid, a water-soluble antioxidant renowned for effectively neutralizing reactive oxygen species [89]. Overall, the polyphenol-rich extracts showed a better antioxidant activity than the EO. Naringenin, the main flavonoid present in both extracts, is a powerful antioxidant that can destroy free radicals and attenuate their formation [90]. However, the GF-EO antioxidant activity could be attributed to the high % of limonene [47]. Citrus polyphenols, which are mainly flavonoids, display exceptional health-promoting protection against oxidative stress-related diseases, including antioxidant, anti-inflammatory, antimicrobial, and other activities [13].

**Table 4.** Percentage (%) inhibition of the GF-EO extracted from grapefruit peels in terms of DPPH and nitric oxide free radicals at different concentrations (μg/mL) and their $IC_{50}$ values compared to vitamin C (standard). Values represent the mean ± standard deviation. Different letters mean significant differences between concentrations ($p < 0.05$).

| | DPPH | | Nitric Oxide | |
|---|---|---|---|---|
| Concentration (μg/mL) | GP-EO | Vit. C | GP-EO | Vit. C |
| 250 | 21.8 ± 0.69 e | 23.73 ± 0.53 e | 9.85 ± 0.047 e | 43.32 ± 0.32 e |
| 500 | 35.7 ± 0.37 d | 41.08 ± 0.52 d | 14.63 ± 0.18 d | 63.61± 0.42 d |
| 1000 | 46.7 ± 1.27 c | 64.57 ± 0.59 c | 29.37 ± 0.33 c | 77.83 ± 0.51 c |
| 1500 | 54.8 ± 0.46 b | 81.85 ± 0.46 b | 39.86 ± 0.53 b | 85.25 ± 0.32 b |
| 2000 | 65.8 ± 0.28 a | 96.09 ± 0.14 a | 64.71 ± 0.52 a | 92.94 ± 0.28 a |
| $IC_{50}$ | 1271.24 ± 0.85 b | 734.42 ± 0.43 c | 1656.19 ± 0.71 a | 263.60 ± 0.52 h |

**Table 5.** Percentage inhibition (%) of the polyphenol-rich extracts from GF peels in terms of scavenged DPPH and nitric oxide free radicals at different concentrations (μg/mL) and their $IC_{50}$ values. Vitamin C was used as the standard. Values represent the mean ± standard deviation. Different letters mean significant differences between concentrations ($p < 0.05$).

| | DPPH | | | Nitric Oxide | | |
|---|---|---|---|---|---|---|
| Concentration (μg/mL) | GP-CE | GF-PE | Vit. C | GP-CE | GF-PE | Vit. C |
| 31.25 | 31.20 ± 0.72 f | 2.97 ± 0.82 f | 23.73 ± 0.471 f | 27 ± 1.31 f | 2.60 ± 0.25 e | 43.32 ± 0.32 f |
| 62.5 | 46.06 ± 0.53 e | 13.80 ± 0.76 e | 41.08 ± 0.68 e | 45.33 ± 0.88 e | 9.62 ± 0.71 d | 63.61 ± 0.42 e |
| 125 | 62.23 ± 0.28 d | 21.27 ± 0.44 e | 64.57 ± 0.53 d | 65.67 ± 0.33 d | 10.5 ± 0.40 d | 77.83 ± 0.51 d |
| 250 | 71.13 ± 0.33 c | 25.83 ± 0.75 c | 71.85 ± 0.52 c | 75.33 ± 0.67 c | 28.86 ± 0.52 c | 85.25 ±0.32 c |
| 500 | 88.96 ± 0.52 b | 31.42 ± 1.47 b | 86.09 ± 0.59 b | 84.33 ± 0.33 b | 44.04 ± 0.76 b | 88.12 ± 0.64 b |
| 1000 | 92.26 ± 0.54 a | 42.87 ± 0.62 a | 96.09 ± 46 a | 90 ± 0.58 a | 55.30 ± 0.51 a | 94.89 ± 0.78 a |
| $IC_{50}$ | 75.69 ± 0.81 e | 1069.00 ± 0.56 a | 118.16 ± 0.78 c | 113.45 ± 0.71 d | 791.40 ± 0.52 b | 59.61 ± 0.65 f |

### 3.4. Antimicrobial Activity

Annually, contaminated food results in 600 million illnesses and 420,000 deaths. Moreover, microbiological spoilage stands as the primary contributor to food waste [91]. Hence, novel and efficient strategies are required to inhibit and eradicate contamination. This research focuses on natural food preservatives endowed with antimicrobial attributes poised as safer substitutes for synthetic counterparts [92]. Therefore, the antimicrobial activities of the GF-EO and GF polyphenol-rich extracts were tested against a panel of several Gram+ and Gram− bacteria and a fungus, specifically selected based on their importance to public health.

#### 3.4.1. Grapefruit Essential Oil (GF-EO)

Plant-derived EOs are one group of natural food preservatives. The effectiveness of EOs is associated with the presence of multiple volatile compounds with antimicrobial activity reported since ancient times. This results in the effectiveness and specificity of each EO against various microorganisms. The results of the antimicrobial activity measured by the disc diffusion method for the EO extracted from GF and miconazole (standard) are shown in Figure 4. The GF-EO showed a good antibacterial activity against Gm+ bacteria: *Bacillus cereus, Micrococcus leutus, Staphylococcus epidermidis,* and *Staphylococcus aureus*. The highest antibacterial activity was demonstrated by the EO against *S. aureus* (15 mm), followed by the inhibition zone against *B. cereus* (13 mm). Regarding *S. epidermidis* and *M. leutus,* the inhibition zone was 10 mm. *S. aureus* and *S. epidermidis* contain teichoic acid in the peptidoglycan layer and are inhibited by different citrus peel EOs and extracts [93]. In addition, the GF-EO also had a great inhibition spectrum toward the pathogenic yeast

*Candida albicans* (16 mm), showing a better potential than amoxicillin (standard). For the Gm− bacteria, there was no inhibition.

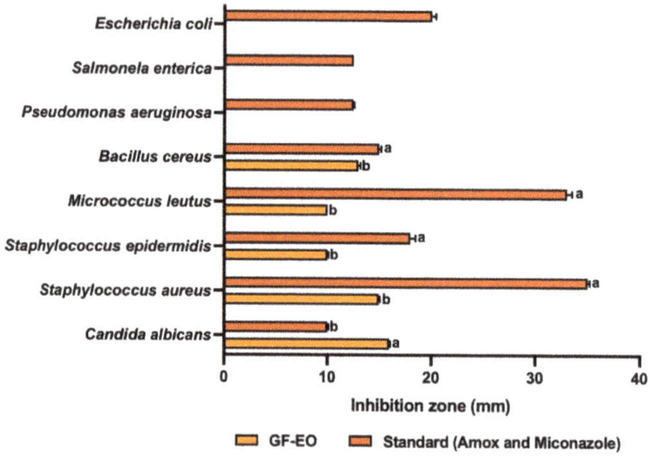


**Figure 4.** Antimicrobial activities of the GF-EO and the standards (amoxicillin and miconazole) determined by the disc diffusion assay. Values represent the mean ± standard deviation. Different letters mean significant differences between the GF-EO and the standard ($p < 0.05$).

Our results align with most of the studies investigating the efficacy of EOs extracted from grapefruit peels against food spoilage organisms and foodborne pathogens. However, some studies reported different findings, mainly regarding Gram− bacteria inhibition. For instance, Deng et al. [94] reported the inhibition for *E. coli* (26.86 mm) and *P. aeruginosa* (8.57 mm). In addition, another recent study from Luciardi et al. [95] demonstrated that the GG-EO at 0.1 mg/mL could not inhibit *P. aeruginosa* growth but inhibited its biofilm production in the range of 52–55%. The differences observed may be related to the method of EO extraction. This study used hydrodistillation, whereas other comparative studies used cold-pressing extraction. However, the previously reported studies agree that EOs are generally slightly more active against Gram+ than Gram− bacteria. Gram+ bacteria typically exhibit a higher susceptibility due to their thick peptidoglycan cell wall layer. Conversely, Gram− bacteria possess a thinner peptidoglycan layer along with an additional outer membrane composed of phospholipids and lipopolysaccharides [96]. The presence of non-oxygenated compounds and oxygenated sesquiterpenes, such as D-Limonene, β-pinene, γ-Terpinene, α-pinene, and nootkatone, contributes significantly to the antimicrobial activity. Although the precise mechanism of action of EOs remains unclear, it is hypothesized that they disrupt cell membrane function and structure, consequently interfering with electron chain transport, enzyme activity, nutrient uptake, and the synthesis of nucleic acids and proteins [80].

3.4.2. Grapefruit Polyphenol-Rich Extracts (GF-CE and GF-PE)

In addition to GF-EOs, extracts rich in phenolic compounds, another group of secondary metabolites in plants, are important BCs due to their bioactive properties. They demonstrate remarkable effectiveness as inhibitors against numerous foodborne pathogenic and spoilage bacteria. Moreover, they are excellent natural alternatives [97]. The results of the antimicrobial activity measured by the disc diffusion method of the polyphenol-rich extracts (GF-CE and GF-PE) are shown in Figure 5. The results show that the GF peel extract has a good activity against both Gm+ and Gm− and is also active against *Candida albicans*. However, the GF-PE shows a better activity than the GF-CE against *P. aeruginosa* (7 mm) and *L. monocytogenes* (7 mm). Although extracts rich in phenolic compounds can effectively inhibit pathogens, they are less potent than EOs.

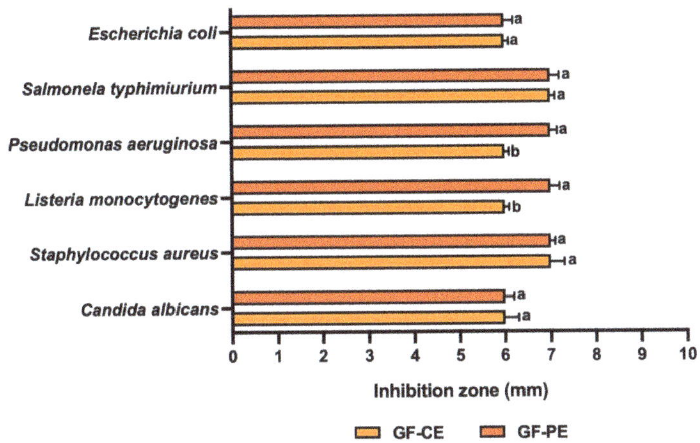


**Figure 5.** Antimicrobial activities of the polyphenol-rich extracts (GF-CE and GF-PE) determined by the disc diffusion assay. Values represent the mean ± standard deviation. Different letters mean significant differences between the polyphenol-rich extracts ($p < 0.05$).

To the best of our knowledge, there is limited literature available on the antimicrobial activity of GF peel extracts rich in polyphenols. Generally, GF polyphenol-rich extracts are obtained from seeds. A recent study from Arsène et al. [98] demonstrated that the antibacterial activity of a hydroethanolic extract (80%, $v/v$) was dose-dependent, and a higher inhibition diameter was observed for the Gm+ bacteria *S. aureus* (13 mm) and the Gm− bacteria *E. coli* (11 mm). Overall, the results are in line with those observed in our study. To the best of our knowledge, the antifungal activity of the polyphenol-rich extracts obtained from GF peels against *Candida albicans* is reported for the first time in this study. However, a study from Yaldiz et al. [99] showed that a treatment with mixtures of polyphenol-rich extracts and EO from GF peels (25:75%) for 8 h had an effective antifungal activity against *C. albicans*. Therefore, EOs are generally considered to be particularly effective in this context. The research indicates that phenolic compounds, such as rutin, quercetin, and naringenin, interact with bacterial cell membranes, increasing permeability, reducing ATP production, binding to metabolic enzymes, and disrupting membrane integrity, ultimately resulting in bacterial cell membrane destruction [100]. Additionally, flavonoids, a large group of phenolic compounds found in GF, demonstrate the ability to inhibit bacterial metabolism and the synthesis of DNA and RNA in bacteria [101,102].

### 3.5. Anti-Inflammatory Activity

The relationship between anti-inflammatory activity and the enzymes COX-1 (cyclooxygenase-1) and COX-2 (cyclooxygenase-2) is closely tied to the inflammatory response and the production of prostaglandins [103]. COX-1 and COX-2 are enzymes involved in synthesizing prostaglandins from arachidonic acid, a fatty acid found in cell membranes. COX1 is constitutively expressed in many tissues and maintains normal physiological functions, such as protecting the stomach lining and regulating blood platelets. COX-1 produces prostaglandins that are involved in these housekeeping functions [103]. Meanwhile, COX2 is typically produced during inflammation and is responsible for producing the prostaglandins that contribute to the inflammatory response. COX-2 is often upregulated in response to various stimuli, including injury or infection [103]. Non-steroidal anti-inflammatory drugs (NSAIDs), including indomethacin and celecoxib, exert their anti-inflammatory effects by suppressing the activity of both COX-1 and COX-2 enzymes [104]. Nevertheless, there is considerable interest in natural extracts or compounds with anti-inflammatory properties that may act as inhibitors of COX-1 and COX-2 [105–107]. One of these groups may be polyphenols, more precisely, flavonoids [108].

Figure 6 shows the anti-inflammatory effects of the GF peel polyphenol-rich extracts (GF-CE and GF-PE) and GF-EO against the biomarkers COX1 and COX2.

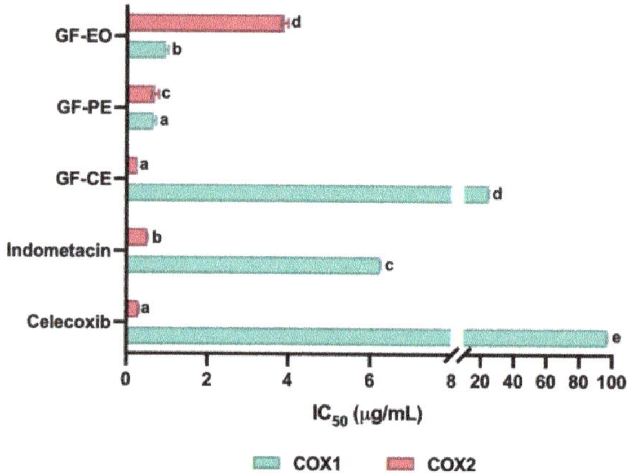


**Figure 6.** Anti-inflammatory activity in vitro for the biomarkers COX1 and COX2 expressed as $IC_{50}$ (ug/mL) for the polyphenol-rich extracts (GF-CE and GF-PE) and essential oil (GF-EO) obtained from grapefruit peels. Indomethacin and celecoxib were used as the standards. Values represent the mean ± standard deviation. Different letters mean significant differences between the bioactive extracts and standards for each cyclooxygenase (same color in the bar).

The GF-EO, GF-CE, and GF-PE exhibited activities as COX1 and COX2 inhibitors, with $IC_{50}$ values of 0.97 ± 0.1, 25 ± 0.1, and 0.67 ± 0.058 ug/mL for COX1 and 3.9 ± 0.1, 0.28 ± 0.0006, and 0.7 ± 0.1 ug/mL for COX2, compared to the reference standards celecoxib and indomethacin, with values of 97.5 ± 0.1 and 6.25 ± 0.002, and 0.31 ± 0.01 and 0.52 ± 0.01, respectively. It was noticed from the results that EO has a weak activity as a COX2 inhibitor and good activity as an inhibitor of COX1, while the GF-CE has selective activity as a COX2 inhibitor and mild inhibitor of COX1, but the GF-PE has a weak activity as a COX2 inhibitor compared to the extract obtained from the crude GF peels and good activity as a COX1 inhibitor. The bioactive extracts obtained from GF peels showed a significantly better activity ($p < 0.05$) than synthetic drugs. Our results are consistent with recent studies, which demonstrate the inhibitory effect of grapefruit peel extracts on the production of inflammatory mediators, including prostaglandin E2 and nitric oxide (NO), in LPS-activated RAW 264.7 cells. Flavanone glycosides have been identified as significant contributors to the anti-inflammatory activity of citrus peels due to their higher abundance compared to polymethoxyflavones [109]. Furthermore, in vivo investigations have indicated that the EO extracted from grapefruit peels exhibits anti-inflammatory properties compared to the negative control group over the entire 5 h post-induction assessment period [110].

The correlation between inhibiting COX-1 and COX-2 and anti-inflammatory activity is rooted in the role of these enzymes in the inflammatory process. Inflammation involves the production of prostaglandins, which COX-1 and COX-2 synthesize. By inhibiting these enzymes, the production of prostaglandins is reduced, leading to a decreased inflammatory response. The GF-PE had a higher content of valuable phytochemicals, such as naringenin, narirutin, hesperidin, ferulic acid, and chlorogenic acid. These BCs exhibit diverse anti-inflammatory properties by inhibiting various pathways, including regulatory enzyme inhibition, the alteration of arachidonic acid metabolism, the modulation of gene expression, and targeting transcription factors crucial for regulating inflammatory mediators [111–113].

Additionally, it can modulate the activity of human macrophages and, in turn, reduce inflammation [114]. Inflammation is a biological response to noxious stimuli, such as pathogens, which cause tissue and cell damage [115]. It may be acute or chronic, depending on the reaction of the body to the stimuli, with an either short or prolonged response.

## 4. Conclusions

The current study presented a comparative study of the valorization of GF peels to obtain a single BCs (EO or polyphenol-rich extracts) or a multi-cascade BCs from the same by-product. A sustainable and integrated approach using Egyptian GF peel, a by-product from the juice industry, by producing consecutive GF-EO and polyphenol-rich extracts (GF-PE) using the same by-product was assessed. The method of extraction and pretreatment applied before the polyphenol extraction significantly influences the content of phenolic compounds and the biological activity of the final ingredient. The TPC and TFC increased 2-fold, while naringenin, narirutin, and hydroxycinnamic acids increased by 6.2-, 1.9-, and 3.0-fold, respectively. However, the crude GF extract (GF-CE) exhibited a stronger free radical scavenging activity, mainly in DPPH ($IC50 = 75.69 \pm 0.81$ μg/mL), than the GF-PE ($113.45 \pm 0.85$ μg/mL). However, the GF-PE showed a higher anti-inflammatory activity than the GF-CE ($0.67 \pm 0.058$ ug/mL for COX1 and $0.7 \pm 0.1$ ug/mL for COX2). Lastly, both the GF-CE and GF-PE showed antimicrobial activity ($p > 0.05$) against Gram− and Gram+ bacteria; however, the GF-PE showed a better inhibitory capacity against *P. aeruginosa* (7 mm) and *L. monocytogenes* (7 mm). Nevertheless, the GF-EO demonstrated a higher antimicrobial activity than the polyphenol-rich extracts, mainly against Gram+ bacteria compared to the reference standard (amoxicillin) and a strong activity against the yeast *Candida albicans* (inhibition zone of 16 mm). The major compounds in the GF-EO included D-limonene (25%), nootkatone (24%), and β-pinene (8%). Although the GF-EO demonstrates anti-inflammatory capacity, its ability to inhibit COX 1 ($0.97 \pm 0.1$ μg/mL) is greater than that to inhibit COX 2 ($3.9 \pm 0.1$ μg/mL).

This study highlights the potential of GF peels as a sustainable source of valuable ingredients/extracts for functional foods and their potential to establish synergies with the cosmetic and pharmaceutical industries. In addition, it proves the efficiency of the consecutive extraction of BCs from GF peels. By using these new upcycled ingredients, we can contribute to the achievement of the SDGs and zero waste, a set of principles focused on waste prevention that encourage finding new uses for resources instead of throwing them away.

**Author Contributions:** Conceptualization, F.M.I., R.S.M. and E.S.E.H.; methodology, F.M.I., R.S.M., E.A. and W.E.S.A.; software, R.S.M.; validation, R.S.M. and F.M.I.; formal analysis, F.M.I., R.S.M. and E.A.; investigation, F.M.I., R.S.M., E.A. and W.E.S.A.; resources, E.S.E.H.; data curation, F.M.I.; writing—original draft preparation, A.A.V.-B., F.M.I., R.S.M. and E.A.; writing—review and editing, A.A.V.-B., F.M.I. and M.P.; project administration, E.S.E.H. All authors have read and agreed to the published version of the manuscript.

**Funding:** This research was financial supported by the Science, Technology and Innovation Funding Authority (STDF) from Egypt and PRIMA Program through the MEDISMART project—Mediterranean Citrus innovative soft-processing solutions for S.M.A.R.T (Sustainable, Mediterranean, Agronomically evolved nutritionally enriched Traditional) products (reference number: PRIMA-EU-2019-SECTION2; http://doi.org/10.54499/PRIMA/0014/2019).

**Institutional Review Board Statement:** Not applicable.

**Informed Consent Statement:** Not applicable.

**Data Availability Statement:** Data will be available upon request.

**Acknowledgments:** The authors gratefully acknowledge the Science, Technology and Innovation Funding Authority (STDF) from Egypt and, the scientific collaboration of the Escola Superior de Biotecnologia—Universidade Católica Portuguesa through CBQF under the FCT project UIDB/50016/2020 and http://doi.org/10.54499/PRIMA/0014/2019. Lastly, the author Ana A. Vilas-Boas would like to acknowledge the FCT for the individual PhD grant (reference: 2020.05655.BD).

**Conflicts of Interest:** The authors declare no conflicts of interest.

## References

1. Wedamulla, N.E.; Fan, M.; Choi, Y.-J.; Kim, E.-K. Citrus Peel as a Renewable Bioresource: Transforming Waste to Food Additives. *J. Funct. Foods* **2022**, *95*, 105163. [CrossRef]
2. Shahbandeh, M. Grapefruit Production Worldwide from 2012/2013 to 2022/2023. Available online: https://www.statista.com/statistics/577836/world-grapefruit-production/ (accessed on 17 March 2024).
3. UN Food and Agriculture Organization, C.S.D. (FAOSTAT). Grapefruit Production in 2022, Crops/Regions/World List/Production Quantity (Pick Lists). Available online: https://www.fao.org/faostat/en/#data/QCL (accessed on 17 March 2024).
4. Khalil, M.N.A.; Farghal, H.H.; Farag, M.A. Outgoing and Potential Trends of Composition, Health Benefits, Juice Production and Waste Management of the Multi-Faceted Grapefruit Citrus X Paradisi: A Comprehensive Review for Maximizing Its Value. *Crit. Rev. Food Sci. Nutr.* **2022**, *62*, 935–956. [CrossRef] [PubMed]
5. Chavan, P.; Singh, A.K.; Kaur, G. Recent Progress in the Utilization of Industrial Waste and By-products of Citrus Fruits: A Review. *J. Food Process Eng.* **2018**, *41*, e12895. [CrossRef]
6. Vilas-Boas, A.A.; Gómez-García, R.; Campos, D.A.; Correia, M.; Pintado, M. Integrated Biorefinery Strategy for Orange Juice By-Products Valorization: A Sustainable Protocol to Obtain Bioactive Compounds. In *Food Waste Conversion*; Springer: Berlin/Heidelberg, Germany, 2023; pp. 113–124.
7. Sharma, H.P.; Patel, H. Sugandha Enzymatic Added Extraction and Clarification of Fruit Juices—A Review. *Crit. Rev. Food Sci. Nutr.* **2017**, *57*, 1215–1227. [CrossRef]
8. Saini, R.K.; Ranjit, A.; Sharma, K.; Prasad, P.; Shang, X.; Gowda, K.G.M.; Keum, Y.-S. Bioactive Compounds of Citrus Fruits: A Review of Composition and Health Benefits of Carotenoids, Flavonoids, Limonoids, and Terpenes. *Antioxidants* **2022**, *11*, 239. [CrossRef] [PubMed]
9. Sir Elkhatim, K.A.; Elagib, R.A.A.; Hassan, A.B. Content of Phenolic Compounds and Vitamin C and Antioxidant Activity in Wasted Parts of Sudanese Citrus Fruits. *Food Sci. Nutr.* **2018**, *6*, 1214–1219. [CrossRef] [PubMed]
10. Saleem, M.; Durani, A.I.; Asari, A.; Ahmed, M.; Ahmad, M.; Yousaf, N.; Muddassar, M. Investigation of Antioxidant and Antibacterial Effects of Citrus Fruits Peels Extracts Using Different Extracting Agents: Phytochemical Analysis with in Silico Studies. *Heliyon* **2023**, *9*, e15433. [CrossRef]
11. Alsakhawy, S.A.; Baghdadi, H.H.; El-Shenawy, M.A.; El-Hosseiny, L.S. Comparative Phytochemical Composition and Antimicrobial Activity of Citrus Peel Essential Oils and Phenolic Compounds. *Antiinfect. Agents* **2023**, *21*, 57–68. [CrossRef]
12. Liu, N.; Li, X.; Zhao, P.; Zhang, X.; Qiao, O.; Huang, L.; Guo, L.; Gao, W. A Review of Chemical Constituents and Health-Promoting Effects of Citrus Peels. *Food Chem.* **2021**, *365*, 130585. [CrossRef]
13. Addi, M.; Elbouzidi, A.; Abid, M.; Tungmunnithum, D.; Elamrani, A.; Hano, C. An Overview of Bioactive Flavonoids from Citrus Fruits. *Appl. Sci.* **2022**, *12*, 29. [CrossRef]
14. Kaur, S.; Panesar, P.S.; Chopra, H.K. Citrus Processing By-Products: An Overlooked Repository of Bioactive Compounds. *Crit. Rev. Food Sci. Nutr.* **2023**, *63*, 67–86. [CrossRef] [PubMed]
15. Maqbool, Z.; Khalid, W.; Atiq, H.T.; Koraqi, H.; Javaid, Z.; Alhag, S.K.; Al-Shuraym, L.A.; Bader, D.M.D.; Almarzuq, M.; Afifi, M.; et al. Citrus Waste as Source of Bioactive Compounds: Extraction and Utilization in Health and Food Industry. *Molecules* **2023**, *28*, 1636. [CrossRef] [PubMed]
16. Jeong, D.; Park, H.; Jang, B.K.; Ju, Y.; Shin, M.H.; Oh, E.J.; Lee, E.J.; Kim, S.R. Recent Advances in the Biological Valorization of Citrus Peel Waste into Fuels and Chemicals. *Bioresour. Technol.* **2021**, *323*, 124603. [CrossRef] [PubMed]
17. Deng, Y.; Tu, Y.; Lao, S.; Wu, M.; Yin, H.; Wang, L.; Liao, W. The Role and Mechanism of Citrus Flavonoids in Cardiovascular Diseases Prevention and Treatment. *Crit. Rev. Food Sci. Nutr.* **2022**, *62*, 7591–7614. [CrossRef]
18. Razavi, B.M.; Hosseinzadeh, H. A Review of the Effects of *Citrus paradisi* (Grapefruit) and Its Flavonoids, Naringin, and Naringenin in Metabolic Syndrome. In *Bioactive Food as Dietary Interventions for Diabetes*, 2nd ed.; Academic Press: Cambridge, MA, USA, 2019; pp. 515–543.
19. Chao, C.-L.; Weng, C.-S.; Chang, N.-C.; Lin, J.-S.; Kao, S.-T.; Ho, F.-M. Naringenin More Effectively Inhibits Inducible Nitric Oxide Synthase and Cyclooxygenase-2 Expression in Macrophages than in Microglia. *Nutr. Res.* **2010**, *30*, 858–864. [CrossRef] [PubMed]
20. Nwosu, O.K.; Keskin, M.; Lohani, H.; Egbuna, C.; Haider, S.Z. Bioactive Lead Compounds and Molecular Targets for the Development of Antiinflammatory Drugs. In *Phytochemicals as Lead Compounds for New Drug Discovery*; Elsevier: Amsterdam, The Netherlands, 2020; pp. 317–331.
21. Deng, W.; Liu, K.; Cao, S.; Sun, J.; Zhong, B.; Chun, J. Chemical Composition, Antimicrobial, Antioxidant, and Antiproliferative Properties of Grapefruit Essential Oil Prepared by Molecular Distillation. *Molecules* **2020**, *25*, 217. [CrossRef]
22. Nikolic, D.; Bosco, L.; Moschetti, M.; Tinnirello, V.; Pucci, M.; Corleone, V.; Raimondo, S.; Alessandro, R.; Fontana, S. Anti-inflammatory Properties of an Aldehydes-enriched Fraction of Grapefruit Essential Oil. *J. Food Sci.* **2023**, *88*, 1172–1187. [CrossRef] [PubMed]
23. Osaili, T.M.; Dhanasekaran, D.K.; Zeb, F.; Faris, M.E.; Naja, F.; Radwan, H.; Ismail, L.C.; Hasan, H.; Hashim, M.; Obaid, R.S. A Status Review on Health-Promoting Properties and Global Regulation of Essential Oils. *Molecules* **2023**, *28*, 1809. [CrossRef] [PubMed]

24. Hassoun, A.; Bekhit, A.E.-D.; Jambrak, A.R.; Regenstein, J.M.; Chemat, F.; Morton, J.D.; Gudjónsdóttir, M.; Carpena, M.; Prieto, M.A.; Varela, P.; et al. The Fourth Industrial Revolution in the Food Industry—Part II: Emerging Food Trends. *Crit. Rev. Food Sci. Nutr.* **2024**, *64*, 407–437. [CrossRef]
25. Guo, Y.; Li, Z.; Chen, F.; Chai, Y. Polyphenols in Oral Health: Homeostasis Maintenance, Disease Prevention, and Therapeutic Applications. *Nutrients* **2023**, *15*, 4384. [CrossRef]
26. Kim, J.H.; Hong, W.S.; Oh, S.W. Effect of Layer-by-Layer Antimicrobial Edible Coating of Alginate and Chitosan with Grapefruit Seed Extract for Shelf-Life Extension of Shrimp (*Litopenaeus vannamei*) Stored at 4 °C. *Int. J. Biol. Macromol.* **2018**, *120*, 1468–1473. [CrossRef] [PubMed]
27. Durmus, M. The Effects of Nanoemulsions Based on Citrus Essential Oils (Orange, Mandarin, Grapefruit, and Lemon) on the Shelf Life of Rainbow Trout (*Oncorhynchus mykiss*) Fillets at 4 ± 2 °C. *J. Food Saf.* **2020**, *40*, e12718. [CrossRef]
28. Olszewska, M.A.; Gędas, A.; Simões, M. Antimicrobial Polyphenol-Rich Extracts: Applications and Limitations in the Food Industry. *Food Res. Int.* **2020**, *134*, 109214. [CrossRef] [PubMed]
29. Cioni, E.; Migone, C.; Ascrizzi, R.; Muscatello, B.; De Leo, M.; Piras, A.M.; Zambito, Y.; Flamini, G.; Pistelli, L. Comparing Metabolomic and Essential Oil Fingerprints of *Citrus australasica* F. Muell (Finger Lime) Varieties and Their in Vitro Antioxidant Activity. *Antioxidants* **2022**, *11*, 2047. [CrossRef] [PubMed]
30. Choudhary, P.; Khade, M.; Savant, S.; Musale, A.; Chelliah, M.S.; Dasgupta, S. Empowering Blue Economy: From Underrated Ecosystem to Sustainable Industry. *J. Environ. Manag.* **2021**, *291*, 112697. [CrossRef] [PubMed]
31. Aschemann-Witzel, J.; Asioli, D.; Banovic, M.; Perito, M.A.; Peschel, A.O.; Stancu, V. Defining Upcycled Food: The Dual Role of Upcycling in Reducing Food Loss and Waste. *Trends Food Sci. Technol.* **2023**, *132*, 132–137. [CrossRef]
32. Some, S.; Roy, J.; Chatterjee, J.S.; Butt, M.H. Low Demand Mitigation Options for Achieving Sustainable Development Goals: Role of Reduced Food Waste and Sustainable Dietary Choice. *J. Clean. Prod.* **2022**, *369*, 133432. [CrossRef]
33. Lemaire, A.; Limbourg, S. How Can Food Loss and Waste Management Achieve Sustainable Development Goals? *J. Clean. Prod.* **2019**, *234*, 1221–1234. [CrossRef]
34. Adams, R.P. *Identification of Essential Oil Components by Gas Chromatography/Mass Spectrometry*, 4.1. ed.; Allured Pub Corp: Carol Stream, IL, USA, 2017; ISBN 978-1-932633-21-4.
35. Farid, M.M.; Ibrahim, F.M.; Ragheb, A.Y.; Mohammed, R.S.; Hegazi, N.M.; Shabrawy, M.O.E.L.; Kawashty, S.A.; Marzouk, M.M. Comprehensive Phytochemical Characterization of *Raphanus raphanistrum* L.: In Vitro Antioxidant and Antihyperglycemic Evaluation. *Sci. Afr.* **2022**, *16*, e01154. [CrossRef]
36. Ibrahim, F.M.; Fouad, R.; EL-Hallouty, S.; Hendawy, S.F.; Omer, E.A.; Mohammed, R.S. Egyptian *Myrtus communis* L. Essential Oil Potential Role as Invitro Antioxidant, Cytotoxic and α-Amylase Inhibitor. *Egypt. J. Chem.* **2021**, *64*, 3005–3017. [CrossRef]
37. Mostafa, F.A.; Abd El Aty, A.A.; Hamed, E.R.; Eid, B.M.; Ibrahim, N.A. Enzymatic, Kinetic and Anti-Microbial Studies on Aspergillus Terreus Culture Filtrate and Allium Cepa Seeds Extract and Their Potent Applications. *Biocatal. Agric. Biotechnol.* **2016**, *5*, 116–122. [CrossRef]
38. Visan, D.-C.; Oprea, E.; Radulescu, V.; Voiculescu, I.; Biris, I.-A.; Cotar, A.I.; Saviuc, C.; Chifiriuc, M.C.; Marinas, I.C. Original Contributions to the Chemical Composition, Microbicidal, Virulence-Arresting and Antibiotic-Enhancing Activity of Essential Oils from Four Coniferous Species. *Pharmaceuticals* **2021**, *14*, 1159. [CrossRef] [PubMed]
39. McFarland, J. The Nephelometer: An Instrument for Estimating the Number of Bacteria in Suspensions Used for Calculating the Opsonic Index and for Vaccines. *J. Am. Med. Assoc.* **1907**, *49*, 1176–1178. [CrossRef]
40. El-Liethy, M.A.; Hemdan, B.A.; El-Taweel, G.E. Phenotyping Using Semi-Automated BIOLOG and Conventional PCR for Identification of Bacillus Isolated from Biofilm of Sink Drainage Pipes. *Acta Ecol. Sin.* **2018**, *38*, 334–338. [CrossRef]
41. Blobaum, A.L.; Marnett, L.J. Structural and Functional Basis of Cyclooxygenase Inhibition. *J. Med. Chem.* **2007**, *50*, 1425–1441. [CrossRef] [PubMed]
42. Tran, T.H.; Dao, T.P.; Le, X.T.; Huynh, B.L.; Minh, L.T.N. Volatile Compounds of Grapefruit (*Citrus grandis* (L.) Osbeck) Peel Essential Oil by Cold Pressing and Hydrodistillation Methods. In *IOP Conference Series: Earth and Environmental Science*; Institute of Physics; IOP Publishing: Bristol, UK, 2023; Volume 1241.
43. Kim, Y.W.; Kim, M.J.; Chung, B.Y.; Bang, D.Y.; Lim, S.K.; Choi, S.M.; Lim, D.S.; Cho, M.C.; Yoon, K.; Kim, H.S. Safety Evaluation and Risk Assessment of D-Limonene. *J. Toxicol. Environ. Health Part. B* **2013**, *16*, 17–38. [CrossRef] [PubMed]
44. Ahmed, S.; Rattanpal, H.S.; Gul, K.; Dar, R.A.; Sharma, A. Chemical Composition, Antioxidant Activity and GC-MS Analysis of Juice and Peel Oil of Grapefruit Varieties Cultivated in India. *J. Integr. Agric.* **2019**, *18*, 1634–1642. [CrossRef]
45. Matera, R.; Lucchi, E.; Valgimigli, L. Plant Essential Oils as Healthy Functional Ingredients of Nutraceuticals and Diet Supplements: A Review. *Molecules* **2023**, *28*, 901. [CrossRef]
46. Kirbaşlar, F.G.; Boz, I. Composition of Turkish Lemon and Grapefruit Peel Oils. *J. Essent. Oil Res.* **2006**, *18*, 525–543. [CrossRef]
47. Kirbaşlar, F.G.; Tavman, A.; Dülger, B.; Türker, G. Antimicrobial Activity of Turkish Citrus Peel Oils. *Pak. J. Bot.* **2009**, *41*, 3207–3212.
48. Anandakumar, P.; Kamaraj, S.; Vanitha, M.K. D-Limonene: A Multifunctional Compound with Potent Therapeutic Effects. *J. Food Biochem.* **2021**, *45*, e13566. [CrossRef] [PubMed]
49. Murali, R.; Karthikeyan, A.; Saravanan, R. Protective Effects of D-limonene on Lipid Peroxidation and Antioxidant Enzymes in Streptozotocin-induced Diabetic Rats. *Basic. Clin. Pharmacol. Toxicol.* **2013**, *112*, 175–181. [CrossRef] [PubMed]

50. Yoon, W.-J.; Lee, N.H.; Hyun, C.-G. Limonene Suppresses Lipopolysaccharide-Induced Production of Nitric Oxide, Prostaglandin E2, and pro-Inflammatory Cytokines in RAW 264.7 Macrophages. *J. Oleo Sci.* **2010**, *59*, 415–421. [CrossRef] [PubMed]
51. Li, X.; Ren, J.-N.; Fan, G.; Zhang, L.-L.; Pan, S.-Y. Advances on (+)-Nootkatone Microbial Biosynthesis and Its Related Enzymes. *J. Ind. Microbiol. Biotechnol.* **2021**, *48*, kuab046. [CrossRef] [PubMed]
52. Bezerra Rodrigues Dantas, L.; Silva, A.L.M.; da Silva Júnior, C.P.; Alcântara, I.S.; de Oliveira, M.R.; Oliveira Brito Pereira Bezerra Martins, A.; Ribeiro-Filho, J.; Coutinho, H.D.M.; Rocha Santos Passos, F.; Quintans-Junior, L.J.; et al. Nootkatone Inhibits Acute and Chronic Inflammatory Responses in Mice. *Molecules* **2020**, *25*, 2181. [CrossRef] [PubMed]
53. Zhang, L.-L.; Chen, Y.; Li, Z.-J.; Fan, G.; Li, X. Production, Function, and Applications of the Sesquiterpenes Valencene and Nootkatone: A Comprehensive Review. *J. Agric. Food Chem.* **2022**, *71*, 121–142. [CrossRef] [PubMed]
54. Barra, L.; Schulz, B.; Dickschat, J.S. Pogostol Biosynthesis by the Endophytic Fungus Geniculosporium. *ChemBioChem* **2014**, *15*, 2379–2383. [CrossRef] [PubMed]
55. Bora, H.; Kamle, M.; Mahato, D.K.; Tiwari, P.; Kumar, P. Citrus Essential Oils (CEOs) and Their Applications in Food: An Overview. *Plants* **2020**, *9*, 357. [CrossRef] [PubMed]
56. Singh, B.; Singh, J.P.; Kaur, A.; Singh, N. Phenolic Composition, Antioxidant Potential and Health Benefits of Citrus Peel. *Food Res. Int.* **2020**, *132*, 109114. [CrossRef]
57. Antony, A.; Farid, M. Effect of Temperatures on Polyphenols during Extraction. *Appl. Sci.* **2022**, *12*, 2107. [CrossRef]
58. Chaaban, H.; Ioannou, I.; Chebil, L.; Slimane, M.; Gérardin, C.; Paris, C.; Charbonnel, C.; Chekir, L.; Ghoul, M. Effect of Heat Processing on Thermal Stability and Antioxidant Activity of Six Flavonoids. *J. Food Process Preserv.* **2017**, *41*, e13203. [CrossRef]
59. Chaves, J.O.; De Souza, M.C.; Da Silva, L.C.; Lachos-Perez, D.; Torres-Mayanga, P.C.; Machado, A.P.D.F.; Forster-Carneiro, T.; Vázquez-Espinosa, M.; González-de-Peredo, A.V.; Barbero, G.F.; et al. Extraction of Flavonoids from Natural Sources Using Modern Techniques. *Front. Chem.* **2020**, *8*, 507887. [CrossRef]
60. Babaoğlu, A.S.; Ainiwaer, T.; Özkan, H.; Karakaya, M. Grapefruit and Pomelo Peel Extracts as Natural Antioxidants for Improved Storage Stability of Turkey Patties during Refrigerated Storage. *J. Food Sci. Technol.* **2022**, *59*, 4067–4074. [CrossRef] [PubMed]
61. Islam, M.; Malakar, S.; Dwivedi, U.; Kumar, N.; Prabakar, P.K.; Kishore, A.; Kumar, A. Impact of Different Drying Techniques on Grapefruit Peels and Subsequent Optimization of Ultrasonic Extraction Conditions for Bioactive Compounds. *J. Food Process Eng.* **2023**, *46*, e14331. [CrossRef]
62. Garcia-Castello, E.M.; Rodriguez-Lopez, A.D.; Mayor, L.; Ballesteros, R.; Conidi, C.; Cassano, A. Optimization of Conventional and Ultrasound Assisted Extraction of Flavonoids from Grapefruit (*Citrus paradisi* L.) Solid Wastes. *LWT-Food Sci. Technol.* **2015**, *64*, 1114–1122. [CrossRef]
63. Singh, V.; Chahal, T.S.; Grewal, S.K.; Gill, P.S. Effect of Fruit Development Stages on Antioxidant Properties and Bioactive Compounds in Peel, Pulp and Juice of Grapefruit Varieties. *J. Food Meas. Charact.* **2021**, *15*, 2531–2539. [CrossRef]
64. Bağdatli, İ.; Khalily, F. Changes in Total Phenolic Content and Antioxidant Activity of Grapefruit and Mandarin Peels Extracted with Different Solvents. *Eurasian J. Food Sci. Technol.* **2022**, *6*, 88–93.
65. Rafiq, S.; Kaul, R.; Sofi, S.A.; Bashir, N.; Nazir, F.; Nayik, G.A. Citrus Peel as a Source of Functional Ingredient: A Review. *J. Saudi Soc. Agric. Sci.* **2018**, *17*, 351–358. [CrossRef]
66. Ademosun, A.O.; Oboh, G.; Passamonti, S.; Tramer, F.; Ziberna, L.; Boligon, A.A.; Athayde, M.L. Phenolics from Grapefruit Peels Inhibit HMG-CoA Reductase and Angiotensin-I Converting Enzyme and Show Antioxidative Properties in Endothelial EA. *Hy* 926 Cells. *Food Sci. Hum. Wellness* **2015**, *4*, 80–85. [CrossRef]
67. Castro-Vazquez, L.; Alañón, M.E.; Rodríguez-Robledo, V.; Pérez-Coello, M.S.; Hermosín-Gutierrez, I.; Díaz-Maroto, M.C.; Jordán, J.; Galindo, M.F.; Arroyo-Jimenez, M.D.M. Bioactive Flavonoids, Antioxidant Behaviour, and Cytoprotective Effects of Dried Grapefruit Peels (*Citrus paradisi* Macf.). *Oxidative Med. Cell. Longev.* **2016**, *2016*, 8915729. [CrossRef]
68. Gómez-Mejía, E.; Rosales-Conrado, N.; León-González, M.E.; Madrid, Y. Citrus Peels Waste as a Source of Value-Added Compounds: Extraction and Quantification of Bioactive Polyphenols. *Food Chem.* **2019**, *295*, 289–299. [CrossRef] [PubMed]
69. Peng, G.; Gan, J.; Dong, R.; Chen, Y.; Xie, J.; Huang, Z.; Gu, Y.; Huang, D.; Yu, Q. Combined Microwave and Enzymatic Treatment Improve the Release of Insoluble Bound Phenolic Compounds from the Grapefruit Peel Insoluble Dietary Fiber. *LWT* **2021**, *149*, 111905. [CrossRef]
70. Awad, A.M.; Kumar, P.; Ismail-Fitry, M.R.; Jusoh, S.; Ab Aziz, M.F.; Sazili, A.Q. Green Extraction of Bioactive Compounds from Plant Biomass and Their Application in Meat as Natural Antioxidant. *Antioxidants* **2021**, *10*, 1465. [CrossRef] [PubMed]
71. Lim, K.J.A.; Cabajar, A.A.; Lobarbio, C.F.Y.; Taboada, E.B.; Lacks, D.J. Extraction of Bioactive Compounds from Mango (*Mangifera indica* L. Var. Carabao) Seed Kernel with Ethanol–Water Binary Solvent Systems. *J. Food Sci. Technol.* **2019**, *56*, 2536–2544. [CrossRef] [PubMed]
72. Dibacto, R.E.K.; Tchuente, B.R.T.; Nguedjo, M.W.; Tientcheu, Y.M.T.; Nyobe, E.C.; Edoun, F.L.E.; Kamini, M.F.G.; Dibanda, R.F.; Medoua, G.N. Total Polyphenol and Flavonoid Content and Antioxidant Capacity of Some Varieties of *Persea americana* Peels Consumed in Cameroon. *Sci. World J.* **2021**, *2021*, 8882594. [CrossRef] [PubMed]
73. Strugała, P.; Tronina, T.; Huszcza, E.; Gabrielska, J. Bioactivity in Vitro of Quercetin Glycoside Obtained in Beauveria Bassiana Culture and Its Interaction with Liposome Membranes. *Molecules* **2017**, *22*, 1520. [CrossRef] [PubMed]
74. Chiriac, E.R.; Chiţescu, C.L.; Geană, E.-I.; Gird, C.E.; Socoteanu, R.P.; Boscencu, R. Advanced Analytical Approaches for the Analysis of Polyphenols in Plants Matrices—A Review. *Separations* **2021**, *8*, 65. [CrossRef]

75. Kubra, I.R.; Jagan Mohan Rao, L. Microwave Drying of Ginger (*Zingiber Officinale Roscoe*) and Its Effects on Polyphenolic Content and Antioxidant Activity. *Int. J. Food Sci. Technol.* **2012**, *47*, 2311–2317. [CrossRef]
76. Mukherjee, S.; Mandal, N.; Dey, A.; Mondal, B. An Approach towards Optimization of the Extraction of Polyphenolic Antioxidants from Ginger (*Zingiber Officinale*). *J. Food Sci. Technol.* **2014**, *51*, 3301–3308. [CrossRef]
77. da Silveira, T.F.F.; Meinhart, A.D.; Ballus, C.A.; Godoy, H.T. The Effect of the Duration of Infusion, Temperature, and Water Volume on the Rutin Content in the Preparation of Mate Tea Beverages: An Optimization Study. *Food Res. Int.* **2014**, *60*, 241–245. [CrossRef]
78. Volf, I.; Ignat, I.; Neamtu, M.; Popa, V.I. Thermal Stability, Antioxidant Activity, and Photo-Oxidation of Natural Polyphenols. *Chem. Pap.* **2014**, *68*, 121–129. [CrossRef]
79. M'hiri, N.; Ioannou, I.; Ghoul, M.; Mihoubi Boudhrioua, N. Phytochemical Characteristics of Citrus Peel and Effect of Conventional and Nonconventional Processing on Phenolic Compounds: A Review. *Food Rev. Int.* **2017**, *33*, 587–619. [CrossRef]
80. Li, X.; Ren, J.-N.; Fan, G.; Pan, S.-Y. Changes of Aroma Compounds and Qualities of Freshly-Squeezed Orange Juice during Storage. *J. Food Sci. Technol.* **2018**, *55*, 4530–4543. [CrossRef] [PubMed]
81. Bozkir, H.; Kola, O.; Duran, H.; Şimşek, M.; Kelebek, H. Effect of Thermal Processing on Carotenoids of Some Orange Juices. *J. Food Agric. Environ.* **2015**, *13*, 52–57.
82. Oikeh, E.I.; Oriakhi, K.; Omoregie, E.S. Proximate Analysis and Phytochemical Screening of Citrus Sinensis Fruit Wastes. *Biosci. J.* **2013**, *1*, 164–170.
83. Oikeh, E.I.; Omoregie, E.S.; Oviasogie, F.E.; Oriakhi, K. Phytochemical, Antimicrobial, and Antioxidant Activities of Different Citrus Juice Concentrates. *Food Sci. Nutr.* **2016**, *4*, 103–109. [CrossRef]
84. Moran, E.P.; Wang, Z.; Chen, J.; Sapieha, P.; Smith, L.E.H.; Ma, J. Neurovascular Cross Talk in Diabetic Retinopathy: Pathophysiological Roles and Therapeutic Implications. *Am. J. Physiol.-Heart Circ. Physiol.* **2016**, *311*, H738–H749. [CrossRef]
85. Xu, G.; Ye, X.; Chen, J.; Liu, D. Effect of Heat Treatment on the Phenolic Compounds and Antioxidant Capacity of Citrus Peel Extract. *J. Agric. Food Chem.* **2007**, *55*, 330–335. [CrossRef]
86. Zhang, M.; Duan, C.; Zang, Y.; Huang, Z.; Liu, G. The Flavonoid Composition of Flavedo and Juice from the Pummelo Cultivar (*Citrus grandis* (L.) Osbeck) and the Grapefruit Cultivar (*Citrus paradisi*) from China. *Food Chem.* **2011**, *129*, 1530–1536. [CrossRef]
87. Al-Ogaili, N.A.; Yasin, Z. Qualitative and Quantitative Investigation of Iraqi Grapefruit (*Citrus padisi*) Flavonoids From Peel and Seeds and Comparing Their Aqueous Extracts for Antimicrobial Activity. *Iraqi J. Sci.* **2016**, *57*, 2627–2633.
88. Can, Z.; Keskin, B.; Üzer, A.; Apak, R. Detection of Nitric Oxide Radical and Determination of Its Scavenging Activity by Antioxidants Using Spectrophotometric and Spectrofluorometric Methods. *Talanta* **2022**, *238*, 122993. [CrossRef] [PubMed]
89. Rekha, C.; Poornima, G.; Manasa, M.; Abhipsa, V.; Devi, J.P.; Kumar, H.T.V.; Kekuda, T.R.P. Ascorbic Acid, Total Phenol Content and Antioxidant Activity of Fresh Juices of Four Ripe and Unripe Citrus Fruits. *Chem. Sci. Trans.* **2012**, *1*, 303–310. [CrossRef]
90. Stabrauskiene, J.; Kopustinskiene, D.M.; Lazauskas, R.; Bernatoniene, J. Naringin and Naringenin: Their Mechanisms of Action and the Potential Anticancer Activities. *Biomedicines* **2022**, *10*, 1686. [CrossRef] [PubMed]
91. World Health Organization. Food Safety. Available online: https://www.who.int/news-room/fact-sheets/detail/food-safety (accessed on 5 February 2024).
92. Faustino, M.; Veiga, M.; Sousa, P.; Costa, E.M.; Silva, S.; Pintado, M. Agro-Food Byproducts as a New Source of Natural Food Additives. *Molecules* **2019**, *24*, 1056. [CrossRef] [PubMed]
93. Bozkurt, T.; Gülnaz, O.; Kaçar, Y.A. Chemical Composition of the Essential Oils from Some Citrus Species and Evaluation of the Antimicrobial Activity. *J. Environ. Sci. Toxicol. Food Technol.* **2017**, *11*, 29–33.
94. Deng, Q.; Penner, M.H.; Zhao, Y. Chemical Composition of Dietary Fiber and Polyphenols of Five Different Varieties of Wine Grape Pomace Skins. *Food Res. Int.* **2011**, *44*, 2712–2720. [CrossRef]
95. Luciardi, M.C.; Blázquez, M.A.; Alberto, M.R.; Cartagena, E.; Arena, M.E. Grapefruit Essential Oils Inhibit Quorum Sensing of Pseudomonas Aeruginosa. *Food Sci. Technol. Int.* **2020**, *26*, 231–241. [CrossRef] [PubMed]
96. Jackson, M.; Stevens, C.M.; Zhang, L.; Zgurskaya, H.I.; Niederweis, M. Transporters Involved in the Biogenesis and Functionalization of the Mycobacterial Cell Envelope. *Chem. Rev.* **2020**, *121*, 5124–5157. [CrossRef] [PubMed]
97. Takó, M.; Kerekes, E.B.; Zambrano, C.; Kotogán, A.; Papp, T.; Krisch, J.; Vágvölgyi, C. Plant Phenolics and Phenolic-Enriched Extracts as Antimicrobial Agents against Food-Contaminating Microorganisms. *Antioxidants* **2020**, *9*, 165. [CrossRef]
98. Arsène, M.M.J.; Podoprigora, I.V.; Davares, A.K.L.; Razan, M.; Das, M.S.; Senyagin, A.N. Antibacterial Activity of Grapefruit Peel Extracts and Green-Synthesized Silver Nanoparticles. *Vet. World* **2021**, *14*, 1330. [CrossRef]
99. Yaldiz, B.; Saglam-Metiner, P.; Cakmak, B.; Kaya, E.; Deliogullari, B.; Yesil-Celiktas, O. Essential Oil and Supercritical Carbon Dioxide Extract of Grapefruit Peels Formulated for Candida Albicans Infections: Evaluation by an in Vitro Model to Study Fungal–Host Interactions. *ACS Omega* **2022**, *7*, 37427–37435. [CrossRef] [PubMed]
100. Côté, H.; Pichette, A.; Simard, F.; Ouellette, M.-E.; Ripoll, L.; Mihoub, M.; Grimard, D.; Legault, J. Balsacone C, a New Antibiotic Targeting Bacterial Cell Membranes, Inhibits Clinical Isolates of Methicillin-Resistant Staphylococcus Aureus (MRSA) without Inducing Resistance. *Front. Microbiol.* **2019**, *10*, 2341. [CrossRef] [PubMed]
101. Wen, Q.-H.; Wang, R.; Zhao, S.-Q.; Chen, B.-R.; Zeng, X.-A. Inhibition of Biofilm Formation of Foodborne Staphylococcus Aureus by the Citrus Flavonoid Naringenin. *Foods* **2021**, *10*, 2614. [CrossRef] [PubMed]
102. Li, A.-P.; He, Y.-H.; Zhang, S.-Y.; Shi, Y.-P. Antibacterial Activity and Action Mechanism of Flavonoids against Phytopathogenic Bacteria. *Pestic. Biochem. Physiol.* **2022**, *188*, 105221. [CrossRef] [PubMed]

103. Ricciotti, E.; FitzGerald, G.A. Prostaglandins and Inflammation. *Arterioscler. Thromb. Vasc. Biol.* **2011**, *31*, 986–1000. [CrossRef] [PubMed]
104. Vishwakarma, R.K.; Negi, D.S. The Development of COX-1 and COX-2 Inhibitors: A Review. *Int. J. Pharm. Sci. Res.* **2020**, *11*, 3544.
105. Cui, J.; Jia, J. Natural COX-2 Inhibitors as Promising Anti-Inflammatory Agents: An Update. *Curr. Med. Chem.* **2021**, *28*, 3622–3646. [CrossRef] [PubMed]
106. Ambati, G.G.; Jachak, S.M. Natural Product Inhibitors of Cyclooxygenase (COX) Enzyme: A Review on Current Status and Future Perspectives. *Curr. Med. Chem.* **2021**, *28*, 1877–1905. [CrossRef]
107. Ahmadi, M.; Bekeschus, S.; Weltmann, K.-D.; von Woedtke, T.; Wende, K. Non-Steroidal Anti-Inflammatory Drugs: Recent Advances in the Use of Synthetic COX-2 Inhibitors. *RSC Med. Chem.* **2022**, *13*, 471–496. [CrossRef]
108. Ribeiro, D.; Freitas, M.; Tomé, S.M.; Silva, A.M.S.; Laufer, S.; Lima, J.L.F.C.; Fernandes, E. Flavonoids Inhibit COX-1 and COX-2 Enzymes and Cytokine/Chemokine Production in Human Whole Blood. *Inflammation* **2015**, *38*, 858–870. [CrossRef]
109. Huang, Y.-S.; Ho, S.-C. Polymethoxy Flavones Are Responsible for the Anti-Inflammatory Activity of Citrus Fruit Peel. *Food Chem.* **2010**, *119*, 868–873. [CrossRef]
110. Miya, G.; Nyalambisa, M.; Oyedeji, O.; Gondwe, M.; Oyedeji, A. Chemical Profiling, Toxicity and Anti-Inflammatory Activities of Essential Oils from Three Grapefruit Cultivars from KwaZulu-Natal in South Africa. *Molecules* **2021**, *26*, 3387. [CrossRef]
111. Arafah, A.; Rehman, M.U.; Mir, T.M.; Wali, A.F.; Ali, R.; Qamar, W.; Khan, R.; Ahmad, A.; Aga, S.S.; Alqahtani, S. Multi-Therapeutic Potential of Naringenin (4′, 5, 7-Trihydroxyflavonone): Experimental Evidence and Mechanisms. *Plants* **2020**, *9*, 1784. [CrossRef]
112. Dayarathne, L.A.; Ranaweera, S.S.; Natraj, P.; Rajan, P.; Lee, Y.J.; Han, C.-H. Restoration of the Adipogenic Gene Expression by Naringenin and Naringin in 3T3-L1 Adipocytes. *J. Vet. Sci.* **2021**, *22*, e55. [CrossRef]
113. Kampschulte, N.; Alasmer, A.; Empl, M.T.; Krohn, M.; Steinberg, P.; Schebb, N.H. Dietary Polyphenols Inhibit the Cytochrome P450 Monooxygenase Branch of the Arachidonic Acid Cascade with Remarkable Structure-Dependent Selectivity and Potency. *J. Agric. Food Chem.* **2020**, *68*, 9235–9244. [CrossRef]
114. Cheng, L.; Zheng, W.; Li, M.; Huang, J.; Bao, S.; Xu, Q.; Ma, Z. Citrus Fruits Are Rich in Flavonoids for Immunoregulation and Potential Targeting ACE2. *Nat. Prod. Bioprospect.* **2020**, *12*, 4.
115. Ferrero-Miliani, L.; Nielsen, O.H.; Andersen, P.S.; Girardin, S. Chronic Inflammation: Importance of NOD2 and NALP3 in Interleukin-1β Generation. *Clin. Exp. Immunol.* **2007**, *147*, 227–235. [CrossRef]

applied sciences

MDPI

*Article*

# Assessment of the Fungistatic Properties of *Calendula officinalis* L. Water Extract and the Effect of Its Addition on the Quality of Wheat Bread

**Izabela Podgórska-Kryszczuk and Urszula Pankiewicz ***

Department of Analysis and Food Quality Assessment, University of Life Sciences in Lublin, Skromna 8, 20-704 Lublin, Poland; izabela.podgorska-kryszczuk@up.lublin.pl
* Correspondence: urszula.pankiewicz@up.lublin.pl

**Abstract:** The potential of *Calendula officinalis* water extract against fungi *Aspergillus niger* and *Penicillium* sp. and the effect of extract addition on the quality of wheat bread were investigated. In vitro, the extract reduced the mycelial growth and biomass production of *A. niger*, but there was no inhibitory effect on *Penicillium* sp. Enriched bread showed significantly higher total phenolic content, by about 77% and 95% in the bread, in which 10% and 15% of the water was replaced with extract, respectively. The antioxidant potential against DPPH• was significantly higher (compared to the control) in both variants used in the experiment, and the level of antioxidant activity increased with the addition of extract. The enriched bread had good quality characteristics—lower baking losses and higher volume than the control. The moisture content and acidity of the crumb of the extract-enriched bread were also higher. The extract additive used did not affect the sensory properties of the bread.

**Keywords:** *Aspergillus niger*; *Penicillium* sp.; marigold; biopreservatives; fungal control; bread fortification

## 1. Introduction

Bread and other bakery products are important diet components in many parts of the world. The baking mainly uses white wheat flour obtained after removing wheat (*Triticum aestivum* L.) bran and germ, which are sources of fibre, phytochemicals, and important nutrients. Because of this, the final product contains less dietary fibre and phenolic compounds. Foods enriched with natural antioxidants are being developed to recompense for this loss. Functional foods consumed as part of a regular diet provide biologically active ingredients that provide health benefits [1]. Unfortunately, bread is a perishable food that goes through physical, chemical, sensory, and microbiological changes during storage. One of the biggest problems in baking is the growth of toxigenic fungi, mainly from the genera *Aspergillus* and *Penicillium* [2]. The growth of fungi leads to quality changes in food, but it can also have negative health effects, as some of these microorganisms produce mycotoxins. They are a public health problem due to their widespread occurrence in the world's food supply [3]. To avoid spoilage of food products by fungi, preservatives are used. However, they have some disadvantages, including altering the aftertaste of light flour products, delaying dough fermentation [4], and not being inert to the consumer. There are reports in the literature of the occurrence of urticaria and contact dermatitis after consuming products with sorbic acid, which is approved for preserving bread [5]. In addition, chemical preservatives are ineffective against some common fungal species in bread, such as *Penicillium paneum* and *P. roqueforti* [2].

The threat to food health safety posed by the presence of toxigenic fungi and the demand for natural products has prompted the search for new methods of preserving baked goods and eliminating harmful microorganisms and their metabolites. There is great interest in natural ingredients with multifunctional properties with potential uses in food production. Plant extracts have been widely studied as potential biopreservatives, as plants contain many important antimicrobial compounds, including phenolic compounds, glucosinolates, alkaloids, cyanogenic glycosides, and oxylipins [6,7].

In this context, *Calendula officinalis* (common marigold), a member of the *Asteraceae* family, is a valuable product with rich medicinal and functional properties. Many secondary metabolites are responsible for the biological activity of marigolds, such as flavonoids, triterpenoids, and polyphenols, found mainly in flower extract [8,9]. These compounds are responsible for antioxidant activity, exhibit neuroprotective effects, prevent degenerative diseases such as diabetes and cardiovascular diseases, and have anticancer, anti-inflammatory, and antimicrobial effects [10]. Extracts from *Calendula officinalis* petals have proven antimicrobial activity against many pathogens, including *Pseudomonas aeruginosa*, *Staphylococcus aureus* [11], *Escherichia coli*, *Bacillus subtilis*, *Klebsiella aerogenes*, *Enterococcus faecalis*, *Candida albicans*, *C. glabrata*, *Aspergillus flavus*, *A. fumigatus*, and *A. niger* [12].

Reports from other researchers indicate that, depending on the plant species, the best solvents for extracting bioactive compounds are acetone, ethanol, and methanol [13,14]. However, according to current trends, analytical procedures should strive to be minimal, as simple as possible to perform, low-cost, and environmentally friendly. Therefore, the study presented in this paper decided to use deionized water as a solvent to extract bioactive compounds from marigolds. Water extract preparation is a relatively easy and inexpensive process, which is important from the consumer's point of view.

This study aimed to evaluate the fungistatic properties of the *Calendula officinalis* water extract against fungi *Aspergillus niger* and *Penicillium* sp. and to determine the effect of extract addition on the quality of wheat bread.

## 2. Materials and Methods

### 2.1. Fungal Pathogens

Pure cultures of *Aspergillus niger* and *Penicillium* sp., isolated from mouldy bread, were obtained from the Department of Analysis and Food Quality Assessment of the University of Life Sciences in Lublin, Poland. The cultures of fungi were kept fresh and viable by periodical transfers on YPD medium (10 g yeast extract, 20 g peptone, 20 g glucose, 20 g agar per 1 L of distilled water; pH 6.0 $\pm$ 0.2) under aseptic conditions throughout the study. Strains were stored at 4 °C for routine cultivation.

### 2.2. Plant Material

Dried petals of *Calendula officinalis* originating from organic farming (Dary Natury, Grodzisk, Poland) were purchased at a local pharmacy and stored at room temperature.

### 2.3. Preparation of Plant Water Extract

*Calendula officinalis* flower water extract was prepared using the method described by Gonelimali et al. [15], with some modifications. The extraction was done using deionised water with a sample-to-solvent ratio of 1:20 ($w/v$). The dried material was powdered in a laboratory mill (IKA A11 basic, IKA-Werke GmbH & Co. KG, Staufen, Germany) until it passed through a 0.35 mm sieve. Twenty g of powder was soaked in distilled water in a round bottom flask, heated for 30 min at 90 °C, and then incubated for 24 h at 37 °C and 150 rpm in a shaking incubator. The obtained extracts were centrifuged for 10 min at 10,000 rpm and filtered using Whatman No. 1 filter.

### 2.4. Qualitative Phytochemical Screening

The water extract of *Calendula officinalis* flowers was evaluated for qualitative determination of major phytochemicals, including alkaloids, saponins, tannins, flavonoids, glycosides, terpenoids, and reducing sugars according to the methodology described below [16–18]. Specific reagents were used for each chemical group to induce chemical reactions that developed distinct colours and/or precipitates characteristic of each class of substances.

Test for alkaloids: A few drops of sulfuric acid were added to the 3 mL of extract, and the formation of orange colour indicated the presence of flavonoids.

Test for saponins: 3 mL of extract was shaken vigorously with 3 mL of distilled water in a test tube, and the mixture was warmed. The formation of stable foam was taken as an indication of the presence of saponins.

Test for tannins: 3 mL of the extract was mixed with 3 mL of water and heated in a water bath. Then ferric chloride was added to the mixture. The dark green colour indicated the presence of tannins.

Test for flavonoids: A few drops of lead acetate solution were added to the 3 mL extract. A yellow-colour precipitate indicated the presence of flavonoids.

Test for glycosides: To 2 mL of the extract, 3 mL of glacial acetic acid, one drop of 5% $FeCl_3$ and concentrated $H_2SO_4$ were added. A reddish-brown colour appeared at the junction of the two liquid layers, and the top layer was blue-green, indicating the presence of glycosides.

Test for terpenoids: 3 mL of the extract was mixed with 2 mL of chloroform and 3 mL of concentrated sulfuric acid added carefully to form a layer. A reddish-brown colour indicated the presence of terpenoids.

Test for reducing sugars: To 0.5 mL of extract, 5 mL of Benedict's reagent was added. Then the mixture was boiled for 5 min. The presence of a bluish-green precipitate indicated the presence of reducing sugars.

### 2.5. *Phenolic Content of Calendula officinalis Water Extract*

The total phenolic content (TPC) of the extract was determined using Folin-Ciocalteau Reagent (FCR) as described by Singleton and Rossi [19]. To 0.1 mL of the extract were added 3 mL of distilled water and 0.4 mL FCR. After 3 min, 1.5 mL of sodium carbonate (100 g/L) was added, and the contents were mixed and allowed to stand for 30 min. The absorbance of the resulting solutions was measured at 760 nm using a UV–VIS spectrophotometer (Cary 50 Scan, Varian, USA) against a reagent blank. The concentration of TPC was calculated from a standard curve constructed using gallic acid solutions following the method described above. The range of gallic acid standard concentration was between 0 and 1 g/L. The TPC in the extract was expressed as gallic acid equivalents (GAE) in mg/g of dry weight (DW). The experiment was performed in triplicate.

### 2.6. *Antifungal Activity of Calendula officinalis Water Extract In Vitro*

#### 2.6.1. Direct Contact Assay

The effect of the *Calendula officinalis* water extract on mycelial growth was studied according to the methodology described by Gleń and Boligłowa [20]. Twenty millilitres of YPD medium supplemented with *Calendula officinalis* water extract (5 mg/mL) at 5%, 10%, and 15% (*v*/*v*) were poured into sterilized Petri dishes (inner diameter 90 mm). After the medium had solidified, it was inoculated centrally with a 10-day-old culture of fungi *A. niger* and *Penicillium* sp. discs (5 mm in diameter). Control was the Petri dishes with pure YPD medium inoculated similarly. The Petri dishes were incubated at 28 °C. The experiment was performed in triplicate.

The effect of *Calendula officinalis* water extract on the mycelial growth of the tested fungi after 3, 5, and 7 days was calculated from the formula:

$$\text{Inhibitory activity } \% = (C - T)/C \times 100\%$$

where C and T are the average diameters (mm) of fungal mycelia in the control and the treatment, respectively.

#### 2.6.2. Influence of *Calendula officinalis* Water Extract on Fungal Biomass Production

Preparation of Spore Suspension of *Aspergillus niger* and *Penicillium* sp.

*Aspergillus niger* and *Penicillium* sp. were kept on YPD medium plates at 28 °C. Spores of fungus were harvested from 10-day-old cultures in Ringer's solution (peptone 1 g, NaCl 8 g per 1 L of distilled water; pH 6.9 ± 0.2) and loosened by 15 min shaking. The suspension was adjusted to $1 \times 10^6$ spores/mL.

Fungal Biomass Production

The biomass growth of tested fungi was carried out in 200 mL Erlenmeyer flasks per 100 mL of YPG medium (10 g yeast extract, 20 g peptone, 20 g glucose per 1 L of distilled water; pH 6.0 ± 0.2), in which part (5%, 10%, and 15%) was replaced with a previously prepared *Calendula officinalis* water extract, according to the methodology of Gleń and Boligłowa [20] with some modifications. One mL of the prepared spore suspension of the tested fungi was added to each flask. The control was the flask with pure YPG medium inoculated in the same way. The flasks were incubated at 28 °C for 7 days and at 150 rpm in a shaking incubator. After the incubation period, the contents of each flask were filtered through standard filter paper (each filter had been weighed). The harvested mycelium was dried at 80 °C until a constant weight was obtained. The fungal biomass of each treatment was weighed and compared with the control. The experiment was performed in triplicate.

The influence of *Calendula officinalis* water extract on fungal biomass production was calculated by using the formula:

$$\text{Inhibitory activity } \% = ((\text{Wbc} - \text{Wbt})/\text{Wbc}) \times 100\%$$

where: Wbc is the weight of biomass control, and Wbt is the weight of biomass treatment.

### 2.7. *Effect of the Calendula officinalis Water Extract Addition on Wheat Bread Quality*

#### 2.7.1. Laboratory Baking

The control bread was prepared from wheat flour type 750 (150 g), water (90 mL), compressed baker's yeast (4.5 g), saccharose (1.5 g), and salt (3 g). The study material was wheat bread made of dough, in which part of the water (10% and 15%) prescribed by the recipe was replaced with a previously prepared *Calendula officinalis* water extract (5 mg/mL). The ingredients were mixed, and the dough was formed for 50 min at 35 °C. Bread loaves were baked for 25 min at 210 °C, then immediately transferred to the laminar flow cabinet and cooled under sterile conditions. Bread loaves were weighted, transferred into plastic bags (one loaf per bag), sealed, and stored in room conditions (until the next day) for further analysis.

#### 2.7.2. Bread Quality Characteristics

The prepared bread dough was weighed, and the dough yield (Yd) was calculated according to the formula:

$$\text{Yd } (\%) = (\text{Wd} \times 100\%)/\text{Fw}$$

where: Wd is the weight of dough (g), and Fw is the weight of used flour (g).

The cooled bread (24 h after baking) was weighed, and the bread yield (Yb) was calculated according to the formula:

$$\text{Yb } (\%) = (\text{Wb} \times \text{Yd})/\text{Wd}$$

where: Wb is the weight of bread after cooling (g), Yd is the dough yield (%), and Wd is the weight of dough (g).

The total baking loss (Bl) was calculated according to the formula:

$$\text{Bl } (\%) = ((\text{Wd} - \text{Wb}))/\text{Wd} \times 100\%$$

where: Wd is the weight of dough (g), and Wb is the weight of bread after cooling (g).

Bread loaf volume was measured 24 h after baking using the millet (*Cenchrus americanus* L.) seed displacement method. Specific volume (SV) was calculated as the loaf volume divided by weight. The Polish standard determined bread acidity and crumb moisture (PN-A-74108: 1996). To determine acidity, 25 g of bread was weighed into flasks each and 250 mL of distilled water at about 60 °C was added, then closed tightly with a rubber stopper. The sample was shaken for 3 min, then allowed to stand for another 3 min and shaken again for 1 min. The suspension was filtered through cotton wool, and 50 mL of filtrate was taken from each sample into a conical flask. It was titrated with 0.1 M NaOH solution in the presence of a few drops of 1% phenolphthalein until the pink color of the solution appeared and persisted for 30 s. The result was calculated according to the formula:

$$x_5 = 2a$$

where: $x_5$—bread acidity [°], a—volume of 0.1 M NaOH solution used for titration [mL].

### *2.8. Phenolics Content and Antioxidant Potential of Bread*

#### 2.8.1. Extraction of Bioactive Compounds

The bread samples (1 g) were ground, placed in Falcon tubes, and shaken with 10 mL of a 4:1 (*v*/*v*) ethanol/water mixture for 120 min in a laboratory shaker. Then, the samples were centrifuged at 3000 rpm for 10 min. The supernatant was stored at −18 °C for further analysis.

#### 2.8.2. Phenolics Content

The TPC of the bread extract was determined using FCR as described in Section 2.5. and the results are expressed as gallic acid equivalents in mg/g of bread.

#### 2.8.3. DPPH (2-Diphenyl-1-Picrylhydrazyl) Radical Scavenging Activity

The DPPH assay was determined according to Shahidi et al. [21], with some modifications. A 0.15 mL of the prepared extract was mixed with 2.85 mL of a 0.1 mM solution of DPPH• in 75% methanol. The absorbance was measured after 30 min of reaction at 517 nm. The scavenging percentage was calculated using the formula:

$$\text{Scavenging activity } \% = (\text{Acontrol} - \text{Asample})/\text{Acontrol} \times 100\%$$

where Asample is the absorbance of the mixture of sample and DPPH•; Acontrol is the absorbance of the control (DPPH• solution).

#### 2.8.4. ABTS (2,2′-Azobis(3-Etylobenzotiazolino-6-Sulfonianu)) Radical Scavenging Activity

The ABTS assay was determined according to Re et al. [22], with some modifications. The radical solution was prepared with ABTS and potassium persulfate, diluted in water to a final concentration of 2.45 mM, and left in the dark for 16 h to allow for radical development. The solution was diluted to reach the absorbance measures around 0.7 at 734 nM. Then, 2.85 mL of the ABTS•+ solution was mixed with 0.15 mL of each sample. The absorbance was measured after 30 min of the reaction at 734 nm. The scavenging percentage was calculated using the formula:

$$\text{Scavenging activity } \% = (\text{Acontrol} - \text{Asample})/\text{Acontrol} \times 100\%$$

where Asample is the absorbance of the sample and ABTS•+; A control is the absorbance of the control (ABTS•+ solution).

### 2.9. Microbiological Quality of Bread

The number of yeasts and moulds per 1 g of bread was determined using the plate technique on YGC medium (5 g yeast extract, 20 g glucose, 0.1 g chloramphenicol, 15 g agar per 1 L of distilled water; pH 6.6 ± 0.2). The assay was performed on the first day (after cooling) and after 7 days of storage in sterile Petri dishes at 28 °C. To isolate yeasts and moulds from bread, 10 g of crumbs were collected under sterile conditions under a laminar chamber into 300 mL conical flasks containing 90 mL of sterile Ringer's solution. The flasks were shaken at 180 rpm for 10 min. Then serial dilutions were made, and 1 mL was transferred to sterile Petri dishes containing YGC medium with chloramphenicol. The flasks were incubated for 5 days at 28 °C. After incubation, the colonies were counted and expressed as colony-forming units per gram (cfu/g) of samples. The test was performed in triplicate for each sample.

### 2.10. Consumer Evaluation

Consumer evaluation of bread was conducted on 29 consumers (18–25 years old). The bread was cut into slices about 1.5 cm thick the day after baking and coded with a three-digit number. The external appearance, crust (appearance and connection with the crumb; colour; thickness), crumb (appearance; colour; porosity; elasticity), smell, and taste were assessed. The bread was evaluated using the 5-point quality scale method. Grade 5 meant very good quality, grade 4—good quality, grade 3—satisfactory quality, grade 2—unsatisfactory quality, and grade 1—bad quality. The evaluation was based on a table with assigned definitions to each scale of scores for individual bread characteristics. Each quality discriminant has assigned an appropriate weighting factor (external appearance 0.1; crust 0.1; crumb 0.2; smell 0.2; taste 0.4). Plain water was used for mouth rinsing before and after each sample test.

### 2.11. Statistical Analysis

The results were analyzed statistically using Statistica 13.3 (StatSoft, Cracow, Poland) and Excel 2019 (Microsoft, Washington, DC, USA). A one-way analysis of variance (ANOVA) was carried out to compare the results, and the significance of differences between group means was determined using Tukey's post hoc test. All statistical hypotheses were verified at the significance level of $p < 0.05$.

## 3. Results and Discussion

### 3.1. Qualitative Phytochemical Screening

The results of qualitative phytochemical screening of the Calendula officinalis water extract are shown in Table 1. The preliminary phytochemical test results indicated the presence of alkaloids, saponins, flavonoids, glycosides and terpenoids in the tested extract.

**Table 1.** Phytochemical screening of the Calendula officinalis flower water extract.

| Phytochemicals | Alkaloids | Saponins | Tannins | Flavonoids | Terpenoids | Glycosides | Reducing Sugars |
|---|---|---|---|---|---|---|---|
| Test result | + | + | − | + | + | + | − |

"+" presence and "−" absence of phytochemicals in the *Calendula officinalis* water extract.

The results agree with the literature data [23,24] and indicate that marigold can be considered a potential raw material for the extraction of bioactive components. Calendula officinalis is an important medicinal plant with diverse phytochemicals and biological activities, such as antioxidant, anti-inflammatory, and antimicrobial. Among the phytochemicals in marigold petals, triterpenes, flavonoids, phenolic acids, quinones and coumarins, and carotenoids are reported to play the most important role [25]. These compounds are responsible for antimicrobial activity, and their mechanisms of action are diverse, including inhibition of nucleic acid synthesis, interference with cytoplasmic membrane function, and slowing of microbial metabolism [26].

### 3.2. Total Phenolic Content of *Calendula officinalis* Water Extract

The TPC determined in the *Calendula officinalis* water extract was 12.8 ± 0.007 mg GAE/g DM.

Using various solvents and extraction conditions, many researchers have attempted to extract phenolic compounds from *Calendula officinalis*. Sytar et al. [27] showed that among the tested methanolic extracts of leaves of the representative family *Asteraceae*, marigold leaf extracts have been shown to have the highest total content of these compounds (1.125 ± 0.153 mg/g DM). A study by Olennikov and Kashchenko [28] reported that the TPC in the ethanolic extract of marigold leaves ranged from 29.21 to 50.24 mg/g, depending on the variety. In a study by Piekut [29], the ethanolic extract of marigold flowers had a low TPC content of 6.19 mg/g DM. Petkova et al. [30] found that calendula infusion showed a TPC of 16.49 ± 0.20 mg/g DM, while the decoction had a higher content of these compounds (19.44 ± 0.04 mg/g DM). Water extracts of *Calendula officinalis* flowers and leaves were studied by Mubashar Sabir et al. [31], showing that the TPC was 72.91 ± 2.1 mg/g and 23.1 ± 0.8 mg/g, respectively. Krochmal-Marczak and Kiełtyka-Dadasiewicz [32] found a significantly lower polyphenols content ranging from 0.06 to 3.64 mg/100 g DW in an infusion of marigold flowers, depending on brewing time and temperature.

Comparing the obtained results with the above reports by other authors, it can be concluded that the TPC in the plant extract is affected by many factors. These include the type of solvents used, their ratio, extraction time and temperature, the chemical composition and physical properties of the plant materials, and even the part of the plant used (flowers, leaves). Data available in the literature state that the chemical compounds found in plants are also influenced by the region where the plant is grown, climatic conditions (e.g., rainy and dry seasons), growing season (beginning or end of flowering), and genetic factors [27,33]. Therefore, all these factors should be considered when comparing the TPC in an extract from a given plant. The repeatability of chemical composition is one of the main problems in using plant extracts as food preservative ingredients. Extracts of natural origin, or different batches of extracts produced by different manufacturers, can exhibit quantitative and qualitative variability.

### 3.3. Antifungal Activity of *Calendula officinalis* Water Extract In Vitro

#### 3.3.1. Direct Contact Assay

The study found that adding *Calendula officinalis* water extract to the medium limited the mycelial growth of *A. niger* after 3, 5, and 7 days of incubation on Petri dishes (Figure 1). The greatest inhibition was found in a Petri dish with the addition of 15% ($v/v$) of water extract in the medium, and it ranged from 49.81% on the 7th day of cultivation to 62.98% on the 3rd day. On the other hand, the tested extract did not affect inhibiting the mycelial growth of *Penicillum* sp. (Figure 2).

Numerous literature reports indicate the effectiveness of extracts prepared from various plant species, including marigolds, against pathogenic microorganisms [6,12,27,34,35]. Efstratiou et al. [12] found that methanolic and ethanolic extracts of *Calendula officinalis* petals showed excellent antifungal activity against 9 different species from genera *Candida, Aspergillus*, and *Exophiala*. Similar to the present study, the researchers proved that marigold extract reduced the growth of *A. niger* on Petri dishes [12]. Rigane et al. [36] also showed such activity for water-methanol extracts from calendula flowers and leaves. Ikeura et al. [37] screened the antifungal activity of extracts from 16 plants (dichloromethane and diethyl ether were the solvents) against *Penicillium expansum* and showed that marigold extract strongly reduced the growth of the pathogen by the steam contact method.

The antimicrobial properties of *Calendula officinalis* are due to the presence of numerous bioactive compounds; among them, phenolic compounds play an important role in the control of fungi [38]. The literature shows that their antimicrobial activity is based on different mechanisms due to the wide variation in the structure of these substances. Phenolic compounds can interact with the cytoplasmic membrane, cell wall, nucleic acids, and energy transport, altering or inhibiting their functions. They also denature enzymes or bind vitamins, minerals, or carbohydrates, making these compounds unavailable to microorganisms [39].

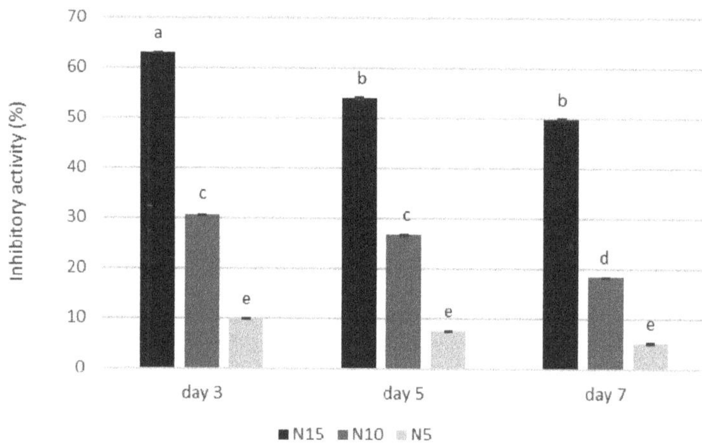


**Figure 1.** Effect of adding *Calendula officinalis* water extract on the mycelial growth of *Aspergillus niger*. N15, N10, N5—addition of *Calendula officinalis* water extract to the medium in doses of 15%, 10%, and 5% (*v*/*v*), respectively; Values marked with the same letters do not differ significantly at $p < 0.05$ (Tukey's post hoc test); mean value ($n = 3$) ± SD.

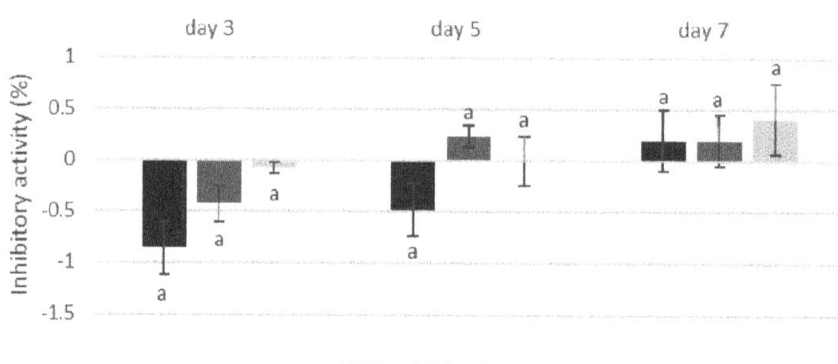


**Figure 2.** Effect of adding *Calendula officinalis* water extract on the mycelial growth of *Penicillium* sp. N15, N10, N5—addition of *Calendula officinalis* water extract to the medium in doses of 15%, 10%, and 5% (*v*/*v*), respectively; Negative values denote stimulation of fungal biomass production; Values marked with the same letters do not differ significantly at $p < 0.05$ (Tukey's post hoc test); mean value ($n = 3$) ± SD.

### 3.3.2. Influence of *Calendula officinalis* Water Extract on Fungal Biomass Production

There was a significant inhibition of *A. niger* biomass production after 7 days of incubation when 15% and 10% *Calendula officinalis* water extract were added to the culture medium. The reduction in pathogen biomass was 15.50% and 9.17%, respectively. On the other hand, as in the previous experiment on Petri dishes, the study extract did not significantly affect the inhibition of *Penicillium* sp. biomass production.

Studies on the inhibitory effect of plant extract on the growth of pathogenic fungi biomass have been reported by some researchers. Bajwa et al. [40] evaluated the potential of water extracts of *Parthenium hysterophorus*, a plant in the *Asteraceae* family, against three pathogenic fungi: *Drechslera tetramera*, *Phoma glomerata*, and *A. niger*. The lower extract concentrations proposed in the experiment caused a decrease in fungal biomass production, and this response was species-specific. In contrast, increasing extract concentrations increased fungal biomass production at all harvest intervals. Gleń and Boligłowa [20] showed that water extracts of walnut (*Juglans regia*) leaves, birch (*Betula verrucosa Ehrh.*) bark, and nettle (*Utrica dioica* L.) herb strongly reduced the biomass production of *Fusarium culmorum* (29.60–53.70%) and *Botrytis cinerea* (17.90–53.80%), while in some variants of the experiment tested extracts showed a stimulating effect on the biomass growth of *Alternaria alternata*, *Phoma exiqua*, and *Sclerotina sclerotiorum*. El–Mohamedy and Abdallah [41] tested the antifungal activity of *Moringa oleifera* Lam. seed oil and extract against *Fusarium oxysporum*, *F. solani*, *Alternaria solani*, *A. alternate*, *Rhizoctonia solani*, *Sclerotium rolfsii* and *Macrophomina phaseolina*, and the researchers showed that study oil and extract had an inhibitory effect on the growth of pathogen biomass.

The results presented in the paper and reports by other researchers indicate that a particular plant extract can reduce the biomass production of one pathogen and have no effect or even stimulate the growth of another. Therefore, testing a broad spectrum of product-specific microorganisms is appropriate when designing new foods with plant extracts. This will make it possible to create foods of high quality and high safety for the consumer.

### 3.4. Bread Quality Characteristics

Due to the low fungistatic activity obtained with the 5% addition of *Calendula officinalis* water extract in the culture medium, it was decided to use only the higher addition in bread production and replace the water recommended by the recipe with 10% and 15% extract. The evaluation of the control and bread with the addition of extract quality parameters is shown in Table 2. The appearance of the prepared bread is shown in Figure 3.

**Table 2.** Evaluation of bread quality parameters.

| Parameters | Sample | | |
|---|---|---|---|
| | **Control** | **N10** | **N15** |
| Bread yield (%) | 139.54 ± 2.78 [a] | 139.42 ± 2.60 [a] | 140.49 ± 0.73 [a] |
| Baking loss (%) | 11.14 ± 1.11 [a] | 10.68 ± 0.39 [b] | 9.16 ± 1.46 [c] |
| Specific volume ($cm^3/g$) | 2.12 ± 0.19 [b] | 2.67 ± 0.17 [a] | 2.61 ± 0.23 [a] |
| Crumb moisture (%) | 31.64 ± 3.29 [b] | 37.26 ± 5.40 [a] | 37.49 ± 5.89 [a] |
| Acidity [°] | 1.73 ± 0.12 [c] | 2.00 ± 0.00 [b] | 2.23 ± 0.03 [a] |

N10, N15—bread, in which 10% and 15% of the water have been replaced with *Calendula officinalis* water extract, respectively. Within each row, values with the same letter do not differ significantly at $p < 0.05$ (Tukey's post hoc test); mean value ($n = 3$) ± SD.

The addition of *Calendula officinalis* water extract used in the study did not affect bread yields, and the results were similar to the control. The experiments found that bread with the extract showed significantly lower baking losses compared to the control. The total baking loss depends on numerous factors, including the size, shape, volume, and type of bread, dough management and baking method, and even the type of oven used [42]. Alotaibi et al. [43] report that the lower baking losses in bread with the addition of marigold may be due to the presence of a large amount of lutein, which binds to water and thus reduces moisture loss. Accordingly, it was found that the enriched bread had a significantly higher (about 18%) moisture content compared to the control. The bread with extract also had significantly higher volume than the control, by about 23% and 26% in the bread, in which 15% and 10% of the water was replaced with *Calendula officinalis* extract, respectively. The specific bread volume is considered one of the main characteristics defining consumer acceptability [43], so the results obtained in this paper are significant. In contrast to the present study, Alotaibi et al. [43] found that enriching bread with dried marigold flower powder reduced volume. In the experiments presented here, it was also noted that the acidity of the bread increased as the addition of extract increased.

**Figure 3.** The appearance of control bread (K) and with the addition of *Calendula officinalis* water extract, 10% (N10) and 15% (N15), respectively.

Dough expansion depends on dough rheological characteristics, primarily due to the gluten and starch components. Adding bioactive ingredients may physically modify the development of bread dough [44] since antioxidants can form complexes with proteins and/or polysaccharides. Several studies have shown an adverse effect of the powdered plant material addition on bread volume [45]. This can be explained by the presence of fibre, which interacts with proteins and weakens the gluten network of the dough, resulting in a reduction in the amount of $CO_2$ retained in the dough during fermentation [46]. Jakubczyk et al. [47] report that *Calendula officinalis* has a high total fibre content ($62.33 \pm 9.17$ g/100 g), but most of the fraction is insoluble ($57.54 \pm 8.32$ g/100 g). In general, insoluble fibres reduce the technological quality of baked goods, while soluble fibres have a positive effect [48], so using water extract was reasonable in the present study.

### 3.5. Phenolics Content and Antioxidant Potential of Bread

The effect of adding *Calendula officinalis* water extract on the TPC and antioxidant activity of wheat bread is presented in Table 3. As expected, marigold extract positively affected fortified bread's phenolic content and antioxidant status. Compared to the control, the TPC was significantly higher by about 77% and 95% in the bread, in which 10% and 15% of the water was replaced with *Calendula officinalis* extract, respectively. In addition, the experiment observed significantly higher free radical scavenging activity for ABTS•+ than DPPH•, which may be related to their solubility in water and organic solvents. The same relation was found in other studies for marigold flower extract [23,30]. In the present study, the antioxidant potential against DPPH• was significantly higher in both enriched bread than bread without extract. In this case, the level of antioxidant activity increased with the addition of marigold water extract. In contrast, the antiradical activity against ABTS•+ was higher only in bread, in which 15% of the water was replaced by *Calendula officinalis* extract.

**Table 3.** Phenolics content and antioxidant activity of wheat bread with *Calendula officinalis* water extract.

| | Sample | | |
|---|---|---|---|
| | **Control** | **N10** | **N15** |
| TPC (mg GAE/g DW) | 0.44 ± 0.00 [b] | 0.78 ± 0.01 [a] | 0.86 ± 0.01 [a] |
| DPPH (%) | 6.87 ± 0.01 [c] | 8.47 ± 0.01 [b] | 10.40 ± 0.01 [a] |
| ABTS (%) | 20.24 ± 0.01 [b] | 20.53 ± 0.01 [b] | 33.59 ± 0.02 [a] |

N10, N15—bread, in which 10% and 15% of the water have been replaced with *Calendula officinalis* water extract, respectively. TPC—total phenolic contents; ABTS—the ability to quench ABTS radicals; DPPH—the ability to quench DPPH radicals; Within each row and for the selected assays, values with the same letter do not differ significantly at $p < 0.05$ (Tukey's post hoc test); mean value ($n$ = 3) ± SD.

Food fortification is of increasing interest to consumers. Many products are enriched to improve their phenolic levels and antioxidant potential, thereby increasing their health-promoting properties. This is particularly important for bread derived from white wheat flour. Edible flowers are a good source of compounds with antioxidant activity [10], and *Calendula officinalis* has been extensively studied in this aspect [23,27,30]. Measuring the antioxidant capacity can provide a wealth of information on the oxidation resistance, the quantitative contribution of antioxidants, or the antioxidant effects that may occur in the body during consumption [49]. The radical-scavenging properties of natural compounds are often associated with their ability to quench free radicals. Many literature reports confirm that adding plant extracts increases bread's polyphenol content and antioxidant activity [50,51], and our results indicate the potential of *Calendula officinalis* in food fortification.

### 3.6. Microbiological Quality of Bread

After 7 days of bread storage in Petri dishes at 28 °C, there was no visual evidence of microbial spoilage. However, after the isolation of yeasts and moulds from the bread, their presence was found in all samples, but their amount was within acceptable standards for bread (Table 4). The highest number of yeast and mould was found in the control, both on the baking day ($1.50 \times 10^2$ cfu/g) and after 7 days of storage ($2.30 \times 10^2$ cfu/g). Similar numbers of these microorganisms ($2.20 \times 10^2$ cfu/g) were found after storage in bread with 15% of the prepared extract. Conversely, the lowest number of yeast and mould on a baking day ($1.23 \times 10^2$ cfu/g) as well as after 7 days of storage ($1.90 \times 10^2$ cfu/g) was observed in bread in which 10% of the water was replaced with the *Calendula officinalis* extract.

**Table 4.** Number of yeasts and moulds in bread.

| Sample | Number of Yeasts and Moulds in Bread (cfu/g) | |
|---|---|---|
| | **Baking Day** | **After 7 Days of Storage** |
| Control | $1.50 \times 10^2 \pm 0.05$ [a] | $2.30 \times 10^2 \pm 0.04$ [a] |
| N10 | $1.23 \times 10^2 \pm 0.06$ [b] | $1.90 \times 10^2 \pm 0.06$ [b] |
| N15 | $1.30 \times 10^2 \pm 0.03$ [b] | $2.20 \times 10^2 \pm 0.05$ [a] |

N10, N15—bread, in which 10% and 15% of the water have been replaced with *Calendula officinalis* water extract, respectively; Within each column and for the selected day, values with the same letter do not differ significantly at $p < 0.05$ (Tukey's post hoc test); mean value ($n$ = 3) ± SD.

One of the difficult challenges facing the bakery industry is to reduce microbial spoilage, especially fungi, thereby extending shelf life and guaranteeing product safety. It has been proven that some plant extracts can inhibit the growth of pathogenic microorganisms, so enriching bread is reasonable. Pinilla et al. [52] found that bread fortified with garlic extract was more microbiologically stable compared to the control, and the addition of the extract inhibited the growth of the fungi *A. flavus*, *Penicillium herquei*, and *F. graminearum* in the bread. A study by Doudi et al. [53] proved that cinnamon (*Cinnamomum verum*) extract improves the shelf life of bread, delays spoilage, and reduces the growth of *Aspergillus* sp. fungi. Numerous works by other researchers have prompted the search for natural bread additives to extend the shelf life of the finished product. Several reports in the literature about the effectiveness of marigold extract against fungi of the *Aspergillus* and *Penicillium* genera [12,36,37] pose a threat in baking. The results presented in this paper prove that the appropriate addition of *Calendula officinalis* water extract can be successfully applied to bread and positively affect the microbiological quality of the product.

### 3.7. Consumer Evaluation

The average consumer ratings of the bread from the individual parameters were similar in each sample and were not statistically significantly different from each other (Table 5). The control and enriched bread received similar overall ratings and were of good quality, according to consumers. The control bread received an overall score of 3.96, the bread in which 10% of the water was replaced with prepared *Calendula officinalis* water extract received 4.02, and in the case of 15% replacement of water with extract 3.94. The results show that the addition of *Calendula officinalis* water extract used in the experiment had no significant effect on the organoleptic properties of the bread.

**Table 5.** Sensory evaluation of bread prepared with *Calendula officinalis* water extract.

| Evaluation Parameters | | Control | N10 | N15 |
|---|---|---|---|---|
| External appearance | | $4.14 \pm 0.64$ a | $4.10 \pm 0.72$ a | $4.00 \pm 0.76$ a |
| Crust | Appearance and connection with the crumb | $4.10 \pm 0.62$ a | $4.09 \pm 0.70$ a | $4.00 \pm 0.89$ a |
| | Color | $4.10 \pm 0.77$ a | $4.07 \pm 0.88$ a | $3.86 \pm 0.95$ a |
| | Thickness | $4.14 \pm 0.74$ a | $3.93 \pm 0.88$ a | $4.14 \pm 0.79$ a |
| Crumb | Appearance | $4.07 \pm 0.70$ a | $4.00 \pm 0.80$ a | $4.17 \pm 0.80$ a |
| | Color | $3.93 \pm 0.80$ a | $4.14 \pm 0.83$ a | $4.14 \pm 0.95$ a |
| | Porosity | $3.97 \pm 0.73$ a | $3.83 \pm 1.04$ a | $4.14 \pm 0.83$ a |
| | Elasticity | $3.83 \pm 0.85$ a | $3.86 \pm 0.83$ a | $3.86 \pm 0.89$ a |
| Aroma | | $4.07 \pm 0.84$ a | $4.07 \pm 0.80$ a | $4.07 \pm 0.88$ a |
| Taste | | $3.83 \pm 0.71$ a | $4.00 \pm 0.85$ a | $3.79 \pm 0.90$ a |
| Overall | | $3.96 \pm 0.12$ a | $4.02 \pm 0.10$ a | $3.94 \pm 0.14$ a |

N10, N15—bread, in which 10% and 15% of the water have been replaced with *Calendula officinalis* water extract, respectively; Within each row and for each factor, values with the same letter do not differ significantly at $p < 0.05$ (Tukey's post hoc test); mean value ($n = 29$) $\pm$ SD.

The attractiveness of bread consists of many different characteristics, but the first to attract the consumer's attention is external appearance and colour. Edible flowers or extracts from them can be an interesting alternative to artificial colours used in food technology [54]. The obtained *Calendula officinalis* extract was brown, but the amount of additive used was low enough that it did not significantly change the colour of the finished product. The addition of plant extracts rich in phenolic compounds, which aim to increase the antioxidant capacity of wheat bread, can also affect certain rheological properties of the dough, such as gumminess, strength, elasticity, adhesiveness, and chewiness, among others [1]. As a result, the sensory properties of the final product are also modified, which is not always acceptable to consumers. For example, in a study by Czaja et al. [51], according to consumers, bread enriched with onion extract had a less acceptable smell and taste than the control bread. In the development of functional foods, a key role is played by finding ingredients that allow specific beneficial effects to be achieved without modifying the sensory properties of the enriched product. Therefore, the results obtained in the present study are significant and encourage further research into using *Calendula officinalis* water extract for bread fortification.

## 4. Conclusions

The demand for natural, chemical-free, "clean label" products induces the search for new methods to extend food shelf life and eliminate harmful microorganisms from food products. Plants extracts are an active area for research due to their rich phytochemical composition. *Calendula officinalis* is widely used in cosmetics, while the present study demonstrated the potential of this plant in food fortification. The studied marigold extract showed antifungal properties against *A. niger*, and the enriched bread had satisfactory quality characteristics, exhibited antiradical activity, and had unchanged sensory properties. The results encourage further research into using *Calendula officinalis* in food technology.

**Author Contributions:** Conceptualization, I.P.-K. and U.P.; methodology, I.P.-K. and U.P.; software, I.P.-K.; validation I.P.-K.; formal analysis, I.P.-K.; investigation, I.P.-K.; data curation, I.P.-K.; writing—original draft preparation, I.P.-K.; writing—review and editing, U.P.; supervision, U.P. All authors have read and agreed to the published version of the manuscript.

**Funding:** This research received no external funding.

**Institutional Review Board Statement:** Not applicable.

**Informed Consent Statement:** Not applicable.

**Data Availability Statement:** The data presented in this study are available on request from the corresponding author.

**Conflicts of Interest:** The authors declare no conflict of interest.

## References


1. Czajkowska–González, Y.A.; Alvarez–Parrilla, E.; del Rocío Martínez-Ruiz, N.; Vázquez–Flores, A.A.; Gaytán–Martínez, M.; de la Rosa, L.A. Addition of Phenolic Compounds to Bread: Antioxidant Benefits and Impact on Food Structure and Sensory Characteristics. *Food Prod. Process. Nutr.* **2021**, *3*, 25. [CrossRef]
2. Garcia, M.V.; Bernardi, A.O.; Copetti, M.V. The Fungal Problem in Bread Production: Insights of Causes, Consequences, and Control Methods. *Curr. Opin. Food Sci.* **2019**, *29*, 1–6. [CrossRef]
3. Daou, R.; Joubrane, K.; Maroun, R.G.; Khabbaz, L.R.; Ismail, A.; El Khoury, A.; Daou, R.; Joubrane, K.; Maroun, R.G.; Khabbaz, L.R.; et al. Mycotoxins: Factors Influencing Production and Control Strategies. *AIMS Agric. Food* **2021**, *6*, 416–447. [CrossRef]
4. Fik, M. Czerstwienie Pieczywa i Sposoby Przedłużania Jego Świeżości. *ŻYWNOŚĆ. Nauk. Technol. Jakość* **2004**, *2*, 5–22.
5. Abdulmumeen, H.A.; Risikat, A.N.; Sururah, A.R. Food: Its Preservatives, Additives and Applications. *Int. J. Chem. Biochem. Sci.* **2012**, *1*, 36–47. [CrossRef]
6. Pires, T.C.S.P.; Dias, M.I.; Barros, L.; Calhelha, R.C.; Alves, M.J.; Oliveira, M.B.P.P.; Santos-Buelga, C.; Ferreira, I.C.F.R. Edible Flowers as Sources of Phenolic Compounds with Bioactive Potential. *Food Res. Int.* **2018**, *105*, 580–588. [CrossRef]
7. Rahman, M.; Islam, R.; Hasan, S.; Zzaman, W.; Rana, M.R.; Ahmed, S.; Roy, M.; Sayem, A.; Matin, A.; Raposo, A.; et al. A Comprehensive Review on Bio-Preservation of Bread: An Approach to Adopt Wholesome Strategies. *Foods* **2022**, *11*, 319. [CrossRef] [PubMed]
8. Givol, O.; Kornhaber, R.; Visentin, D.; Cleary, M.; Haik, J.; Harats, M. A Systematic Review of *Calendula officinalis* Extract for Wound Healing. *Wound Repair Regen.* **2019**, *27*, 548–561. [CrossRef]
9. Martin, D.; Navarro Del Hierro, J.; Villanueva Bermejo, D.; Fernández-Ruiz, R.; Fornari, T.; Reglero, G. Bioaccessibility and Antioxidant Activity of *Calendula officinalis* Supercritical Extract as Affected by in Vitro Codigestion with Olive Oil. *J. Agric. Food Chem.* **2016**, *64*, 8828–8837. [CrossRef]
10. Prabawati, N.B.; Oktavirina, V.; Palma, M.; Setyaningsih, W. Edible Flowers: Antioxidant Compounds and Their Functional Properties. *Horticulturae* **2021**, *7*, 66. [CrossRef]
11. Hamad, M.N.; Mohammed, H.; Merdaw, M. Antibacterial Activity of *Calendula officinalis* Flowers in vitro. *Ibn Al-Haitham J. Pure Appl. Sci.* **2011**, *24*, 3. [CrossRef]

12. Efstratiou, E.; Hussain, A.I.; Nigam, P.S.; Moore, J.E.; Ayub, M.A.; Rao, J.R. Antimicrobial Activity of *Calendula officinalis* Petal Extracts against Fungi, as Well as Gram-Negative and Gram-Positive Clinical Pathogens. *Complement. Ther. Clin. Pract.* **2012**, *18*, 173–176. [CrossRef] [PubMed]
13. Do, T.H.; Truong, H.B.; Nguyen, H.C. Optimization of Extraction of Phenolic Compounds from *Ocimum basilicum* Leaves and Evaluation of Their Antioxidant Activity. *Pharm. Chem. J.* **2020**, *54*, 162–169. [CrossRef]
14. Sepahpour, S.; Selamat, J.; Abdul Manap, M.Y.; Khatib, A.; Abdull Razis, A.F. Comparative Analysis of Chemical Composition, Antioxidant Activity and Quantitative Characterization of Some Phenolic Compounds in Selected Herbs and Spices in Different Solvent Extraction Systems. *Molecules* **2018**, *23*, 402. [CrossRef] [PubMed]
15. Gonelimali, F.D.; Lin, J.; Miao, W.; Xuan, J.; Charles, F.; Chen, M.; Hatab, S.R. Antimicrobial Properties and Mechanism of Action of Some Plant Extracts against Food Pathogens and Spoilage Microorganisms. *Front. Microbiol.* **2018**, *9*, 1639. [CrossRef]
16. Kannan, V.; Fahad, S.M.; Arumugam, C.S.; Vinothkumar, D.; Ramesh Babu, N.G. Phytochemical Screening of *Bauhinia purpurea* L.: An Important Medicinal Plant. *Int. Res. J. Pharm.* **2015**, *6*, 802–804. [CrossRef]
17. Nortjie, E.; Basitere, M.; Moyo, D.; Nyamukamba, P. Extraction Methods, Quantitative and Qualitative Phytochemical Screening of Medicinal Plants for Antimicrobial TTextiles: A Review. *Plants* **2022**, *11*, 2011. [CrossRef]
18. Shaikh, J.R.; Patil, M. Qualitative Tests for Preliminary Phytochemical Screening: An Overview. *Int. J. Chem. Stud.* **2020**, *8*, 603–608. [CrossRef]
19. Singleton, V.L.; Rossi, J.A. Colorimetry of Total Phenolics with Phosphomolybdic-Phosphotungstic Acid Reagents. *Am. J. Enol. Vitic.* **1965**, *16*, 144–158. [CrossRef]
20. Gleń, K.; Boligłowa, E. Ocena Aktywności Fungistatycznej Wyciągów Roślinnych w Testach in Vitro. *J. Res. Appl. Agric. Eng.* **2012**, *57*, 104–109.
21. Shahidi, F.; Liyana-Pathirana, C.M.; Wall, D.S. Antioxidant Activity of White and Black Sesame Seeds and Their Hull Fractions. *Food Chem.* **2006**, *99*, 478–483. [CrossRef]
22. Re, R.; Pellegrini, N.; Proteggente, A.; Pannala, A.; Yang, M.; Rice-Evans, C. Antioxidant Activity Applying an Improved ABTS Radical Cation Decolorization Assay. *Free Radic. Biol. Med.* **1999**, *26*, 1231–1237. [CrossRef] [PubMed]
23. Ak, G.; Zengin, G.; Sinan, K.I.; Mahomoodally, M.F.; Picot-Allain, M.C.N.; Cakir, O.; Bensari, S.; Yilmaz, M.A.; Gallo, M.; Montesano, D. A Comparative Bio-Evaluation and Chemical Profiles of *Calendula officinalis* L. Extracts Prepared via Different Extraction Techniques. *Appl. Sci.* **2020**, *10*, 5920. [CrossRef]
24. Ghaima, K.K.; Rasheed, S.F.; Ahmed, E.F. Antibiofilm, Antibacterial and Antioxidant Activities of Water Extract of *Calendula officinalis* Flowers. *Int. J. Biol. Pharm. Res.* **2013**, *4*, 465–470.
25. Escher, G.B.; Borges, L.D.C.C.; Santos, J.S.; Cruz, T.M.; Marques, M.B.; Do Carmo, M.A.V.; Azevedo, L.; Furtado, M.M.; Sant'ana, A.S.; Wen, M.; et al. From the Field to the Pot: Phytochemical and Functional Analyses of *Calendula officinalis* L. Flower for Incorporation in an Organic Yogurt. *Antioxidants* **2019**, *8*, 559. [CrossRef]
26. Krzepiłko, A.; Kordowska-Wiater, M.; Sosnowska, B.; Pytka, M. *Oddziaływanie Ekstraktów Roślinnych Na Drobnoustroje*; University of Life Sciences in Lublin: Lublin, Poland, 2020; ISBN 9788372593252.
27. Sytar, O.; Hemmerich, I.; Zivcak, M.; Rauh, C.; Brestic, M. Comparative Analysis of Bioactive Phenolic Compounds Composition from 26 Medicinal Plants. *Saudi J. Biol. Sci.* **2018**, *25*, 631–641. [CrossRef]
28. Olennikov, D.N.; Kashchenko, N.I. Componential Profile and Amylase Inhibiting Activity of Phenolic Compounds from *Calendula officinalis* L. Leaves. *Sci. World J.* **2014**, *2014*, 654193. [CrossRef]
29. Piekut, J. Ocena Wybranych Ekstraktów Roślin Przyprawowych Pod Względem Ich Właściwości Przeciwdrobnoustrojowych Oraz Zawartości Fenolokwasów. *Zesz. Probl. Postępów Nauk Rol.* **2018**, *588*, 103–111. [CrossRef]
30. Petkova, D.; Mihaylova, D.; Denev, P.; Krastanov, A. Antioxidant Activity of Some Edible Flowers Water Extracts from Bulgaria. *Bull. Univ. Agric. Sci. Vet. Med. Cluj-Napoca. Food Sci. Technol.* **2020**, *77*, 54. [CrossRef] [PubMed]
31. Mubashar Sabir, S.; Khan, M.F.; Rocha, J.B.T.; Boligon, A.A.; Athayde, M.L. Phenolic Profile, Antioxidant Activities and Genotoxic Evaluations of *Calendula officinalis*. *J. Food Biochem.* **2015**, *39*, 316–324. [CrossRef]
32. Krochmal-Marczak, B.; Kiełtyka-Dadasiewicz, A. Wpływ Temperatury Wody i Czasu Parzenia Na Właściwości Antyoksydacyjne Naparów z Nagietka Lekarskiego (*Calendula officinalis* L.). *Herbalism* **2018**, *4*, 43–51. [CrossRef]
33. Avci, A.B.; Inan, M. Comparing of Cultivated Annual and Perennial *Calendula officinalis* L. Species. *J. Agric. Nat.* **2021**, *4*, 579–585. [CrossRef]
34. Ali, E.M.; Abd El-Moaty, H. Antifungal Activity of *Achillea santolina* L. and *Calendula officinalis* L. Essential Oils and Their Constituents against Fungal Infection of Liver as Complication of Cyclophosphamide Therapy. *J. Essent. Oil-Bear. Plants* **2017**, *20*, 1030–1043. [CrossRef]
35. Chaleshtori, S.H.; Kachoie, M.A.; Pirbalouti, A.G. Phytochemical Analysis and Antibacterial Effects of *Calendula offcinalis* Essential Oil. *Biosci. Biotechnol. Res. Commun.* **2016**, *9*, 517–522. [CrossRef]
36. Rigane, G.; Ben Younes, S.; Ghazghazi, H.; Ben Salem, R. Investigation into the Biological Activities and Chemical Composition of *Calendula officinalis* L. Growing in Tunisia. *Int. Food Res. J.* **2013**, *20*, 3001–3007.
37. Ikeura, H.; Somsak, N.K.F.; Kanlayanara, S.H.Y. Application of Selected Plant Extracts to Inhibit Growth of *Penicillium expansum* on Apple Fruits. *Plant Pathol. J.* **2011**, *10*, 79–84. [CrossRef]
38. El-Shahir, A.A.; El-Wakil, D.A.; Latef, A.A.H.A.; Youssef, N.H. Bioactive Compounds and Antifungal Activity of Leaves and Fruits Methanolic Extracts of *Ziziphus Spina-christi* L. *Plants* **2022**, *11*, 746. [CrossRef]

39. Skroza, D.; Šimat, V.; Smole Možina, S.; Katalinić, V.; Boban, N.; Generalić Mekinić, I. Interactions of Resveratrol with Other Phenolics and Activity against Food-Borne Pathogens. *Food Sci. Nutr.* **2019**, *7*, 2312–2318. [CrossRef] [PubMed]
40. Bajwa, R.; Khalid, A.; Shahid Cheem, T. Antifungal Activity of Allelopathic Plant Extracts III: Growth Response of Some Pathogenic Fungi to Aqueous Extract of *Parthenium hysterophorus*. *Plant Pathol. J.* **2003**, *2*, 145–156. [CrossRef]
41. El–Mohamedy, R.S.R.; Abdallah, A.M. Antifungal Activity of Moringa Oleifera Oil and Seed Extract against Some Plant Pathogenic Fungi. *Middle East J. Agric. Res.* **2014**, *3*, 42–249.
42. Wirkijowska, A.; Zarzycki, P.; Sobota, A.; Nawrocka, A.; Blicharz-Kania, A.; Andrejko, D. The Possibility of Using By-Products from the Flaxseed Industry for Functional Bread Production. *LWT* **2020**, *118*, 108860. [CrossRef]
43. Alotaibi, H.N.; Anderson, A.K.; Sidhu, J.S. Influence of Lutein Content of Marigold Flowers on Functional Properties of Baked Pan Bread. *Ann. Agric. Sci.* **2021**, *66*, 162–168. [CrossRef]
44. Sivam, A.S.; Sun-Waterhouse, D.; Waterhouse, G.I.N.; Quek, S.; Perera, C.O. Physicochemical Properties of Bread Dough and Finished Bread with Added Pectin Fiber and Phenolic Antioxidants. *J. Food Sci.* **2011**, *76*, H97–H107. [CrossRef]
45. Wójcik, M.; Różyło, R.; Cacak-, G. Textural and Sensory Properties of Wheat Bread Fortified with Nettle (*Urtica dioica* L.) Produced by the Scalded Flour Method. *J. Food Process. Preserv.* **2021**, *45*, e15851. [CrossRef]
46. Cacak-Pietrzak, G.; Dziki, D.; Gawlik-Dziki, U., Sułek, A.; Kalisz, S.; Sujka, K. Effect of the Addition of Dried Dandelion Roots (*Taraxacum officinale* F. H. Wigg.) on Wheat Dough and Bread Properties. *Molecules* **2021**, *26*, 7564. [CrossRef]
47. Jakubczyk, K.; Koprowska, K.; Gottschling, A.; Janda-Milczarek, K. Edible Flowers as a Source of Dietary Fibre (Total, Insoluble and Soluble) as a Potential Athlete's Dietary Supplement. *Nutrients* **2022**, *14*, 2470. [CrossRef]
48. Sciarini, L.S.; Bustos, M.C.; Vignola, M.B.; Paesani, C.; Salinas, C.N.; Pérez, G.T. A Study on Fibre Addition to Gluten Free Bread: Its Effects on Bread Quality and *in Vitro* Digestibility. *J. Food Sci. Technol.* **2017**, *54*, 244. [CrossRef]
49. Zielińska, E.; Pankiewicz, U. Nutritional, Physiochemical, and Antioxidative Characteristics of Shortcake Biscuits Enriched with Tenebrio Molitor Flour. *Molecules* **2020**, *25*, 5629. [CrossRef]
50. Baiano, A.; Viggiani, I.; Terracone, C.; Romaniello, R.; Del Nobile, M.A. Physical and Sensory Properties of Bread Enriched with Phenolic Aqueous Extracts from Vegetable Wastes. *Czech J. Food Sci.* **2015**, *33*, 247–253. [CrossRef]
51. Czaja, A.; Czubaszek, A.; Wyspiańska, D.; Sokół-Łętowska, A.; Kucharska, A.Z. Quality of Wheat Bread Enriched with Onion Extract and Polyphenols Content and Antioxidant Activity Changes during Bread Storage. *Int. J. Food Sci. Technol.* **2020**, *55*, 1725–1734. [CrossRef]
52. Pinilla, C.M.B.; Thys, R.C.S.; Brandelli, A. Antifungal Properties of Phosphatidylcholine-Oleic Acid Liposomes Encapsulating Garlic against Environmental Fungal in Wheat Bread. *Int. J. Food Microbiol.* **2019**, *293*, 72–78. [CrossRef] [PubMed]
53. Doudi, M.; Setorki, M.; Rezayatmand, Z. Effects of Aqueous Extract of *Cinnamomum verum* on Growth of Bread Spoilage Fungi. *Int. J. Med. Res. Health Sci.* **2016**, *5*, 162–171.
54. Takahashi, J.A.; Rezende, F.A.G.G.; Moura, M.A.F.; Dominguete, L.C.B.; Sande, D. Edible Flowers: Bioactive Profile and Its Potential to Be Used in Food Development. *Food Res. Int.* **2020**, *129*, 108868. [CrossRef] [PubMed]

applied sciences

MDPI

*Review*

# Functional Foods, Gut Microbiome and Association with Obesity and Metabolic Syndrome: A Literature Review

**Despoina Koumpouli [1], Varvara Koumpouli [2] and Antonios E. Koutelidakis [1,*]**

[1] Unit of Human Nutrition, Laboratory of Nutrition and Public Health, Department of Food Science and Nutrition, University of the Aegean, Leoforos Dimokratias 66, 81400 Myrina, Greece; dkoub@hotmail.com
[2] Laboratory of Hematology, University General Hospital of Ioannina, Stavrou Niarchou Avenue, 45500 Ioannina, Greece; vana_koumpouli@yahoo.gr
* Correspondence: akoutel@aegean.gr; Tel.: +30-225-408-3123; Fax: +30-225-408-3109

**Abstract:** The human gastrointestinal gut consists of about 100 trillion microorganisms, including up to 5000 different types of bacteria, as well as Archaea, Eukarya, parasites, viruses and bacteriophages that together are called the "gut microbiome". Changes in gut microorganism composition (dysbiosis) can cause various diseases. The present study aims to investigate if diet, and more specifically, functional foods have an impact on the intestinal microbiome, and whether the intestinal microbiome has an influence on metabolic syndrome (MetS) and obesity. This systematic review was accomplished according to PRISMA guidelines, mostly using the key words functional foods, microbiome, obesity, MetS, and Mediterranean diet. The search focused on recent scientific articles from the Pubmed, Scopus, and Google Scholar databases. Most of the studies discussed showed a potential therapeutic effect of the Mediterranean diet, which is rich in beneficial nutrients, on body weight and fat deposition, through reshaping of the gut microbiome's synthesis. This literature review showed a possible relationship between microflora metabolites, endotoxemia, obesity and MetS. The role of probiotics, prebiotics, and polyphenols in the prevention of obesity and MetS is of high importance in promoting healthy aging. The future challenge is to comprehend how different dietary patterns could regulate the gut microflora's composition and whether these changes could be long term.

**Keywords:** functional foods; microbiome; obesity; metabolic syndrome; polyphenols; prebiotics

## 1. Introduction

The human gastrointestinal gut contains approximately 100 trillion microorganisms, called the "gut microbiome", which weighs about 2 kg and contains up to 5000 different species of bacteria that correspond to 5 million genes [1–4]. Furthermore, there are also members of the kingdom Archaea present, such as Methanobrevibacter species that produce methane in the gut; Eukarya, in the form of the yeast Candida; parasites such as Entamoeba; and macro parasites such as helminths. In addition, viruses and bacteriophages play a crucial role in the gut's ecology and the maintenance of a healthy balance. These populations differ between individuals, demonstrating the heterogeneity of human intestine microflora [5] and are renewed every 3 days. Through competition for resources and colonization sites, they offer protection to the host from colonization by pathogenic microorganisms [3,6,7]. Most have a symbiotic relationship with the host that is critical for the human body, due to their possible role in nutrition, metabolism, and the pathogenesis of diseases. From birth, the gut microbiome is constantly evolving, and factors, such as diet, lifestyle, age, and genetics, have an important influence on the microbial composition of the human gut [1].

The gut microbiome population contains mainly the bacterial phyla of Firmicutes (mainly genus Clostridia), Bacteroidetes (mainly Bacteroides), Actinobacteria and Proteobacteria [1,5,8]; 90% of the human gut microbiome belong almost to only two phyla, Firmicutes and Bacteroidetes [7].

The most widespread, non-communicable diseases, such as obesity and MetS, are increasing worldwide. They are both distinguished by a state of inflammation and reshaping of the intestinal microflora [5]. The World Health Organization (WHO) defines obesity as an excessive accumulation of body fat, mainly visceral fat, which can induce health damage [9–12]. Worldwide, 2.1 billion people are overweight or obese [13]. The major concern about obesity is its relationship with chronic metabolic diseases such as insulin resistance (IR), cardiovascular diseases, type II diabetes and MetS [9,10,14]. WHO defines MetS as a pathological condition characterized by obesity, IR, hyperlipidemia, and hypertension [14,15].

Several studies have shown that dysbiosis leads to low-grade inflammation, obesity, and, consequently, MetS [2,16]. Dysbiosis is a condition of imbalance between the various microorganisms in the human gut. It can be caused by hereditary factors, antibiotics, radiotherapy, chemotherapy, and age; however, it is mainly related to imbalances in nutrition, physical and psychological stress, resulting in dysregulation of the immune system [17].

Recent scientific research associates MetS with intestinal dysbiosis [14]. Intestinal dysbiosis increases intestinal permeability, which can cause the translocation of lipopolysaccharide (LPS), a component of the external membrane of Gram-negative bacteria, resulting in metabolic endotoxemia, low-grade inflammation, IR, and weight gain [2,5,7,10,18–20]. The gut microbiota are considered a metabolic gateway between the external environment and the host, in terms of energy metabolism, bodyweight homeostasis and inflammation regulation. Obesity has been shown to be related to the modification of gut microbiota and obesity phenotypes and can be transmitted through gut microbiota in rodent models. Probiotic bacteria, as part of the fermented traditional functional foods of the Mediterranean diet, may contribute to reducing dysbiosis and promoting human health [16,21].

Diet is considered a key factor for the modification of microbial gut diversity. Some food ingredients have possible beneficial effects on human health beyond the provision of basic nutritional needs, and are known as functional foods. Functional foods are foods that offer potential benefits to the human organism other than basic diet, when consumed as part of a regular, balanced diet [11] and they are defined as industrially processed or natural foods. Functional foods contain a plethora of bioactive compounds with possible effects on promoting human health and contributing to disease prevention [22]. Phenolic compounds, and especially flavonoids, are of particular interest, as part of many traditional foods of the Mediterranean diet, such as herbs, greens, fruit, and vegetables, because of their possible effects as antioxidant and anti-inflammatory compounds. Eating foods rich in flavanols has been shown to reshape gut microflora, exerting prebiotic effects. Furthermore, microflora alteration contributes to improving IR, glycemic control, and glucose tolerance [11].

The aim of this literature review was to investigate the most recent scientific information about the relationship between nutritional attitudes, functional foods, probiotics, prebiotics, and polyphenols consumption with the modification of the gut microbiome, as well as the association of the gut microbiome with obesity and MetS. The study focuses on whether the gut microbiome predisposes the occurrence of obesity and MetS and the optimal eating habits for developing a healthy intestinal microbiome. The future challenge is to fully understand how different dietary patterns could modulate gut microbiota composition and whether these changes are long-lasting.

## 2. Materials and Methods

The literature search was conducted according to PRISMA (Preferred Reporting Items for Systematic Reviews and Meta-Analyses) guidelines via the online academic search engines PubMed, Scopus, and Google Scholar. The keywords used were "functional foods", and "microbiome", and "obesity", or "metabolic syndrome", or "polyphenols", or "prebiotics", or "Mediterranean diet". The search was limited to research articles that were published in reviewed journals between 2014 and 2024 and were written in English. From this research, 120 articles were initially studied and evaluated and the articles that included the specific keywords in their title or abstract were selected for further analysis. The selection criteria were primarily contemporary research conducted in the last decade, including analyses mainly of nutritional and clinical studies, in which there was an evaluation of the influence of functional foods on obesity and MetS through their effect on intestinal microbiome. Systematic reviews, articles in another language except English, articles with a year of publication before 2014, articles that did not include information about the microbiome and functional foods in correlation with obesity and MetS, studies with unreliable study designs and incompletely documented findings and evidence were excluded. Finally, 19 article were selected for the literature review.

## 3. Results

### 3.1. Effects of the Mediterranean Diet on the Intestinal Microbiome, Obesity, and Metabolic Syndrome

In a recent case–control study, 18 overweight obese subjects and 18 normal-weight volunteers followed a Mediterranean diet enriched with 40 g/day high-quality extra virgin olive oil (HQ-EVOO) for 3 months. All subjects received a typical Mediterranean diet (55–60% carbohydrates, 25–30% polyunsaturated and monounsaturated fats, 15–20% proteins) and cases followed a low-calorie Mediterranean diet (1552 $\pm$ 160 Kcal). Both cases and controls consumed 40 g/day of HQ-EVOO for 3 months as the exclusive cooking and dressing fat. Anthropometric assessments, questionnaires, blood, and fecal samples were collected at the beginning and after 3 months of the Mediterranean diet rich in HQ-EVOO (T1). There was a significant reduction in inflammation and oxidative stress in both controls and cases, while IL-10 and adiponectin were increased in the cases. Beneficial gut bacteria, mostly *Lactobacillus* strains, increased in the cases after a Mediterranean diet enriched with HQ-EVOO [23].

In another study, 82 healthy overweight and obese subjects that followed a sedentary lifestyle, participated in a parallel 8-week randomized controlled trial. Forty-three participants consumed a Mediterranean diet and 39 consumed their regular diet. Individual conformity to the protocol was estimated every 2 weeks by self-recorded 7-day food diaries and physical activity questionnaires. Visits and sample collections were carried out at baseline, 4 weeks and 8 weeks. The dietary modifications in the group that followed the Mediterranean diet led to a reduction in carnitine in plasma and urine, as well as a decrease in plasma cholesterol, an increase in insulin sensitivity and a reshaping of gut microbiome composition. A significant inverse correlation was observed between the variations in gut microbial gene abundance and inflammatory status estimated by serum C-reactive protein (CRP) variations. During the increasing Mediterranean diet adherence phase, *Ruthenibacterium lactatiformans, Flavonifractor plautii, Parabacteroides merdae, Ruminococcus torques* and *Ruminococcus gnavus* were significantly reduced [24]. The latter species are proinflammatory species due to the secretion of a polysaccharide that induces tumor necrosis factor alpha (TNF-a) [25]. In parallel, there was an increase in potentially beneficial species, including the fiber-degrading *F. prausnitzii*, Roseburia and members of the Clostridiales and Lachnospiraceae taxa, related to the butyrate precursor functional pathways. These potentially beneficial species were previously documented for their anti- inflammatory properties [26] and were found to be increased in the participants who consumed the Mediterranean diet [24].

In a 12-week single-arm pilot trial, nine participants followed nutritional counseling sessions according to a Mediterranean diet. The participants were men and women aged 40–80 years, non-smokers, with type II diabetes. Biochemical parameters, body composition, gut microbiota, and blood pressure were measured at the beginning of the study, 4 weeks, and 12 weeks after the intervention. Bacterial diversity and richness were increased. Bacterial variety was negatively associated with glycated hemoglobin (HbA1c) and bacterial abundance was negatively correlated with fasting glucose levels. An increase in the Prevotella/Bacteroides ratio was observed and there was a reduction in HbA1c. An evaluation of Alkaline phosphatase activity in fecal samples was conducted as a parameter of intestinal permeability and inflammation; this was positively related to bacterial diversity and negatively correlated with HbA1c [27]. Consuming a Mediterranean diet increased the activity of intestinal ALP, while the activity of this enzyme induced an increase in bacterial richness and a lowering of HbA1c. This can be ascribed to the ability of intestinal ALP to reduce inflammation by modifying gut microbiota synthesis and to dephosphorylate LPS, reducing intestinal permeability and increasing insulin sensitivity [28,29].

Another transversal study investigated the relationship between the Mediterranean diet and gut microbiota in a group of 31 adults with non-declared pathology. Only those individuals who had not used antibiotics, glucocorticoids, immunosuppressive drugs, monoclonal antibodies, or other immunotherapies participated in this study. Dietary intake was evaluated using an annual semi-quantitative food frequency questionnaire (FFQ), with 160 items. Expert dieticians asked about cooking methods, the number and amounts of ingredients used in each recipe (e.g., type of oil or milk used), and other relevant information such as whether the participants ate the fruit peels. During an interview, participants were asked, food by food, whether they normally consumed that food and how much they usually ate. For this purpose, three different serving sizes of each cooked food were presented in pictures to the participants so that they could choose from up to seven serving sizes. Food quantity was recorded in household units, by volume, or by measuring with a ruler. As for the alcoholic beverages, each participant was asked about the type and amount, for which household measures, such as a glass, a bottle, etc., were used. At the end of the study, there was an increase in the population of Bacteroidetes, Prevotella and a decrease in the population of Firmicutes and Lachnospiraceae. In addition, the Mediterranean diet increased faucal propionate and butyrate concentrations [30].

Table 1 summarizes the effects of the Mediterranean diet on the gut microbiome, in relation to the recent studies described previously.

**Table 1.** Summary results of the effect of Mediterranean diet on human microbiome.

| Study Type | Study Sample/Duration | Participants | Protocol | Summary of Results | Study Reference |
|---|---|---|---|---|---|
| Clinical trial | 36 participants/3 months | 18 overweight/obese subjects, 18 normal-weight controls | Participants followed Mediterranean diet enriched with 40 g/day HQ-EVOO. | Mediterranean diet rich in HQ-EVOO induced an increase in Lactobacillus strains in gut microbiota. | [23] |
| Randomized controlled trial | 82 Subjects/8 weeks | 82 overweight and obese subjects | 43 participants consumed a Mediterranean diet and 39 maintained their regular diets. | Reductions in carnitine and plasma cholesterol and increased levels of *Faecalibacterium prausnitzii* and insulin sensitivity, rearrangement of the gut microbiome composition. | [24] |
| 12-week single-arm pilot trial | 9 participants/12 weeks | Men and women 40–80 y, nonsmokers, with diagnosis of type 2 diabetes | Participants followed nutritional counseling sessions according to Mediterranean diet. | Increase in microbiota variety, Pre-votella/Bacteroides ratio and reduction in HbA1c. | [27] |
| Transversal study | 31 participants | 31 adults, 23 women, 8 men, 42 y with non-declared pathology | Following of Mediterranean diet. | Mediterranean diet increased the population of Bacteroidetes, Prevotella and decreased the population of Firmicutes and Lachnospiraceae. | [30] |

### 3.2. Effects of Probiotics on the Intestinal Microbiome, Obesity and Metabolic Syndrome

An interventional clinical trial investigated the impact of the combination of a probiotic supplement of 30 g/day of carob and probiotic bacteria *L. helveticus, B. longum, S. Thermophilus, L. Lactis* on body composition and metabolic biomarkers in obese volunteers. Forty-five patients were randomly separated into three groups. The first group was called "diet only" and received a low-calorie diet without any intervention (15 patients). The second group was called "prebiotic group" and 15 patients followed the same diet plan but consumed a prebiotic supplementation (2 carob beans/day of about 30 g). The third group called "probiotic group" and the volunteers consumed the same diet with probiotic supplementation (n = 15). The probiotic constituent used in the study was one tablet containing an association of four microbiological strains and especially *Bifidobacteruim longum, Lactobacillus helveticus, Lactococcus lactis, Streptococcus thermophilus* (1 tablet ($10 \times 10^9$ CFU/capsule)/day). After one month supplementation, in both the prebiotic and probiotic group, there was a significant reduction in fat mass and insulinemia and a beneficial effect on carbohydrate metabolism. In the probiotic group, an important reduction in fasting blood glucose was observed compared to the control group. Microbiome lipopolysaccharides reduction caused by pre- and probiotics may induced appetite reduction by increasing satiety. Prebiotics act on microbiota by increasing the production of SCFAs. SCFA are responsible for a cascade of modifications that may lead to weight reduction and improvement in metabolic parameters [31].

A 3-week, randomized, double-blind, placebo-controlled, parallel pilot study evaluated the effect of a low-calorie diet accompanied by the consumption of a moderate-fat cheese containing the probiotic *Lactobacillus plantarum*. Forty patients compatible with the inclusion criteria were selected. Inclusion criteria were age 30–69 years, and diagnosis of MetS characterized by obesity and accompanied by arterial hypertonia (>130/85 mm Hg). To produce the probiotic cheese, *L. plantarum* was added to the cheese milk in a concentration of $1.5 \times 10^{11}$ CFU/g before renneting. Microbial composition of the test and control cheeses did not vary in the counts and predominance of non-starter microbiota. In probiotic cheese *L. plantarum*, viability was evaluated before the clinical trial. In this study, 25 subjects consumed 50 g/day of probiotic cheese and 15 consumed the same amount of cheese without probiotics. In the probiotic group, there was a significant reduction in body mass index (BMI) compared to the control group. *Lactobacillus plantarum* colonization was positively correlated with the extent of reductions in morning diastolic blood pressure (BP) and systolic BP values. The hypocaloric diet accompanied with the consumption of a probiotic cheese causes a reduction in BMI and arterial blood pressure, both risk factors of MetS. After the consumption of 50 g of cheese containing 26% fat for 3 weeks, a reduction in total cholesterol and low-density lipoprotein (LDL) was also observed. A significant reduction in plasma triglyceride levels was detected only in the probiotic group [32].

In another randomized study, 51 patients with MetS were allocated into two groups, a control group of 25 subjects and a probiotic group of 26 subjects. The probiotic group consumed probiotic-fermented milk for 45 days. The probiotic milk, ready for consumption, exhibited a high probiotics concentration with $3.4 \times 10^8$ CFU/mL of *B. animalis ssp. lactis ssp.*, 0.38% acidity, pH 5.15, 2.80% protein, 7.54% carbohydrates, 0.0% fat, 12.08% total solids and had 33.09 calories per bottle. Pathogens were not found, as requested by legislation; thus, the product was defined as suitable for consumption. Each volunteer was given seven 80 mL bottles every week until they reached 45 d of intake and were asked to consume one bottle every morning. Anthropometric and biochemical parameters were evaluated at the beginning of the study and after 45 days. Waist circumference, body weight, and height were measured, while BMI was calculated (kg/m$^2$). Three blood pressure measurements taken at 1 min intervals were recorded. At the end of the study, beneficial effects of *B. lactis* on lipid profile, BMI, and cytokine levels in patients with MetS were observed. Furthermore, a reduction in the pro-inflammatory cytokines TNF-a and Interleukin (IL-6) was found, contributing to an improvement in lipid metabolism and weight loss, leading to a reduction in the characteristic parameters of MetS and obesity [33].

Another randomized, double-blind, placebo-controlled 8-week study evaluated the effects of a probiotic yogurt fortified with the probiotic strains *Lactobacillus acidophilus* and *Bifidobacterium lactis*. The 44 participants were randomly divided into two groups. Probiotic and regular yogurts contained *lactobacillus bulgaricus* and *Streptococcus thermophilus*, while the probiotic yogurt was supplemented with cultures of *B. lactis* and *L. acidophilus*. The participants consumed 300 g/day of probiotic or regular yogurt for 8 weeks, respectively. As reported by the microbiologic analysis, the average concentrations of *L. acidophilus* and *B. lactis* in the probiotic yogurt was $6.45 \times 10^6$ and $4.94 \times 10^6$ cfu/g, respectively, at the first day after production, which became $2.30 \times 10^6$ and $2 \times 10^6$ cfu/g, respectively, on day 7. Both probiotic bacteria had a constant survival rate during the 7 d storage time at an average concentration of $4.41 \times 10^6$ and $3.55 \times 10^6$ cfu/g, respectively [34]. A significant correlation has been reported among MetS and increased levels of circulating adhesion molecules such as vascular cell adhesion molecule (VCAM)-1, intracellular cell adhesion molecule (ICAM)-1, and plasminogen activator inhibitor (PAI)-1 [35]. Important reductions in VCAM-1 and PAI-1 levels in the serum of participants with MetS were observed. Furthermore, significant reductions were observed in the fasting glucose and insulin levels of the probiotic yogurt group [34].

The aim of another randomized, placebo-controlled, double-blind study was to evaluate the effect of various doses of a multispecies probiotic supplement on LPS levels and cardiometabolic parameters in 81 obese postmenopausal women after 12 weeks of consumption. The participants were randomly divided into three groups that were given a low or high dose of the supplement or a placebo. At the end, 71 participants (placebo group, n = 24, low-dose probiotic (LD) group, n = 24, and high-dose probiotic (HD) group, n = 23) completed the 12-week study. The probiotic group received sachets containing 2 g of freeze-dried powder of the probiotic mixture. The HD group received HD ($1 \times 10^{10}$ CFU per day divided in two equal doses), while the LD group received $2.5 \times 10^9$ CFU per day divided in two equal doses. The probiotic preparation contained nine bacterial strains: *Bifidobacterium bifidum W23*, *Bifidobacterium lactis W51*, *Bifidobacterium lactis W52*, *Lactobacillus acidophilus W37*, *Lactobacillus brevis W63*, *Lactobacillus casei W56*, *Lactobacillus salivarius W24*, *Lactococcus lactis W19*, and *Lactococcus lactis W58*. All strains were present in almost the same amounts, while the viability of the strains was evaluated every 3 months. The placebo group received the same sachets containing only the excipients, i.e., maize starch and maltodextrins. All participants were required to consume two sachets per day, one in the morning and one at the end of the day, after dissolving the contents in a glass of room-temperature water. The participants were asked to continue their habitual physical activity and usual diets and to mention eventual side effects. At the beginning of the study and after 12 weeks of treatment, the anthropometric parameters were measured, and all laboratory tests were performed for each group. After 12 weeks of supplementation in both groups of probiotics, valuable modifications were found in the estimated parameters, but not in the placebo group. High-dose probiotic supplementation for 12 weeks reduced LPS, waist circumference, fat mass, subcutaneous fat, total cholesterol (TC), triglycerides (TG), LDL, and glucose. The enrichment with low-dose probiotics altered the following parameters: waist circumference, fat mass, fat percentage, visceral fat, subcutaneous fat, TC, and LDL. The study concluded that probiotic supplementation ameliorated the intestinal barrier, prevented the translocation of LPS into the circulation, and led to a reduction in inflammation [36].

Table 2 summarizes the actions of probiotics, in relation to the recent studies described previously.

**Table 2.** Summary results of probiotic effects on obesity and MetS.

| Study Type | Study Sample/Duration | Participants | Protocol | Summary of Results | Study Reference |
|---|---|---|---|---|---|
| Prospective interventional study | 45 obese patients/1 month | 42 women, 3 men/33–63 y. | Allocation of patients into three groups: low-calorie diet alone, prebiotic supplementation (2 carob beans/day about 30 g), probiotic supplementation one tablet containing an association of four microbiological strains (1 tablet ($10 \times 10^9$ CFU/capsule) /day). | Prebiotic and probiotic groups significantly decreased fat mass and showed a beneficial effect on carbohydrate metabolism, improvement in insulinemia in the prebiotic group compared to the diet-alone group. The probiotic group showed an improvement in fasting blood glucose compared to the diet group. | [31] |
| Randomized, double-blind, placebo-controlled, parallel pilot study | 40 patients/3 weeks | 30–69 years/with MetS, obesity, and arterial hypertonia (>130/85 mm Hg) | 25 subjects consumed 50 g/day of probiotic cheese and 15 consumed the same amount of cheese without probiotics. | Hypocaloric diet with the addition of 50 g of a probiotic cheese reduced arterial BP, BMI, TC, LDL plasma TG and the risk of MetS in obese patients with hypertension. | [32] |
| A randomized trial. | 51 patients with MetS/45 days | 18–60 years | Control group or untreated patients (n25) and a probiotic group (n26). The probiotic groups were required to consume 80 mL of probiotic milk containing, on average, $3.4 \times 10^8$ CFU/mL of *B. animalis* ssp. lactis ssp. | *B. lactis* significantly ameliorated lipid profile, BMI, and cytokine levels (TNF-a) and (IL-6) leading to an improvement in lipid metabolism and weight loss in patients with MetS. | [33] |
| A randomized, double-blind, placebo-controlled clinical trial | 44 patients/8 weeks | 22 men and 22 women with MetS, 20 to 65 years old. | Participants were randomly divided into two groups, a treatment or control group and consumed 300 g/d of probiotic yogurt containing Bifidobacterium lactis and *Lactobacillus acidophilus* or a regular yogurt for 8 weeks, respectively. | Consumption of probiotic yogurt significantly decreased endothelial function markers VCAM-1 and PAI-1 in the serum, ameliorated fasting blood glucose. | [34] |
| A randomized, placebo-controlled, double-blind intervention | 71 participants/12-week | 71 obese postmenopausal women | Division into three groups that received a placebo (i.e., maize starch and maltodextrins), a low-dose ($2.5 \times 10^9$ colony-forming units (CFU) per day, or a high-dose ($1 \times 10^{10}$ colony-forming units (CFU) per day of lyophilizate powder containing live multispecies probiotic bacteria. | Improvements in waist circumference, lipid profile, visceral fat, glucose metabolism, and LPS concentration in obese postmenopausal women were noted. | [36] |

### 3.3. Effects of Prebiotics on the Intestinal Microbiome, Obesity and Metabolic Syndrome

In a 12-week randomized controlled trial, the aim was to evaluate the influence of prebiotic consumption on metabolic endotoxemia and systemic inflammation in overweight and obese adults. Thirty-seven participants were randomly separated into two groups. One group of 20 subjects consumed 7 g of oligofructose three times a day, and the second group of 17 subjects received an isocaloric maltodextrin placebo. Plasma LPS was reduced by 40% in the oligofructose group while in the placebo group it increased by 48%. Higher levels of LPS and PAI-1 may contribute to the complications of obesity. For this reason prebiotics can help to avoid obesity-related comorbidities [37]. Another triple-blind randomized controlled trial investigated the effects of inulin on inflammatory markers and metabolic endotoxemia in patients with type 2 diabetes. The participants were 49 diabetic women randomly divided into two groups. An intervention group of 24 subjects and a control group of 25 subjects consumed 10 g/d inulin or maltodextrin for 8 weeks, respectively. In the inulin group only, there was a remarkable reduction in body weight and BMI. In the intervention group, in contrast with the control group, energy intake and total fat were significantly decreased. Additionally, in the intervention group there was an important reduction in fasting glucose, HbA1c, fasting insulin and an important reduction in CRP, TNF-a and LPS. These results led to the conclusion that inulin supplementation may ameliorate glycemic status and modulate inflammation and metabolic endotoxemia in women with type 2 diabetes [38].

The aim of another randomized, placebo-controlled, double-blind, 6-week crossover trial was to evaluate the prebiotic effect of inulin-type fructans on SCFAs and gut microbiota in patients with type 2 diabetes. Participants had a BMI $\leq$ 40 kg/m$^2$, HbA1c < 10.0% (86 mmol/mol), and were not treated with insulin or glucagon-like peptide-1 (GLP-1) analogues. For two periods of 6 weeks separated by a 4-week wash-out, the subjects received 16 g per day of inulin-type fructans (a 50/50 mixture of oligofructose and inulin and placebo (maltodextrin 16 g per day) in addition to their usual diet and in a randomized order. The supplements were powdered, were very much alike in colour and taste, and were given in identical and non-transparent portion packages of 8 g. During the first week, the participants were required to consume only 8 g and for the remaining 5 weeks, switched to 16 g per day. Subjects added the supplements to food or drinks and ate them whenever they preferred. Unused packages were returned to measure compliance. Importantly, higher faucal concentrations of bifidobacteria and total SCFA, acetic acid and propionic acid were observed after prebiotic consumption in comparison to placebo. The prebiotic fiber had no effects on the concentration of butyric acid or on the variability of gut microflora. Bifidobacteria are not able to produce butyric acid themselves; however, through cross-feeding, various species metabolize non-digestible carbohydrates through various steps. The bifidobacteria help with their capacity to degrade fructan chains and in this mode prepare for other species to complete the fermentation. Bifidobacteria may also help to prevent endotoxemia and ameliorate regulation of blood glucose. Increased faucal concentrations of SCFA were found, indicating alterations of the microbial metabolic activity in the gut. Total SCFA, acetic acid and propionic acid increased. The results of this study indicate the modest potential of inulin-type fructans to ameliorate gut microbiota composition and increase microbial fermentation in type 2 diabetes [39].

In a single-center, double-blind, placebo-controlled trial conducted in two separate cohorts, the aim was to estimate the impact of prebiotic supplementation on gut microbiota, lipid profile, body composition, insulin concentrations and serum inflammatory markers, in children with overweight and obesity. Subjects were separated to either prebiotic oligofructose-enriched inulin or placebo-control maltodextrin for 16 weeks. The participants consumed either 8 g/day (13.2 kcal/d) of oligofructose-enriched inulin or an isocaloric dose of 3.3 g/day of maltodextrin placebo. The prebiotic and placebo were given as a powder and in pre-weighed individual packets. They were asked to dissolve an entire packet in 250 mL of water in a supplied reusable water bottle. They were told to consume half the dose for the first 2 weeks, for adaptation, and to reduce gastrointestinal symptoms, and to take the full dose for the remaining 14 weeks, 15–20 min before their evening meals. To estimate compliance, unused and empty packets were returned. Supplementation with prebiotics led to significant shaping of the gut microbiota. A quantitative polymerase chain reaction demonstrated a remarkable increase in Bifidobacterium spp. and a reduction in *Bacteroides vulgatus*. There was a normalization in weight gain and a decrease in body fat and trunk fat. Children who received oligofructose-enriched inulin also had an important reduction in IL 6 and there was a significant decrease in serum TG [40].

Table 3 summarizes the effects of prebiotics, consistent with the above recent studies.

**Table 3.** Summary results of prebiotics effects on human microbiome related to metabolic diseases.

| Study Type | Study Sample/Duration | Participants | Protocol | Summary of Results | Study Reference |
|---|---|---|---|---|---|
| A randomized, double-blind, placebo-controlled trial | 37 participants/12 weeks | Adults with overweight and obesity. | 21 g of oligofructose or a maltodextrin placebo. | 40% reduction in LPS concentrations in the oligofructose group compared to a 48% increase in the placebo group. PAI-1 was reduced to a greater extent in the oligofructose group. | [37] |
| Triple-blind randomized controlled study | 54 patients/8 weeks | diabetic females aged 20–65 years. | The intervention group consumed 10 g/d inulin supplement and the control group consumed similar amounts of maltodextrin. | Inulin supplementation modulates metabolic endotoxemia and inflammation in women with type 2. | [38] |
| A placebo-controlled crossover study | 25 patients (15 men)/6 weeks | Adult men and women with type 2 diabetes/41–71 y. | Consumption of 16 g of inulin-type fructans (a mixture of oligofructose and inulin) and 16 g placebo (maltodextrin) in a randomized order. | Inulin-type fructans led to moderate improvement in gut microbiota composition. | [39] |
| Single-center, double-blind placebo-controlled trial | 38 children /16 weeks | Male and female children, 20 in the prebiotic group and 18 in the control Group, aged 7–12 years with overweight or obesity | Supplementation of either oligofructose-enriched inulin 8 g/day or maltodextrin placebo once daily for 16 weeks. | Consumption of Prebiotics normalized weight gain, ameliorated percent body fat and significantly modified gut microbial Composition enriching concentrations of *Bifidobacterium* spp. | [40] |

### 3.4. Effects of Phytochemicals on the Intestinal Microbiome, Obesity and Metabolic Syndrome

A randomized double-blind crossover trial investigated the hypolipidemic activity of curcumin in obese subjects. Each participant consumed curcuminoids or placebo for 30 days and then switched to the other regimen. Sixteen subjects consumed curcuminoids 1 g/day for 30 days and/or a placebo for 30 days. Curcuminoids were given in capsules containing 500 mg curcuminoids and 5 mg bioperine, an extract made of black pepper (*Piper nigrum* L.) or long pepper (*Piper longum* L.) containing 95% piperine, a well-known bioavailability enhancer. Placebo capsules had the same size and shape as the curcuminoid capsules and contained piperine (5 mg). In this study, it was observed that curcuminoid supplementation (1 g/day for 30 days) intensely decreased plasma TG, possibly due to the insulin-sensitizing effects of this phytochemical. Dyslipidemia is an important risk factor for cardiovascular disease and is a common constituent of obesity. Curcumin is a bioactive phytochemical with well-documented antioxidant, anti-inflammatory, and cardioprotective properties. Evidence suggests that obese people have an increased risk of IR syndrome and diabetes mellitus. This increased risk is the result of the pathophysiological modifications in the secretion of adipokines and inflammatory cytokines by the adipose tissue. The amelioration of adipokine status, in addition to the anti-inflammatory effects of curcumin, are possible mechanisms responsible for its beneficial effects on IR [41].

The aim of another double-blind, placebo-controlled, crossover study was to investigate the impact of polyphenol-rich tomato extract (Fruitflow) on trimethylamine N-oxide (TMAO) and gut microbiota. Twenty-two overweight and obese adults were separated into two groups and received 2150 mg of extract per day in one group or placebo (maltodextrin) in the other group for 4 weeks, with a 6-week washout period. Fruitflow decreased urine TMAO, which possesses proinflammatory and proatherogenic action, as well as plasma LPS. Furthermore, there were significant modifications of microbial taxa with Fruitflow, such as a reduction in Bacteroides, Ruminococccus, and Hungatella, as well as an increase in Alistipes, a member of Bacteroidetes phylum known for its involvement in TMA/TMAO metabolism. Considering that the production of TMAO from dietary choline is based on metabolism by the intestinal microbiota, alterations in plasma and urine TMAO are dependent on the modifications of gut microbiota. The study concluded that Fruitflow consumption can influence the synthesis of gut microbiota. This study observed an important reduction in the relative abundance of Bacteroides uniformis, Bacteroides ovatus, Bacteroides acidifaciens, and the associated species Parabacteroides goldsteinii. Moreover, Bacteroides sp. was shown to decrease TMAO to TMA. Thus, it is justifiable to assume that a decrease in Bacteroidetes via Fruitflow is a mechanism that clarifies the observed impact on TMAO. In this study, a reduction in *Hungatella hathewayi* and *Rumonococcus faecis* within the Fruiflow group was also observed. *Hungatella hathewayi* is known to produce TMA. Finally, an increase in *Clostridium carnis* was observed. Furthermore, with Fruitflow, there was a decrease in plasma LPS but not in the placebo group. LPS is a pro-inflammatory factor found in the external membranes of Gram-negative bacteria, and is an important inflammatory factor associated with diseases such as obesity. LPS concentrations are higher in the gut lumen, where gut bacteria live, and lower in plasma, because the intestinal epithelial layer creates an effective barrier against LPS invasion. Thus, the reduction in plasma LPS because of Fruitflow consumption, indicates an improvement in intestinal barrier function [42].

In another double-blind, randomized, parallel-group, placebo-controlled pilot study, 28 obese men with MetS were studied over 35 days. Subjects were randomly separated into two groups and were asked to consume 1 g of trans-resveratrol orally twice a day or placebo and an isocaloric, western-style diet. No other foods or drinks were admitted. In this study, they observed the effects of trans-resveratrol on IR, in obese men with MetS. Trans-resveratrol is a natural product found in grapes, berries, and peanuts. This study observed an increase in insulin sensitivity and a reduction in glucose concentrations of 120 min in the oral glucose tolerance test. An enhancement of *A. municiphila* was also observed to be inversely associated with obesity and low- grade inflammation [43].

In another randomized, placebo-controlled, double-blind, single-center study, middle-aged men between 30 and 60 years old with MetS were divided into either resveratrol or placebo treatment for four months. They received tablets containing placebo, low-dose resveratrol (75 mg twice daily) or high-dose resveratrol (500 mg twice a day). This study presented an intensive metabolomic profile of blood, adipose tissue, skeletal muscle tissue, and urine. This approach allowed for a more detailed investigation of resveratrol's impact on specific intracellular pathways than what is possible by measurements limited to blood. Modifications were noticed in several metabolites in urine derived from essential amino acids, which cannot be produced by the body; however, gut microflora may play a role in their production and degradation. Fermentation of these amino acids by colonic bacteria was found to produce phenols and indoles, which are excreted in the urine. Modifications in urinary amino acids and derivatives observed in this study indicate a resveratrol-induced arrangement of gut microbiota in men with MetS [44].

Another double-blind, crossover, randomized, controlled clinical trial studied the effects of olive oils with different phenolic contents, on urinary tyrosol (T) and hydroxytyrosol (HT) levels and LDL oxidation. The participants were 30 healthy non-smoking volunteers. The clinical trial was organized with three types of olive oils: refined, common, and virgin. During olive oil treatment periods, participants received a raw daily dose of 25 mL of olive oil divided over three meals. Other cooking fats were substituted by refined olive oil throughout the study. After consumption of virgin olive oil, a decrease in LDL oxidation was noticed in vivo. Increases in high-density lipoprotein (HDL) and urinary T and HT were also observed. Urinary T and HT increased in a dose-dependent manner with the phenolic content of the olive oils consumed. There is increasing evidence that the Mediterranean diet, in which olive oil is the main source of fat, has a beneficial impact on diseases associated with oxidative damage. Phenolic compounds are antioxidant agents contained in several foods of the Mediterranean diet, among which virgin olive oil is the most important. HT is considered to be a powerful antioxidant in in vitro studies [45].

A 3-week double-blind, cross-over, randomized controlled trial, estimated the impact of daily consumption (60 mL) of high-polyphenolic extra virgin olive oil (HPOO), in comparison to low-polyphenolic olive oil (LPOO), on oxidative status and on inflammatory biomarkers. Olive oil polyphenols have many benefits for cardiovascular health due to their antioxidant and anti-inflammatory properties. Plasma total antioxidant capacity (TAC) is inversely related to chronic disease risk. The consumption of unrefined extra virgin olive oil with a high content of polyphenols led to an important increase in TAC and to a reduction in plasma-oxidized low-density lipoproteins (ox-LDLs). Both oxidative stress and inflammation may cause damage to endothelial cells, boosting a pro-inflammatory response, as shown by the increased expression of endothelial dysfunction markers such as adhesion molecules and cytokines IL-6 and TNF-α. IL-6 and TNF-α are responsible for the secretion of CRP. Ox-LDL is a very immunogenic particle of great importance to the initiation and progression of atheromatic plaque formation within the arterial wall [46].

Table 4 summarizes the effects of polyphenols in agreement with the above recent studies.

**Table 4.** Summary results of the effects of polyphenols on the human microbiome related to obese and MetS.

| Study Type | Study Sample/Duration | Participants | Protocol | Summary of Results | Study Reference |
|---|---|---|---|---|---|
| Randomized, double-blind, placebo-controlled, crossover trial. | 30 obese Participants /30 days | 18–65 y | Curcuminoids (1 g/day), or placebo. | Curcuminoid supplementation (1 g/day for 30 days) reduced triglycerides concentrations in serum. | [41] |
| A double-blind, placebo-controlled, cross-over study | 22 participants /4 weeks | Overweight and obese adults. | 2150 mg of a water-soluble tomato extract rich in polyphenols per day or placebo (maltodextrin) for 4 weeks with a 6-week wash-out between interventions. | Significantly reduced urine TMAO, related with changes in microbial composition. | [42] |
| A double-blind pilot randomized parallel group design placebo-controlled study | 28 participants/35-days | Obese men with MetS | Resveratrol 1 g orally twice daily or placebo while consuming a western-style diet. | Improvement in insulin sensitivity and glucose tolerance, significant increase in *A. muciniphila*. | [43] |

**Table 4.** *Cont.*

| Study Type | Study Sample/Duration | Participants | Protocol | Summary of Results | Study Reference |
|---|---|---|---|---|---|
| A randomized, placebo-controlled, double-blinded, single-center study. | 66 participants/4 months | Male gender, 30–60 y, and MetS. | Randomized to either resveratrol or placebo treatment. | Urinary derivatives of amino acids, which reflect the synthesis of the gut microbiota, were altered after resveratrol treatment. | [44] |
| Double blind, cross-over, randomized, clinical trial | 30 healthy non-smoking volunteers/3 weeks | Adults/both male and female | Common, virgin, and refined olive oils were sequentially consumed over three periods of 3 weeks. | Reduction in vivo of LDL oxidation and increasement of T and HT in urine. | [45] |
| Double-blind cross-over trial | 43 participants/3 weeks | 20–70 y, 66% females, 44% of study participants were overweight and 4% were obese | 60 mL/day of HPOO (320 mg/kg polyphenols) or LPOO (86 mg/kg polyphenols) for three weeks. | Significant reduction in ox-LDL and CRP in plasma and increase in the TAC of the plasma. | [46] |

## 4. Discussion

Obesity is corelated with an excessive accumulation of fat that is harmful for health. Excessive body weight contributes to MetS pathophysiology parameters such as low-grade chronic systemic inflammation, dyslipidemia, altered glucose metabolism, and hypertension [15]. Obesity also presents an alteration of gut microbiota with a decrease in Bifidobacteria and low bacterial diversity [15]. It has been shown by multiple studies that prebiotics, mostly fructooligosaccharides (FOS) and inulin, reduce hunger and energy intake through the augmentation of GLP-1 and PYY, as a response to higher SCFA production [47,48]. Specifically, GLP-1 peptide and PYY peptide are satiety hormones that impede food intake, contributing in this mode to a reduction in body weight and waist circumference in obese or overweight individuals [15,49].

Low-grade chronic systemic inflammation is associated with an enhancement of inflammatory markers such as TNF-a, IL -6, and CRP. and is correlated with MetS and its characteristics such as obesity, IR, and β cell dysfunction. Inflammation increases gut permeability, and this allows bacteria and toxins entry into the systemic circulation. This status is called metabolic endotoxemia. Endotoxemia leads to increased circulating pro-inflammatory factors, such as LPS [4,15], which is found in the external membranes of Gram-negative bacteria and constitutes a significant inflammatory factor [15,50]. Saccharolytic bacteria, such as Bifidobacteria, play a crucial role in epithelial function and endothelial barrier integrity [48].

A possible mechanism of the role of prebiotics in decreasing low-grade chronic systemic inflammation is the increase in Bifidobacteria, which contribute to the reduction in plasma LPS [48,51]. On the other hand, prebiotics may reduce colon inflammation by increasing the production of SCFA, mostly butyrate, which has an anti-inflammatory effect [11,15,52]. One of the characteristics of MetS is elevated fasting plasma glucose levels, which can also result in IR, the reduced response of cells to insulin. In addition, elevated fasting glucose levels and IR lead to insulin secretion from pancreatic cells and increase fasting insulin levels [15]. People with IR exhibit decreased bacterial diversity and decreased levels of beneficial bacteria. The reduction in Bifidobacteria contributes to a pro-inflammatory state, as previously described. In the circulation, LPS and proinflammatory cytokines alter β-cell function, leading to fasting hyperglycemia and hyperinsulinemia [15]. Prebiotic consumption decreases LPS levels and endotoxemia and improves glucose tolerance [15]. Insoluble prebiotic fiber improves insulin sensitivity and insulin action by increasing SCFA production [11,48]. In addition, prebiotic fibers reduce gastric emptying and glucose absorption and regulate insulin response. Furthermore, prebiotics improve insulin sensitivity due to the production of GLP-1, [14] which enhances insulin secretion and β cell proliferation in the pancreas [15].

Dyslipidemia is a major risk factor for cardiovascular disease. Atherogenic dyslipidemia presents increased levels of LDL and decreased levels of HDL. In addition, increased levels of TG are associated with increased cardiovascular risk [15]. Multiple studies have indicated that metabolic endotoxemia contributes to atherosclerosis and cardiovascular disease, which is caused by an augmentation of TC and a decrease in HDL. A strategy for controlling dyslipidemia is consuming a diet abundant in dietary fiber, fruits, vegetables, and soluble fibers. Soluble prebiotic fibers bind cholesterol and impede its absorption, increasing the excretion of cholesterol in the feces [15]. In addition, prebiotics interact with the intestinal metabolism of bile acids and reduce cholesterol reabsorption. [11] Prebiotics have a lipid-lowering effect, by increasing the production of SCFA, which inhibits the synthesis of TG and cholesterol [11,15,48].

On the other hand, phytochemicals are defined as bioactive non-nutritive plant compounds [53]. Polyphenols are secondary metabolites of plants and represent a great variety of phytochemicals [41,53–57]. They are usually found in plant foods, such as coffee, cocoa, olives, fruits (apples, berries), nuts, some vegetables (lettuce, cabbage), wine and tea, and their consumption as part of the regular diet has been linked to a reduction in the risk of chronic diseases [53,55,58]. The interaction between polyphenols and microflora is well-documented. Only a very small part of polyphenols is absorbed in the small intestine and the rest (90–95%) reach the colon in elevated concentrations, where they are degraded by gut microbiome before being absorbed [7,53,54,59,60] and finally excreted in the urine [41]. Dietary polyphenols are metabolized, as previously mentioned, by the microflora, regulating their composition and generating a series of metabolites [5,7,12,53,59,61]. Polyphenols prevent weight gain, reduce lipid levels in the circulation and reduce the endotoxemia and systemic inflammation induced by high-fat diets. All these results together indicate that polyphenols may have an important role in the control of obesity and MetS [12,54].

Although a decreased intake of polyphenols does not cause specific deficiency diseases, adequate intake presents a lot of possible benefits, specifically for chronic diseases [58]. About 8000 structures of polyphenols have been identified [41,55–58], which are classified into four main groups as follows:

(A) Flavonoids. The first and biggest subgroup of polyphenols, with more than 6000 compounds consisting of 7 main subgroups [41,53,56];

(B) Phenolic Acids (e.g., curcumin). Intestinal microflora metabolize curcumin, which ameliorates the function of intestinal barrier and is considered a potent anti-inflammatory and neuroprotective agent as well as a potential agent for the treatment of obesity [53,61];

(C) Lignans. They are commonly found in oilseeds (sesame), whole grains (wheat, barley), legumes, fruits, and vegetables (mainly berries) and beverages, such as tea, coffee, and wine and, recently, to dairy products, meat and fish. They are related to a possible anti-cancer effect and prevention of cardiovascular diseases [62];

(D) Stilbenoids (e.g., resveratrol). Resveratrol is a natural, non-flavonoid polyphenolic constituent commonly found in grape wines, grape skins (red wine), cranberries, legumes, berries, which may contribute to the French paradox [63]. This phenomenon was initially described by Samuel Black in 1819, and later, in 1992, was named the French paradox by Dr. Renaud. It has been proposed that the primary reason for this phenomenon was the high consumption of red wine in France [64].

Gut microflora participate actively in resveratrol metabolism, increasing its bioavailability [63]. Resveratrol impede effects of LPS in the gut, preventing, in this mode, changes in gut permeability and thereby reducing endotoxemia, low-grade inflammation, and obesity [61,63]. Recent studies show that resveratrol modifies the composition of the gut microflora, leading to lower body weight and fat, improved glucose homeostasis, parameters related to obesity. It seems to be associated with the alteration of bacterial populations by inducing a composition related to a healthy phenotype [5,63]. Studies suggest that resveratrol enhances the population of Bifidobacterium and Lactobacillus [11].

In addition, olive polyphenols have both prebiotic and antimicrobial effects on gut microflora. Olive polyphenols are secondary metabolites of the plant and protect it from parasites and bacterial infections [55,60,65]. High-quality extra virgin olive oils (EVOO) contain a higher concentration of phenolics than refined oils [55,65]. The olive leaf has even greater concentrations of phenolic compounds in comparison to the olive fruit and olive oil. The fresh leaf contains 1350 mg/kg, against 232 ± 15 mg/kg contained in extra virgin olive oil. There are dietary supplements made with olive leaf extract and is being considered the possibility of enriching edible oils with phenolics extracted from olive leaves. The most abundant phenolic compound contained in olive leaves, pulp, seeds, and skin is oleuropein [65]. Furthermore, olive polyphenols impede pro-adipocyte differentiation, decrease adipogenesis, and promote lipolysis and adiponectin secretion [64,65]. Olive polyphenols improve carbohydrate metabolism and decrease hyperglycemia and IR [65,66]. In animal models, it was shown that dietary supplementation with olive polyphenols shows an anti-hyperglycemic effect [57,65].

Dyslipidemia indicates abnormal concentrations of lipids and lipoproteins in the circulation, which leads to dysbiosis of gut microflora and deterioration of lipid disorders. Dyslipidemia, increased fasting triglyceride-rich lipoproteins, reduced HDL, and increased LDL, are indications of MetS [19,65]. Olive oil, as a result of its high MUFA and polyphenols content, decreases TC, LDL and TG in humans, and elevates HDL levels [56,64,65]. On the other hand, the oxidation of LDL caused by free radicals, triggers lipoprotein uptake by macrophages, foam cell development and inflammation. Oxidized LDL contributes to an increased incidence of MetS. The European Food Safety Authority has recognized health claims about the effectiveness of olive oil with high concentrations of HT in controlling lipid peroxidation [5,65,66]. It has been suggested that in vitro protection versus oxidative stress is due to HT and its derivatives, but this has not been proven in humans. In vitro, olive polyphenols neutralize LDL oxidation by both metals and radicals. In vivo antioxidant activities of dietary polyphenols are much lower due to their relatively low bioavailability and fast clearance [55,65].

Probiotics are live microorganisms that can have beneficial effects on the body when administered in sufficient amounts [10,18,22,67–70]. These microorganisms do not colonize the gut permanently and must remain alive all through the digestive system [52,71,72]. They reach the colon temporarily and grow, influencing the composition of the gut microbiota and the production of beneficial metabolites derived from fermentation [72,73]. The most frequent probiotic genera contained in various functional foods and nutritional supplements are Bifidobacterium and Lactobacillus strains [7,18,22,52,70]. The main probiotic bacteria found in dairy products are *Lactobacillus casei, Lactobacillus acidophilus* and *Bifidobacterium* [3]. The biological mechanisms through which they exert their effects on health are discussed in [10,18,22,49,52,67,68,70,71,74,75] and are as follows:

1. Antagonistic adherence to epithelium and mucosa with pro-inflammatory microbes;
2. Adjustment of the gut-associated lymphoid immune system over intestinal cell pattern recognition receptors (Toll-like receptors) or through the release of metabolites;
3. Bile acid deconjugation by *Lactobacillus* strains, which reduce lipid absorption and calorie absorption;
4. Initiation of lipolysis;
5. Obliteration of fat deposition, because of an increase in angiopoietin (inhibitor lipoprotein lipase);
6. Stimulation of genes linked to β-oxidation of fatty acids in the liver and in the muscles;
7. Obliteration of fatty acid synthetase in the liver;
8. Ameliorating insulin sensitivity and glucose tolerance through the production of SCFA and diminishing LPS translocation;
9. Ameliorating the function of the intestinal barrier through SCFA production;
10. Regulation of appetite.

The traditional Mediterranean diet is followed in Mediterranean countries, where the moderate climate supports the production of fruits and vegetables all through the year. It is a nutritionally recommended pattern rich in fruits and vegetables, legumes, nuts, whole grains, and herbs, where olive oil is the main source of fat [5,51,73,76,77]. Furthermore, there is limited consumption of dairy products (mostly fermented products such as cheese and yogurt), fish, and eggs, and red wine is consumed in small proportions along with meals. This type of diet presents a limited meat consumption once a week, mostly lean meats (chicken, turkey, rabbit) [51,56,73,77]. Onion and garlic (abundant in prebiotics) are used, as well as aromatic herbs (oregano, rosemary, basil) and spices (cinnamon, pepper, cumin, saffron), rich in antioxidant and anti-inflammatory components. Greater consumption of plant-origin proteins (mainly from legumes and whole grains), rather than animal proteins, increase the beneficial effects of the Mediterranean diet [4,73,77].

A Mediterranean-type diet present a different microbiome than that associated with a Western-type diet. The microflora, associated with a Mediterranean-type diet, present higher diversity (i.e., a greater number of bacterial species), which exert a positive impact on human health. Mediterranean diet adherence protects against dysbiosis, and is associated with higher SCFA levels, decreased TMAO levels, and the enhancement of fiber-degrading Prevotella and Firmicutes [2,73].

The limited number of studies considered in this review is a limitation; therefore, further analysis of the existed bibliography is needed for more clear results.

## 5. Conclusions

Among the factors that influence the composition of gut microbiota, dietary habits play a key role [5,6]. Several lines of evidence indicate that some dietary patterns are more effective than others in shaping the gut microbiota. In particular, the Mediterranean diet is rich in beneficial nutrients and bioactive compounds, probiotics, prebiotics, and polyphenols and is remarkably effective in modifying the composition of gut microflora and influencing inflammatory processes and subsequent endotoxemia [73]. The future challenge is to completely understand how different dietary patterns can modify gut microbiota composition and whether these changes are long-lasting. Only a robust knowledge will allow the microbial community to be stabilized towards a healthy phenotype and reduce the risk of obesity and MetS.

**Author Contributions:** Conceptualization, A.E.K. and D.K.; methodology, D.K., V.K. and A.E.K.; validation, A.E.K.; resources, D.K. and V.K.; data curation, A.E.K.; writing—original draft preparation, D.K. and V.K.; writing—review and editing, A.E.K.; visualization, A.E.K.; supervision, A.E.K. All authors have read and agreed to the published version of the manuscript.

**Funding:** This research received no external funding.

**Data Availability Statement:** Data are available from the corresponding author on request.

**Conflicts of Interest:** The authors declare no conflicts of interest.

## References

1. Srivastava, A.; Prabhakar, M.R.; Mohanty, A.; Meena, S.S. Influence of gut microbiome on the human physiology. *Syst. Microbiol. Biomanuf.* **2022**, *2*, 217–231. [CrossRef]
2. Croci, S.; D'Apolito, L.I.; Gasperi, V.; Catani, M.V.; Savini, I. Dietary Strategies for Management of Metabolic Syndrome: Role of Gut Microbiota Metabolites. *Nutrients* **2021**, *13*, 1389. [CrossRef] [PubMed]
3. Mazidi, M.; Rezaie, P.; Kengne, A.P.; Mobarhan, M.G.; Ferns, G. Gut microbiome and metabolic syndrome. *Diabetes Metab. Syndr. Clin. Res. Rev.* **2016**, *10*, S150–S157. [CrossRef] [PubMed]
4. Beam, A.; Clinger, E.; Hao, L. Effect of Diet and Dietary Components on the Composition of the Gut Microbiota. *Nutrients* **2021**, *13*, 2795. [CrossRef] [PubMed]
5. Montero, C.G.; Martínez, O.F.; Gómez-Lahoz, A.M.; Pekarek, L.; Castellanos, A.J.; Noguerales-Fraguas, F.; Coca, S.; Guijarro, L.; García-Honduvilla, N.; Asúnsolo, A.; et al. Nutritional Components in Western Diet Versus Mediterranean Diet at the Gut Microbiota–Immune System Interplay. Implications for Health and Disease. *Nutrients* **2021**, *13*, 699. [CrossRef]
6. Wang, B.; Yao, M.; Lv, L.; Ling, Z.; Li, L. The Human Microbiota in Health and Disease. *Engineering* **2017**, *3*, 71–82. [CrossRef]

7. Conlon, M.; Bird, A. The Impact of Diet and Lifestyle on Gut Microbiota and Human Health. *Nutrients* **2015**, *7*, 17–44. [CrossRef]
8. Magne, F.; Gotteland, M.; Gauthier, L.; Zazueta, A.; Pesoa, S.; Navarrete, P.; Balamurugan, R. The Firmicutes/Bacteroidetes Ratio: A Relevant Marker of Gut Dysbiosis in Obese Patients? *Nutrients* **2020**, *12*, 1474. [CrossRef]
9. Siqueira, J.S.; Palacio TL, N.; Vieira, T.A.; Nakandakare-Maia, E.T.; Grandini, N.A.; Togneri Ferron, A.J.; Francisqueti-Ferron, F.V.; Correa, C.R. An overview of the complex interaction between obesity and target organ dysfunction: Focus on redox-inflammatory state. *Nutrire* **2023**, *48*, 21. [CrossRef]
10. Cerdó, T.; García-Santos, J.A.; Bermúdez, M.G.; Campoy, C. The Role of Probiotics and Prebiotics in the Prevention and Treatment of Obesity. *Nutrients* **2019**, *11*, 635. [CrossRef]
11. Laparra, J.M.; Sanz, Y. Interactions of gut microbiota with functional food components and nutraceuticals. *Pharmacol. Res.* **2010**, *61*, 219–225. [CrossRef]
12. Dzah, C.M.; Asante-Donyina, D.; Letsyo, E.; Dzikunoo, J.; Suglo Adams, Z. Dietary Polyphenols and Obesity: A Review of Polyphenol Effects on Lipid and Glucose Metabolism, Mitochondrial Homeostasis, and Starch Digestibility and Absorption. *Plant Foods Hum. Nutr.* **2023**, *78*, 1–12. [CrossRef]
13. Kawadaa, T. Food-derived regulatory factors against obesity and metabolic syndrome. *Biosci. Biotechnol. Biochem.* **2018**, *82*, 547–553. [CrossRef]
14. Li, H.Y.; Zhou, D.D.; Gan, R.Y.; Huang, S.Y.; Zhao, C.N.; Shang, A.; Xiao-Yu Xu, X.Y.; Li, H.B. Effects and Mechanisms of Probiotics, Prebiotics, Synbiotics, and Postbiotics on Metabolic Diseases Targeting Gut Microbiota: A Narrative Review. *Nutrients* **2021**, *13*, 3211. [CrossRef]
15. O'Connora, S.; Chouinard-Castonguaya, S.; Gagnona, C.; Iwona Rudkowskaa, I. Prebiotics in the management of components of the metabolic syndrome. *Maturitas* **2017**, *104*, 11–18. [CrossRef]
16. Green, M.; Arora, K.; Prakash, S. Microbial Medicine: Prebiotic Probiotic Functional Foods to Target Obesity Metabolic Syndrome. *Int. J. Mol. Sci.* **2020**, *21*, 2890. [CrossRef]
17. Tomasello, G.; Mazzola, M.; Leone, A.; Sinagra, E.; Zummo, G.; Farina, F.; Damianic, P.; Cappello, F.; Geagea, A.; Jurjus, A.; et al. Nutrition, oxidative stress and intestinal dysbiosis: Influence of diet on gut microbiota in inflammatory bowel diseases. *Biomed. Pap.* **2016**, *160*, 461–466. [CrossRef]
18. Ferrarese, R.; Ceresola, E.R.; Preti, A.; Canducci, F. Probiotics, prebiotics and synbiotics for weight loss and metabolic syndrome in the microbiome era. *Eur. Rev. Med. Pharmacol. Sci.* **2018**, *22*, 7588–7605.
19. Wang, P.X.; Deng, X.R.; Zhang, C.H.; Yuan, H.J. Gut microbiota and metabolic syndrome. *Chin. Med. J.* **2020**, *133*, 808–816. [CrossRef]
20. Las Heras, V.; Melgar, S.; Mac Sharry, J.; Gahan, C. The Influence of the Western Diet on Microbiota and Gastrointestinal Immunity. *Annu. Rev. Food Sci. Technol.* **2022**, *13*, 489–512. [CrossRef]
21. Harley, I.; Karp, C. Obesity and the gut microbiome: Striving for causality. *Mol. Metab.* **2012**, *1*, 21–31. [CrossRef]
22. Peng, M.; Tabashsum, Z.; Anderson, M.; Truong, A.; Houser, A.; Padilla, J.; Akmel, A.; Bhatti, J.; Rahaman, S.; Biswas, D. Effectiveness of probiotics, prebiotics, and prebiotic-like components in common functional foods. *Compr. Rev. Food Sci. Food Saf.* **2020**, *19*, 1908–1933. [CrossRef]
23. Luisi, M.; Lucarini, L.; Biffi, B.; Rafanelli, E.; Pietramellara, G.; Durante, M.; Vidali, S.; Provensi, G.; Madiai, S.; Francesca Gheri, C.; et al. Effect of Mediterranean Diet Enriched in High Quality Extra Virgin Olive Oil on Oxidative Stress, Inflammation and Gut Microbiota in Obese and Normal Weight Adult Subjects. *Front. Pharmacol.* **2019**, *10*, 1366. [CrossRef]
24. Meslier, V.; Laiola, M.; Munch, H.; De Filippis, F.; Roume, H.; Quinquis, B.; Giacco, R.; Mennella, I.; Ferracane, R.; Pons, N.; et al. Mediterranean diet intervention in overweight and obese subjects lowers plasma cholesterol and causes changes in the gut microbiome and metabolome independently of energy intake. *Gut* **2020**, *69*, 1258–1268. [CrossRef]
25. Henke, M.T.; Kenny, D.J.; Cassilly, C.D.; Vlamakis, H.; Xavier, R.; Clardy, J. Ruminococcus gnavus, a member of the human gut microbiome associated with Crohn's disease, produces an inflammatory polysaccharide. *Proc. Natl. Acad. Sci. USA* **2019**, *116*, 12672–12677. [CrossRef]
26. Vital, M.; Karch, A.; Pieper, D. Colonic butyrate- producing communities in humans: An overview using omics data. *mSystems* **2017**, *2*, e00130-17. [CrossRef]
27. Ismael, S.; Marta, P.; Silvestre, M.; Vasques, M.; Araújo, J.; Morais, J.; Duarte, M.I.; Pestana, D.; Faria, A.; Pereira-Leal, J.; et al. A Pilot Study on the Metabolic Impact of Mediterranean Diet in Type 2 Diabetes: Is Gut Microbiota the Key? *Nutrients* **2021**, *13*, 1228. [CrossRef]
28. Lalle, J. Recent advances in intestinal alkaline phosphatase, inflammation, and nutrition. *Nutr. Rev.* **2019**, *77*, 710–724. [CrossRef]
29. Zaher, D.M.; Aljareh, S.N.; Iqbal, J.; El Gamal, M.I.; Omar, H.A.; Al Shamma, S.A.; Ali, A.J.; Zaib, S. Recent advances with alkaline phosphatase isoenzymes and their inhibitors. *Arch. Pharm.* **2020**, *353*, e2000011. [CrossRef]
30. Gutiérrez-Díaz, I.; Fernández-Navarro, T.; Sánchez, B.; Margollesb, A.; González, S. Mediterranean diet and faecal microbiota: A transversal study. *Food Funct.* **2016**, *7*, 2347–2356. [CrossRef]
31. Othman, R.; Amor, N.; Mahjoub, F.; Berriche, O.; Ghali, C.; Gamoudi, A.; Jamoussi, H. A clinical trial about effects of prebiotic and probiotic supplementation on weight loss, psychological profile and metabolic parameters in obese subjects. *Endocrinol. Diabetes Metab.* **2023**, *6*, e402.

32. Sharafedtinov, K.; Plotnikova, O.; Alexeeva, R.; Sentsova, T.; Songisepp, E.; Stsepetova, J.; Smidt, I.; Mikelsaar, M. Hypocaloric diet supplemented with probiotic cheese improves body mass index and blood pressure indices of obese hypertensive patients—A randomized double-blind placebo-controlled pilot study. *Nutr. J.* **2013**, *12*, 138. [CrossRef]
33. Bernini, L.; Colado Sim~ao, A.; Alfieri, D.; Batisti Lozovoy, M.; Lourenco Mari, N.; Batista de Souza, C.; Dichi, I.; Nobre Costa, G. Beneficial effects of Bifidobacterium lactis on lipid profile and cytokines in patients with metabolic syndrome: A randomized trial. Effects of probiotics on metabolic syndrome. *Nutrition* **2016**, *32*, 716–719. [CrossRef]
34. Rezazadeh, L.; Pourghassem Gargari, B.; Asghari Jafarabadi, M.; Alipour, B. Effects of probiotic yogurt on glycemic indexes and endothelial dysfunction markers in patients with metabolic syndrome. *Nutrition* **2019**, *62*, 162–168. [CrossRef]
35. Blake, G.; Ridker, P. Inflammatory bio-markers and cardiovascular risk prediction. *J. Intern. Med.* **2002**, *252*, 283–294. [CrossRef]
36. Szuli'nska, M.; Łoniewski, I.; van Hemert, S.; Sobieska, M.; Bogda'nski, P. Dose-Dependent Effects of Multispecies Probiotic Supplementation on the Lipopolysaccharide (LPS) Level and Cardiometabolic Profile in Obese Postmenopausal Women: A 12-Week Randomized Clinical Trial. *Nutrients* **2018**, *10*, 773. [CrossRef]
37. Parnell, J.; Klancic, T.; Reimer, R. Oligofructose Decreases Serum Lipopolysaccharide and Plasminogen Activator Inhibitor-1 in Adults with Overweight/Obesity. *Obesity* **2017**, *25*, 510–513. [CrossRef]
38. Dehghan, P.; Pourghassem Gargari, B.; Jafar-Abadi, M.A.; Aliasgharzadeh, A. Inulin controls inflammation and metabolic endotoxemia in women with type 2 diabetes mellitus: A randomized-controlled clinical trial. *Int. J. Food Sci. Nutr.* **2014**, *65*, 117–123. [CrossRef]
39. Birkeland, E.; Gharagozlian, S.; Birkeland, K.; Valeur, J.; Måge, I.; Rud, I.; Aas, A.M. Correction to: Prebiotic efect of inulin-type fructans on faecal microbiota and short-chain fatty acids in type 2 diabetes: A randomised controlled trial. *Eur. J. Nutr.* **2020**, *59*, 3339–3340. [CrossRef]
40. Nicolucci, A.; Hume, M.; Martínez, I.; Mayengbam, S.; Walter, J.; Reimer, R. Prebiotics Reduce Body Fat and Alter Intestinal Microbiota in Children Who Are Overweight or With Obesity. *Gastroenterology* **2017**, *153*, 711–722. [CrossRef]
41. Mohammadi, A.; Sahebkar, A.; Iranshahi, M.; Amini, M.; Khojasteh, R.; Ghayour-Mobarhan, M.; Ferns, G. Effects of Supplementation with Curcuminoids on Dyslipidemia in Obese Patients: A Randomized Crossover Trial. *Phytother. Res.* **2013**, *27*, 374–379. [CrossRef]
42. Rehman, A.; Tyree, S.; Fehlbaum, S.; DunnGalvin, G.; Panagos, C.; Guy, B.; Patel, S.; Dinan, T.; Duttaroy, A.; Duss, R.; et al. A water-soluble tomato extract rich in secondary plant metabolites lowers trimethylamine-n-oxide and modulates gut microbiota: A randomized, double-blind, placebo-controlled cross-over study in overweight and obese adults. *J. Nutr.* **2023**, *153*, 96–105. [CrossRef]
43. Walker, J.; Eckardt, P.; Aleman, J.; Correa da Rosa, J.; Liang, Y.; Iizumi, T.; Etheve, S.; Blaser, M.; Breslow, J.; Holt, P. The effects of trans-resveratrol on insulin resistance, inflammation, and microbiota in men with the metabolic syndrome: A pilot randomized, placebo-controlled clinical trial. *J. Clin. Transl. Res.* **2018**, *4*, 122–135.
44. Korsholm, A.S.; Nordstrøm Kjær, T.; Ornstrup, M.J.; Bønløkke Pedersen, S. Comprehensive Metabolomic Analysis in Blood, Urine, Fat, and Muscle in Men with Metabolic Syndrome: A Randomized, Placebo-Controlled Clinical Trial on the Effects of Resveratrol after Four Months' Treatment. *Int. J. Mol. Sci.* **2017**, *18*, 554. [CrossRef]
45. de la Torre, R. Effects of differing phenolic content in dietary olive oils on lipids and LDL oxidation A randomized controlled trial. *Eur. J. Nutr.* **2004**, *43*, 140–147.
46. Sarapis, K.; George, E.; Marx, W.; Mayr, H.; Willcox, J.; Esmaili, T.; Powell, K.; Folasire, O.; Lohning, A.; Garg, M.; et al. Extra virgin olive oil high in polyphenols improves antioxidant status in adults: A double-blind, randomized, controlled, cross-over study (OLIVAUS). *Eur. J. Nutr.* **2022**, *61*, 1073–1086. [CrossRef]
47. Hul, M.V.; Cani, P.D. Targeting Carbohydrates and Polyphenols for a Healthy Microbiome and Healthy Weight. *Curr. Nutr. Rep.* **2019**, *8*, 307–316.
48. Choque Delgado, G.T.; da Silva Cunha Tamashiro, W.M. Role of prebiotics in regulation of microbiota and prevention of obesity. *Food Res. Int.* **2018**, *113*, 183–188. [CrossRef]
49. Santos-Marcos, J.A.; Perez-Jimenez, F.; Camargo, A. The role of diet and intestinal microbiota in the development of metabolic syndrome. *J. Nutr. Biochem.* **2019**, *70*, 1–27. [CrossRef]
50. Malesza, I.J.; Malesza, M.; Walkowiak, J.; Mussin, N.; Dariusz Walkowiak, D.; Aringazina, R.; Bartkowiak-Wieczorek, J.; Madry, E. High-Fat, Western-Style Diet, Systemic Inflammation, and Gut Microbiota: A Narrative Review. *Cells* **2021**, *10*, 3164. [CrossRef] [PubMed]
51. Campaniello, D.; Corbo, M.R.; Sinigaglia, M.; Speranza, B.; Racioppo, A.; Altieri, C.; Bevilacqua, A. How Diet and Physical Activity Modulate Gut Microbiota: Evidence, and Perspectives. *Nutrients* **2022**, *14*, 2456. [CrossRef] [PubMed]
52. Gomaa, E.Z. Human gut microbiota/microbiome in health and diseases. *Antonie Van Leeuwenhoek* **2020**, *113*, 2019–2040. [CrossRef] [PubMed]
53. Carrera-Quintanar, L.; López Roa, R.; Quintero-Fabián, S.; Sánchez, M.; Vizmanos, B.; Ortuño-Sahagún, D. Phytochemicals That Influence Gut Microbiota as Prophylactics and for the Treatment of Obesity and Inflammatory Diseases. *Mediat. Inflamm.* **2018**, *2018*, 9734845. [CrossRef] [PubMed]
54. Duarte, L.; Gasaly, N.; Poblete-Aro, C.; Uribe, D.; Echeverria, F.; Gotteland, M.; Garcia-Diaz, D.F. Polyphenols and their anti-obesity role mediated by the gut microbiota: A comprehensive review. *Rev. Endocr. Metab. Disord.* **2021**, *22*, 367–388. [CrossRef] [PubMed]

55. Gorzynik-Debicka, M.; Przychodzen, P.; Cappello, F.; Kuban-Jankowska, A.; Marino Gammazza, A.; Knap, N.; Wozniak, M.; Gorska-Ponikowska, M. Potential Health Benefits of Olive Oil Plant Polyphenols. *Int. J. Mol. Sci.* **2018**, *19*, 686. [CrossRef] [PubMed]
56. Nani, A.; Murtaza, B.; Sayed Khan, A.; Akhtar Khan, N.; Hichami, A. Antioxidant and Anti- Inflammatory Potential of Polyphenols Contained in Mediterranean Diet in Obesity: Molecular Mechanisms. *Molecules* **2021**, *26*, 985. [CrossRef] [PubMed]
57. Rigacci, S.; Stefani, M. Nutraceutical Properties of Olive Oil Polyphenols An Itinerary from Cultured Cells through Animal Models to Humans. *Int. J. Mol. Sci.* **2016**, *17*, 843. [CrossRef] [PubMed]
58. Boccellino, M.; D'Angelo, S. Anti-Obesity Effects of Polyphenol Intake: Current Status Future Possibilities. *Int. J. Mol. Sci.* **2020**, *21*, 5642. [CrossRef]
59. Diotallevi, C.; Fava, F.; Gobbetti, M.; Tuohy, K. Healthy dietary patterns to reduce obesity-related metabolic disease: Polyphenol-microbiome interactions unifying health effects across geography. *Curr. Opin. Clin. Nutr. Metab. Care* **2020**, *23*, 437–444. [CrossRef]
60. Clark, J.S.; Simpson, B.; Murphy, K. The role of a Mediterranean diet and physical activity in decreasing age-related inflammation through modulation of the gut microbiota composition. *Br. J. Nutr.* **2022**, *128*, 1299–1314. [CrossRef]
61. Deledda, A.; Annunziata, G.; Tenore, G.C.; Palmas, V.; Manzin, A.; Velluzzi, F. Diet-Derived Antioxidants and Their Role in Inflammation, Obesity and Gut Microbiota Modulation. *Antioxidants* **2021**, *10*, 708. [CrossRef] [PubMed]
62. Durazzo, A.; Lucarini, M.; Souto, E.B.; Cicala, C.; Caiazzo, E.; Izzo, A.; Novelino, E.; Santini, A. Polyphenols: A concise overview on the chemistry, occurrence, and human health. *Phytother. Res.* **2019**, *33*, 2221–2243. [CrossRef] [PubMed]
63. Chaplin, A.; Carpéné, C.; Mercader, J. Resveratrol, Metabolic Syndrome, and Gut Microbiota. *Nutrients* **2018**, *10*, 1651. [CrossRef] [PubMed]
64. Bagetta, D.; Maruca, A.; Lupia, A.; Mesiti, F.; Catalano, R.; Romeo, I.; Moraca, F.; Ambrosio, F.A.; Costa, G.; Artese, A.; et al. Mediterranean products as promising source of multi- target agents in the treatment of metabolic syndrome. *Eur. J. Med. Chem.* **2020**, *186*, 111903. [CrossRef] [PubMed]
65. Saibandith, B.; Spencer, J.P.E.; Rowland, I.R.; Commane, D.M. Olive Polyphenols and the Metabolic Syndrome. *Molecules* **2017**, *22*, 1082. [CrossRef] [PubMed]
66. Peyrol, J.; Riva, C.; Amiot, M.J. Hydroxytyrosol in the Prevention of the Metabolic Syndrome and Related Disorders. *Nutrients* **2017**, *9*, 306. [CrossRef] [PubMed]
67. Sanders, M.; Merenstein, D.; Merrifield, C.; Hutkins, R. Probiotics for human use. *Nutr. Bull.* **2018**, *43*, 212–225. [CrossRef]
68. Song, X.; Wang, L.; Liu, Y.; Zhang, X.; Weng, P.; Liu, L.; Zhang, R.; Wu, Z. The gut microbiota–brain axis: Role of the gut microbial metabolites of dietary food in obesity. *Food Res. Int.* **2022**, *153*, 110971. [CrossRef] [PubMed]
69. Hills, R., Jr.; Benjamin, A.; Pontefract, B.; Hillary, R.; Mishcon, H.; Cody, A. Black C, Sutton S, Theberge C, Gut Microbiome: Profound Implications for Diet and Disease. *Nutrients* **2019**, *11*, 1613. [CrossRef]
70. Adak, A.; Khan, M. An insight into gut microbiota and its functionalities. *Cell. Mol. Life Sci.* **2019**, *76*, 473–493. [CrossRef]
71. Fontané, L.; Benaiges, D.; Goday, A.; Llauradó, G.; Pedro-Botet, J. Influence of the microbiota and probiotics in obesity. *Clin. Investig. Arterioscler.* **2018**, *30*, 271–279. [CrossRef] [PubMed]
72. Leeming, E.; Johnson, A.; Spector, T.; Le Roy, C. Effect of Diet on the Gut Microbiota: Rethinking Intervention Duration. *Nutrients* **2019**, *11*, 2862. [CrossRef] [PubMed]
73. Merra, G.; Noce, A.; Marrone, G.; Cintoni, M.; Tarsitano, M.G.; Capacci, A.; De Lorenzo, A. Influence of Mediterranean Diet on Human Gut Microbiota. *Nutrients* **2021**, *13*, 7. [CrossRef] [PubMed]
74. Abenavoli, L.; Scarpellini, E.; Colica, C.; Boccuto, L.; Salehi, B.; Sharifi-Rad, J.; Aiello, V.; Romano, B.; De Lorenzo, A.; Izzo, A.; et al. Gut Microbiota and Obesity: A Role for Probiotics. *Nutrients* **2019**, *11*, 2690. [CrossRef] [PubMed]
75. Marchesi, J.; Adams, D.; Fava, F.; Hermes, G.; Hirschfield, G.; Hold, G.; Quraishi, M.N.; Kinross, J.; Smidt, H.; Tuohy, K.; et al. The gut microbiota and host health: A new clinical frontier. *Gut* **2016**, *65*, 330–339. [CrossRef] [PubMed]
76. De Filippis, F.; Pellegrini, N.; Vannini, L.; Jeffery, I.B.; La Storia, A.; Laghi, L.; Serrazanetti, D.I.; Cagno, R.; Ferrocino, I.; Lazzi, C.; et al. High-level adherence to a Mediterranean diet beneficially impacts the gut microbiota and associated metabolome. *Gut* **2016**, *65*, 1812–1821. [CrossRef]
77. Mitsou, E.; Aimilia Kakali, A.; Antonopoulou, S.; Mountzouris, K.; Yannakoulia, M.; Panagiotakos, D.; Kyriacou, A. Adherence to the Mediterranean diet is associated with the gut microbiota pattern and gastrointestinal characteristics in an adult population. *Br. J. Nutr.* **2017**, *117*, 1645–1655. [CrossRef]

*applied sciences*

MDPI

*Review*

# High-Temperature Short-Time and Ultra-High-Temperature Processing of Juices, Nectars and Beverages: Influences on Enzyme, Microbial Inactivation and Retention of Bioactive Compounds

**Natalia Polak, Stanisław Kalisz * and Bartosz Kruszewski ***

Department of Food Technology and Assessment, Institute of Food Sciences, Warsaw University of Life Sciences—SGGW, 02-776 Warsaw, Poland; natalia_polak@sggw.edu.pl
* Correspondence: stanislaw_kalisz@sggw.edu.pl (S.K.); bartosz_kruszewski@sggw.edu.pl (B.K.)

**Abstract:** HTST (high-temperature short-time) pasteurization and UHT (ultra-high-temperature) sterilization are techniques commonly used in the dairy industry. Although the use of these methods in fruit and vegetable processing is also well known, the multitude of diverse food matrices determines the need to test and adjust process parameters in order to obtain the best quality of the final product. HTST and UHT are methods that provide effective inactivation of microorganisms and enzymes. Despite the fact that UHT and HTST are thermal processes that cause degradation of bioactive ingredients or color change, in many cases, these two methods are superior to traditional pasteurization, which uses significantly longer exposures to high temperatures. Therefore, this article aims to review the effect of HTST and UHT processing on the quality of juices, nectars and beverages, taking into consideration the quality characteristics, like the presence of microorganisms, pH, titratable acidity, total soluble solids, turbidity, color parameters, contents of bioactive components, antioxidant activity, enzymatic activity and volatile compounds. The impacts of HTST and UHT methods on various food products are discussed, including the food matrix, preservation parameters and the mechanism of interaction. The ability to modify the processing parameters can allow for the selection of adequate preservation parameters for individual products and better results than other unconventional methods, such as HPP (high-pressure processing) or PEF (pulsed electric field). Based on the cited literature, it can be concluded that pH, titratable acidity and TSS most often experience slight changes. As for the other parameters considered, it is extremely important to choose the right temperature and duration for a specific food matrix.

**Keywords:** HTST; UHT; thermal preservation of food; liquid food; bioactive compounds; phenolics; antioxidant capacity; vitamin C; enzyme activity; volatiles profile

## 1. Introduction

Regular consumption of fruits and vegetables has been proven to improve human well-being and reduce incidences of type II diabetes, cancer, obesity and other diet-related diseases [1]. Fruits and vegetables exhibit anti-inflammatory, antihypertensive, and antioxidant properties. This is due to the presence of bioactive components, such as fiber, vitamins, phenolics and carotenoids [2,3]. Because of their short shelf-lives, it is necessary to process them properly and preserve the resulting products with available methods.

The most common method of preserving fruit and vegetable products is pasteurization [4], generally using temperatures of between 80 and 90 °C and a time duration of several minutes. Despite ensuring adequate enzyme inactivation and microbiological stability, it usually leads to negative quality changes [5]. Duhan and Kar [6] indicated that traditional pasteurization of sugarcane juice (70 °C/10 min) resulted in an approximately five-fold reduction in antioxidant activity. Heating at 90 °C for 10 min resulted in a 2–10% decrease in the antioxidant capacity of peach beverage, with negligible changes in total phenolic content and total flavonoid content [7]. Pasteurization (95 °C/3 min) led to reductions in individual anthocyanin contents of unclarified (39–51%) and clarified (11–18%) crude black mulberry juice [8]. Bhagat and Chakraborty [9] indicated absolute color differences in pomegranate juice of ΔE = 4.19 and ΔE = 6.29 after heating at 95 °C for 2 and 3 min, respectively.

Because of changing consumer needs, including the search for products with high contents of bioactive ingredients and highly rated organoleptic quality, recently, intensive studies have been conducted on the use of modern preservation methods [3,10–15]. Also, because of numerous negative reports on the effects of heating on the quality of fruit and vegetable products [12,16–19], further research is needed to optimize preservation parameters in order to obtain the best possible quality. Two modern and still improving methods of food preservation are HTST (high-temperature short-time) pasteurization and UHT (ultra-high-temperature) sterilization. Thanks to the better heat exchange and much shorter exposition time for heating than traditional pasteurization, HTST and UHT allow for obtaining a better quality final product [20].

The aim of the paper is to review the effects of HTST and UHT processing on the quality of juices, nectars and beverages, taking into account quality characteristics such as the presence of microorganisms, pH, titratable acidity, total soluble solids, turbidity, color parameters, content of bioactive components, antioxidant activity, enzymatic activity and volatile compounds. This article can serve as a significant compendium of knowledge for use in work on food processing and scientific research design. Highlighting the different directions of the changes in individual quality characteristics of different food matrices confirms the need for continuous research on the selection of appropriate parameters, also taking into account the food matrix.

When searching for publications, we were guided by the time criterion (40% of all publications used are articles within the last 5 years (i.e., 2020–2024) and 70% from the last 10 years, i.e., 2015–2024) to ensure the timeliness of the review, along with keywords. The use of older publications—prior to 2015—was intended to demonstrate that the discussed methods have been developed over the long term. To ensure the broadest coverage and inclusion of as many relevant publications as possible, we used internationally recognized and reliable scientific databases such as Web of Science and Scopus.

## 2. Applications of UHT and HTST

UHT and HTST are methods of thermally preserving liquid products in a flow-through tubular or plate pasteurizers. They are most often accompanied by bottling into pre-sterilized containers [21]. In the literature, different ways of distinguishing UHT and HTST from each other are encountered. For example, Zvaigzne et al. [22] stated that the UHT method is heating at temperatures above 135 °C for 1–2 s, while Aguiló-Aguayo et al. [21] stated that UHT requires a higher temperature than HTST, above 100 °C, and a shorter time of up to a few seconds. Based on an analysis of the literature on the topic, it can be concluded that the decisive differentiating factor among the analyzed methods is the process temperature. UHT preservation happens at temperatures above 100 °C and a time limited to 10 s, while HTST processing takes place at temperatures up to 100 °C and from few up to several dozen of seconds.

The following stages can be distinguished in both of the preservation techniques: pre-heating, high heating, cooling, and sterile or aseptic packaging. Before these, food products are often homogenized to increase and unify the heat transfer by volume. HTST/UHT systems can work in the direct mode—whereby the product and heating medium are in direct contact with each other. But the most common choice is the use of devices that work in the indirect mode, where the product and heating medium do not come into contact with each other, and heat is transferred via a heat exchanger [23]. In the food industry, UHT and HTST, as continuous flow thermal processes, are mainly applied to milk, soup, honey, stew, soy milk, wine, yogurt and cream [22]. In fruit and vegetable processing, the UHT and HTST methods are used for juices, nectars and beverages.

The most common UHT preservation parameter found in the literature is 110 °C/8.6 s. In the application of the UHT technique for juice preservation, temperatures in a range of 110–138 °C and durations of 2–8 s are used the most often, following the principle that the higher the temperature, the shorter the time. In the case of the application of the HTST technique for heating juices and beverages, the most common temperature chosen is 90 °C, although trials are most often conducted with a temperature range of 71.1–100 °C and times of 2–90 s. Detailed summaries of the temperatures used, depending on the type of technique, as well the type of juice, nectar or beverage, are included in tables in this manuscript.

## 3. Advantages and Disadvantages of UHT and HTST

It is emphasized that the UHT and HTST methods have smaller impacts on the sensory and physicochemical qualities of fruit products compared to traditional pasteurization. This is possible because of the use of high temperatures and shorter processing times. Moreover, direct heating of the product prior to packaging increases the efficiency of the process, because the resistance associated with the packaging material is omitted and the direction of the heat flow during this process changes. These techniques also ensure a high degree of preservation by ensuring the commercial sterility and microbiological safety of the product, allowing for a relatively long shelf-life while maintaining the desired sensory characteristics. However, high temperatures still affect the degradation of bioactive ingredients, such as vitamin C and phenolics, as well as aromatic compounds, and, therefore, reduces antioxidant activity, as well as changes in sensory and nutritional characteristics [21,24–27]. The color changes in HTST and UHT products result not from the activity of enzymes but because high temperatures inactivate them. This is the consequence of nonenzymatic browning reactions (Maillard reaction and caramelization), degradation, or oxidation of pigments due to exposure to heat [27]. It is emphasized that heating at high temperatures may lead to unfavorable changes in aroma, such as an unacceptable off-flavor or odor when cooked [28]. Thermal treatment may also lead to decreases in the quality of proteins and lipid oxidation, as well as depreciation of the biological viabilities of the natural color pigments, influencing consumers' preferences and loss of product marketability [29]. Compared to nonthermal preservation methods, such as high-pressure processing (HPP), high-pressure homogenization (HPH) or ultrasound (US), HTST and UHT may lead to much greater changes in color due to the thermal degradation of color substances or nonenzymatic browning reactions [30,31]. Also, because of thermally induced reactions in the products, the profiles of the aroma compounds may change, as well as consumer acceptance and assessments of quality characteristics, such as taste, smell and color. Samples preserved under high pressure are better in quality in these aspects [25,32]. Therefore, it is important to appropriately select the processing parameters (i.e., time and temperature) in order to maintain product quality as much as possible and, at the same time, ensure sufficient shelf-lives [24]. Over the years, researchers have tried to optimize the process conditions, including by shortening processing times and minimizing unfavorable losses, but this is not always fully possible [21].

## 4. Impacts of HTST and UHT on Food Quality

### 4.1. Microbiological Quality and Shelf-Life

Occurrences of microorganisms in fruit and vegetable products are related to contamination of the raw materials' surfaces during harvesting and further processing. The heat resistance of microorganisms depends on many factors, such as pH, redox potential, water activity, moisture content, salt content, types of targeted microorganisms and activities of the enzymes [21,33]. For selecting the preservation parameters, the most important factor to take into consideration is the pH. Products with a pH up to 4.5 are preserved with temperatures up to 100 °C (pasteurization), inactivating non-spore-forming bacteria, while products with a pH above 4.5 are preserved with temperatures above 100 °C (sterilization), inactivating non-spore-forming and spore-forming microorganisms [21,23]. To ensure their safety, a 5-log reduction in microorganisms is needed [9]. Heat affects various parts of microbial cells and their functioning, ultimately leading to varying degrees of inactivation. The most relevant cellular events that occur in a vegetative bacterial cell associated with heat treatment are DNA alterations; increases in the mutation rate; denaturation and aggregation of proteins; loss of specific protein function; reduction in protein repair capacity; ribosome confirmation loss; outer- and inner-membrane permeabilizations; loss of membrane-associated functions; and loss of intracellular components [34].

A detailed summary of the effects of the UHT and HTST parameters on the preservation efficiency of juices, nectars and beverages is provided in Table 1. According to the literature, the UHT method is very effective in preserving liquid food products. Xu et al. [25] showed that this method resulted in reducing the numbers of the total aerobic bacteria (TAB) and the yeasts and molds (Y&M) to below detection limits and provided a shelf-life for kiwifruit juice of 42 days at 4 °C. Wang et al. [35] also observed TAB and Y&M contents below the detection limits for 12 weeks in purple sweet potato nectar stored at 4 °C and 25 °C. Huang et al. [36] demonstrated complete inactivation of aerobics, psychrotrophs, *Escherichia coli Coliformis* and Y&M in carambola juice after the process, as well as within 40 days of storage. Wang et al. [24] compared different UHT temperatures within the same preservation time, showing that better inactivation occurred at 120 °C and 135 °C than at 110 °C. However, both of these temperatures effectively inactivated microorganisms to safe levels. Chen et al. [37] achieved desirable 90-day shelf-life for pomegranate juice after UHT treatment. Also, other studies have shown the effectiveness of the UHT method in inactivating microorganisms immediately after processing [12,17,38–40], maintaining microbiological safety during storage over 20 days [40,41], 25 days [42], 28 days [27], 90 days [43] and 16 weeks [44,45]. Compared to innovative, emerging methods of preservation like HPP, UHT is much more effective [41].

The HTST method is also effective in ensuring appropriate microbiological stability. Heating peach juice at 72 °C for 15 s resulted in the destruction of *E. coli* O157:H7 by 5 logarithmic cycles [31]. Heating sea buckthorn juice at a configuration of 100 °C/15 s reduced TAB and Y&M counts below detection levels and ensured microbiological safety for 31 days at 4 °C [30]. The good effectiveness of the HTST treatment in inactivating various microorganisms has been demonstrated in many other research works [5,40,46–49]. Deng et al. [5] indicated a shelf-life for cloudy apple juice of 9 weeks. Microbial counts for the HTST-pasteurized orange juice (94 °C/26 s) remained at or below detection limits over the entire study (168 days) [50].

**Table 1.** Effects of UHT and HTST processing on microbial contents.

| Matrix | Parameter | Contents of Microorganisms | Reference |
|---|---|---|---|
| | | **UHT** | |
| Mango nectar | 110 °C/8.6 s | Total aerobic bacteria counts 5.7 log ↓ *<br>Yeasts and molds—nondetected | [45] |
| Watermelon juice | 135 °C/2 s | Total flora count more than 3.8 log ↓ * | [43] |
| Watermelon juice | 110 °C/2 s, 120 °C/2 s, 135 °C/2 s | Total flora count—survival rate below 0.01%, 0.1% and 0.01%, respectively | [24] |
| Açai juice | 138 °C/6 s | Yeast and mold counts 2.2 log ↓ * | [51] |
| Carrot juice | 110 °C/8.6 s | Total plate count 4.9 log ↓ *<br>Yeasts and molds—nondetected | [52] |
| Pepper and orange juice blend | 110 °C/8.6 s | Total aerobic bacteria—nondetected<br>Yeasts and molds—nondetected | [42] |
| Cucumber juice | 110 °C/8.6 s | Total aerobic bacteria 3.6 log ↓ *<br>Yeasts and molds—nondetected | [41] |

**Table 1.** *Cont.*

| Matrix | Parameter | Contents of Microorganisms | Reference |
|---|---|---|---|
| Grapefruit juice | 110 °C/8.6 s | Total plate count—nondetected<br>Yeasts and molds—nondetected | [53] |
| Korla pear juice | 110 °C/8.6 s | Total plate count—nondetected<br>Yeasts and molds—nondetected | [54] |
| Carambola juice | 110 °C/8.6 s | Total aerobics, psychrotrophs, *E. coli*/coliforms, yeasts and molds—nondetected | [36] |
| Mulberry juice | 110 °C/8.6 s | Total aerobic bacteria, yeasts and molds—nondetected | [27] |
| Cloudy ginger juice | 110 °C/8.6 s | Total aerobic bacteria, yeasts and molds—nondetected | [38] |
| Papaya beverage | 110 °C/8.6 s | Total aerobic bacteria, yeasts and molds—nondetected | [39] |
| Clear and cloudy Se-enriched kiwifruit juices | 110 °C/8.6 s | Total aerobic bacteria, yeasts and molds—nondetected | [25] |
| Apricot nectar | 110 °C/8.6 s | Total aerobic bacteria, yeasts and molds—nondetected | [20] |
| Red prickly pear juice | 130 °C/3 s | Total coliforms, yeasts and molds, mesophilic and psychrophilic—nondetected | [40] |
| Black carrot juice | 130 °C/5 s | Total bacterial content, *E. coli* and yeasts and molds—nondetected | [17] |
| Freshly squeezed lettuce juice | 115 °C/5 s | Total aerobic bacteria, yeasts and molds ↓ *<br>(commercial asepsis) | [12] |
| Mulberry juice | 110 °C/8.6 s | Total viable bacteria and yeasts and molds—below 1 log CFU | [55] |
| | | **HTST** | |
| Sea buckthorn juice | 100 °C/15 s | Total aerobic bacteria counts 3.0 log ↓ *<br>Yeast and mold counts—nondetected | [56] |
| Orange and carrot juice | 98 °C/21 s | Total plate counts—nondetected<br>Yeast and mold counts—nondetected | [46] |
| Nonconcentrated and concentrated sea buckthorn juice | 100 °C/15 s | Total plate counts—nondetected<br>Yeast and mold counts—nondetected | [30] |
| Cloudy apple juice | 98 °C/50 s | Total aerobic bacteria counts—nondetected<br>Yeast and mold counts—nondetected<br>*E. coli*—nondetected | [5] |
| Açai juice | 90 °C/6 s | Yeast and mold counts 1.8 log ↓ * | [51] |
| Orange juice | 72 °C/20 s | *E. coli*, *Enterobacteriaceae*, yeasts and molds < 10 cfu/g<br>Total aerobic plate counts < 1000 cfu/g<br>Lactic acid bacteria < 100 cfu/g | [48] |
| Pomegranate juice (MW—Mollar de Elche varietal juice + Wonderful varietal juice; ML—Mollar de Elche + lemon juice; M100—Molar de Elche) | 90 °C/5 s | Total mesophilic aerobic plate count—nondetected | [57] |
| Red prickly pear juice | 80 °C/30 s | Total coliforms, yeasts and molds, mesophilic and psychrophilic—nondetected | [40] |
| Peach juice | 72 °C/15 s | *E. coli* O157:H7 5-log reduction | [31] |
| Pomegranate fermented beverage | 72 °C/15 s | Aerobic mesophilic bacteria 1.3 log ↓ *<br>Yeasts and molds 3.0 log ↓ * | [58] |
| Orange juice | 94 °C/26 s | Total aerobic bacteria—below detection limit | [50] |
| Apple juice with raspberry | 85 °C/6 s | Total aerobic mesophilic and psychrophilic counts, yeasts and molds—nondetected | [13] |
| Apple juice | 72 °C/26 s | Native microorganisms 5 log ↓ | [59] |

**Table 1.** *Cont.*

| Matrix | Parameter | Contents of Microorganisms | Reference |
|---|---|---|---|
| Fruit smoothie-type beverage | 72 °C/15 s | Native microorganisms 3.5 log ↓ *<br>*E. coli* 6.3 log ↓ * | [60] |
| Kale juice | 72 °C/60 s | Total plate count 5.5 log ↓ | [61] |
| Whey–grape juice drink | 72 °C/15 s | Bacterial counts below 1 CFU/mL | [62] |

↓—decrease. * Statistically significant change.

### 4.2. Basic Physicochemical Characteristics

Total soluble solids (TSS) is one parameter that determines the organoleptic characteristics of juices, nectars and beverages. It indicates the content of water-soluble and nonvolatile substances in the studied food matrix. Changes in TSS are associated mainly with changes in the compounds such as sugars and organic acids. A second important parameter influencing the organoleptic characteristics of food is titratable acidity (TA), which corresponds to the total concentration of acidic hydrogen ions. pH is the main indicator for the selection of preservation parameters and determines the microbiological stability of the product. Turbidity and color are the primary qualitative discriminators that a consumer is able to assess during contact with food. The content of macromolecular compounds, as well as chemical reactions, especially thermally induced ones, can lead to an increase or decrease in the turbidity of liquid products.

#### 4.2.1. Total Soluble Solids

The UHT technique, because of the very fast processing time, does not significantly affect the components of the TSS. Generally, many studies confirm the lack of significant changes in TSS after preservation with this method [22,35–37,39,41,45,52–54], as well as during storage [27,41,42,45,53]. However, different observations are found by Jittanit et al. [63]. The significant decrease in TSS they noted was caused by the design of the equipment. The authors assume that because of the need to constantly supply water to the apparatus, the test sample could be diluted and the TSS decreased. Decreases in TSS were also demonstrated by other studies [12,17,42]. Moreover, no significant effects of UHT on the contents of sucrose, glucose and fructose in purple sweet potato nectar were revealed [35]. In another study, there was no effect on the sucrose content, but significant decreases in glucose, fructose and total sugars were observed [45].

Thermal processing of peach juice (72 °C/15 s) did not result in a significant change in TSS following HTST treatment compared to raw juice immediately after processing, as well as during 28 days of storage [31]. Also, other studies have shown no significant effects of HTST on the TSS of juices, nectars and beverages from fruits and vegetables immediately after preservation [5,22,30,32,50,64,65], as well as during storage [5,50,65,66]. Bonilla et al. [67], examining three blackberry–soy–flaxseed-based beverages, also confirmed the lack of significant effects of HTST preservation (71.1 °C/3 s) on TSS. Rivas et al. [46] indicated a 9.5% increase in the TSS in blended orange and carrot juice. In lime juice, a decrease of more than 8% in TSS (85 °C/30 s and 95 °C/30 s) was observed [63].

#### 4.2.2. pH

Many researchers have found no significant changes in pH after UHT [35,38,40,41,53], as well as during storage [27,35,39–41,53]. Only a few researchers found a drop in the pH after the preservation process [17,54,63] or during storage [45].

Typically, researchers indicate that there is no significant effect of HTST heating on pH immediately after preservation [5,13,30,32,40,46,48,50,64–67] and during storage [48,50,65]. However, some researchers point to an increase or decrease in pH after preservation. When preservation conditions of 73 °C/27 s, 80 °C/27 s or 83 °C/27 s were applied to an apple juice, the pH increased by more than 6% [68]. Heating lime juice at 85 °C for 30 s and 95 °C for 30 s resulted in a significant decrease in pH [63]. Conditions of 72 °C/15 s resulted in a significant reduction of 3% in peach juice pH, while the pH dropped by 9.4% over 28 days of storage. However, the dynamics of the changes in pH were less than those in fresh juice, for which the pH decreased by 16.4% [31]. A decrease in pH after the process was also observed by other researchers [69]. There was a decrease in pH of 38.6% during a 50-day storage period, as stated by Zhao et al. [66]. Deng et al. [5] indicated that there was no effect of HTST heating on pH over 9 weeks of storage.

#### 4.2.3. Titratable Acidity

Researchers have mostly determined that there is no significant change in titratable acidity (TA) after UHT preservation [22,25,36,38,41,45,53], as well as during storage [25,35,53,54]. Chen et al. [38], despite showing no significant change in the TA of cloudy ginger juice immediately after preservation, indicated a significant decrease in TA during a 91-day storage period. However, different reports can be found in the literature. Chen et al. [37], in pomegranate juice, observed a significant decrease in TA after UHT preservation. Tian et al. [32] observed an increase in TA of almost 7%, using the same parameters as Chen et al. [37]—110 °C/8.6 s. Bao et al. [17] observed a significant 12.9% increase in acidity after UHT preservation of black carrot juice compared to fresh. Liu et al. [45] observed an increase in acidity within 16 weeks of preservation and Liu et al. [41] within 20 days.

The HTST treatment of peach juice (72 °C/15 s) did not result in a significant change in TA compared to raw juice, immediately after preservation, as well as during a 28-day storage period [31]. Other researchers [5,32,49,56] also observed no significant effects of HTST treatment on the acidity of fruit and vegetable products. In contrast, the acidity of orange juice decreased significantly by 7.6% and 8.4% as a result of heating at 90 °C for 10 and 20 s, while an increase in TA was noted over 180 days. A different observation was made by Rivas et al. [46], whereby TA increased by 10.2% in blended juice after the process.

#### 4.2.4. Turbidity

Tian et al. [32] demonstrated a 110.4% increase in turbidity following heating at 110 °C for 8.6 s. On the other hand, Liu et al. [41] and Zhao et al. [54] showed significant decreases in clarity following UHT preservation of 2.2% and 3.7%, respectively. This was likely due to the degradation of various compounds, such as proteins, phenols or polysaccharide. In addition, liquid products preserved thermally by this method are less transparent than after HPP.

HTST treatment (72 °C/15 s), immediately after the process, did not result in a significant change in the turbidity of peach juice, while there was a significant 25.9% reduction in particle size. During storage, there was a significant increase in turbidity but to a lesser extent than in fresh juice [31]. Bonilla et al. [67] observed an increase in turbidity of 6.4–15.5% in three different formulations of blackberry–soy–flaxseed-based beverages. On the other hand, Tian et al. [32] noted a significant increase in turbidity in cloudy pomegranate juice by 79.4% (110 °C/8.6 s). Rivas et al. [46] observed an increase of as much as 105.7% in blended orange and carrot juice.

#### 4.2.5. Color

Color is affected by the presence of color substances, such as anthocyanins (from red to purple or blue in color), chlorophylls (different intensities of green hues), carotenoids (from yellow to orange or red in color) and betalains (betacyanins—red to violet in color; betaxanthins—yellow to orange in color) [29]. The direction of the change depends on the chemical composition of the food and the preservation conditions. During preservation and storage, the presence of oxygen in the container, enzyme activity or chemical transformations of pigments (for example, degradation, polymerization and Maillard reactions) are also important [39,63]. An increase in the preservation temperature, at the same time, results in a greater color change [24]. Tian et al. [32] showed a significant increase in L*, a* and b* and $\Delta E = 6.6$ compared to raw juice using 110 °C/8.6 s parameters for UHT. Xu et al. [25] showed the difference between clear and cloudy juices—in the case of the former, the L*, a* and b* parameters increased significantly after thermal preservation (ΔE = 2.3), while in the latter juice, the L* and b* parameters decreased (ΔE = 4.6). In addition, their colors continued to change significantly during storage. Visible color changes (ΔE > 3) were also demonstrated by other studies [12,36,38,42,49,53]. An intense color change was demonstrated by Jittanit et al. [63] where ΔE = 11.4 was obtained after preservation of lime juice (136°/4 s), and was associated with decreases in L* of 12.5% and a* of 59.7% and an increase in b* of 85.4%. Smaller color changes were indicated by Liu et al. [41] (ΔE = 3.0), Wang et al. [35] (ΔE = 3.0), Chen et al. [37] (ΔE = 2.4), Chen et al. [39] (ΔE = 1.2) and Zou et al. [27] (ΔE = 0.1), but they confirmed the discoloration process during storage. Wang et al. [35] observed that ΔE = 8.0 at 12 weeks of storage at 4 °C and ΔE = 5.4 at 25 °C, while Liu et al. [41] ΔE = 24.0 at 20 days of storage at 4 °C. In contrast, Gao et al. [53] noted no obvious color changes after the process (ΔE = 2.0) or during a 21-day storage period (ΔE between 1.8 and 2.5). Small ΔEs may be due to the inactivation of enzymes. However, it is emphasized that UHT leads to much larger color changes compared to HPP [37,41]. Yuan et al. [70] compared HTST (85 °C/30 s) with UHT (110 °C/8.6 s). Cloudy pomegranate juice preserved at a higher temperature and for a shorter time achieved a higher value of ΔE = 6.8 than that at a lower temperature and for a longer time, at ΔE= 5.5.

Heating peach juice at 72 °C for 15 s resulted in slight decreases in the L* and a* parameters and a significant increase in the b* value immediately after HTST preservation compared to raw juice. There was a significant change in the brightness and proportion of the yellow color during 28 days of storage. After HTST, the peach juice became more yellow and darker [31]. In contrast, heating at 80 °C for 30 s resulted in increases in the L*, a* and b* values of cloudy pomegranate juice [32]. de Souza et al. [64], on the basis of the thermal processing of two juices and a lemonade, showed different effects for the same HTST preservation conditions on the color parameters. In lemonade, there were significant increases in a* and b* values of 5.6% and 3.9%, respectively. In citrus juice, there was an increase in the a* value of 60.0%, and in green juice there were changes in all three color parameters (L* of 14.1%, a* of 120.0% and b* of 4.9%). The consecutive ΔEs for these products were 0.3, 1.4 and 8.0, respectively. In other studies, for cloudy pomegranate juice, the ΔEs were 5.2 under conditions of 80 °C/30 s [32], 3.6 for sea blackthorn juice preserved at 100 °C/15 s [56] and 1.7–2.6 for blackberry–soy–flaxseed beverages preserved at 71.1 °C/3 s [67]. Small changes in the color of the product (ΔE = 0.6) after HTST processing were observed by Rios-Corripio et al. [58]. Yuan et al. [70] compared the effects of HTST (85 °C/30 s) and UHT (110 °C/8.6 s) on the color of cloudy pomegranate juice—the ΔEs were 5.5 and 6.8, respectively. As can be seen, depending on the heating conditions and food matrix, different visual changes in food can occur. Researchers emphasize that the HTST method exhibits worse retention of the natural color of a product than ultrasound, UV-C radiation, high-pressure homogenization or HPP technique [13,30,31]. However, various observations were made by Lee et al. [71]. In three juices from different apple varieties, the authors indicated that the HTST-treated samples experienced much smaller color changes than the PEF-treated samples. They confirmed significant changes in color during storage. The ΔE of cucumber juice drink after 50 days of storage at 4 °C was as high as 9.2 [66].

### *4.3. Bioactive Compounds*

Detailed summaries of the effects of the applied UHT and HTST preservation temperatures on phenolic compounds, vitamins and other bioactive components in fruit and vegetable juices, nectars and beverages are provided in Tables 2 and 3.

#### 4.3.1. Vitamin C

Vitamin C is a thermolabile, water-soluble antioxidant. It can scavenge singlet oxygen, superoxide radical, hydroxyl radical and hydrogen peroxide [19]. The kinetics of its decomposition depends on numerous factors, such as temperature, pH, light access and the presence of oxygen, metal catalysts or enzymes[71]. Vitamin C contents in products decrease during storage [72]. Degradation is most often associated with the effects of available oxygen (the increase in temperature determines the mobility of the molecules, thus increasing the contact of ascorbic acid with oxygen), causing oxidation of ascorbic acid to dehydroascorbic acid and further degradation to 2,3-diketogulonic acid or the formation of furfural and 3-hydroxy-2-pyrone. Conversion to dehydroascorbic acid is conducted without losses in vitamin C activity, but further degradation results in a loss of these desirable properties [73]. Vitamin C may be partially protected from degradation by the presence of phenolics [74]. Ascorbic acid degradation can reduce the nutritional and sensory quality of food products and cause changes in color and flavor due to nonenzymatic browning reactions [19]. Vitamin C has numerous health-promoting properties and is used in the prevention of numerous diseases, such as heart disease, cancer, the common cold, diabetes, atherosclerosis, macular degeneration, strokes, cataracts, glaucoma and COVID-19 [75]. In addition, it is involved in collagen synthesis and improves iron absorption [72].

UHT heating reduced the vitamin C content by 27.9% in red grapefruit juice [53], 13.4% in korla pear juice [54], and by 38.4% and 22.1% in clear and cloudy kiwifruit juices enriched with selenium, respectively [25] (Table 2). Much greater changes were observed by Zhang et al. [12] in UHT-processed lettuce juice—the vitamin C content dropped by 85.1%. Other researchers have also pointed out the degrading effect of the UHT method in this aspect [18,22,70]. On the other hand, Liu et al. [44] indicated that UHT had no effect on L-ascorbic acid in mango nectar immediately after preservation but showed a significant decreases during 16 weeks of storage at 4 °C and 25 °C, respectively. Huang et al. [36] also showed no significant changes in vitamin C content after preservation. On the other hand, researchers have indicated that by heating at 136 °C for 4.1 s, vitamin C can be extracted and significantly increase in content in acerola juice [76,77]. Researchers highlight that the use of a closed system, which is tubular heat exchange, allows for the protection of this bioactive compound from the degrading effects of oxygen [76].

HTST conditions like 72 °C/15 s resulted in a significant reduction in the ascorbic acid content of peach juice immediately after preservation of 22%. During 28 days of storage, ascorbic acid decreased by 54.3% in HTST-treated juice, while in raw juice, this decrease was only 16.7% [31]. Increased temperatures significantly contribute to changes in the content of bioactive components [31,65]. Amaro et al. [14], in modelling studies on orange juice, indicate that as the HTST preservation temperature increases while maintaining the same processing time, the vitamin C content decreases. Vitamin C retention in orange juice processed at 70, 80, 90 and 100 °C over 10 s were 0.92, 0.87, 0.85 and 0.82, respectively. Other researchers also report decreases in vitamin C contents after HTST preservation, as follows: Mena et al. [57] of 12.9–59.6%, depending on the juice's composition; Deng et al. [5] of 27.2%; Hou et al. [56] of 14.3%; Yuan et al. [70] of 19.8%; and Zvaigzne et al. [22] of about 9%. In lemonade, the ascorbic acid content decreased by 91.7%, while in citrus juice it only decreased by 12.1% under the same preservation conditions [64]. Atuonwu et al. [78] revealed that there was no significant change in vitamin C content as a result of heating orange juice at 76.8 °C/15 s. Torregosa et al. [65] showed the remaining concentration of ascorbic acid to be at a level of 83%, wherein the use of a pulsed electric field preserved as much as 90% of this vitamin. Yuan et al. [70] suggest that HTST is less effective at preserving vitamin C than high-pressure processing. Torregosa et al. [65] and Deng et al. [5] confirmed that the vitamin C contents of HTST-preserved products decreased significantly during storage. The opposite observation was made by Alves Filho et al. [76] and Fonteles et al. [77], with vitamin C contents increased by 35% in acerola juice after preservation at 90 °C for 120 s. The lack of significant changes due to HTST heating is evidenced by Yildiz and Aadil [26] and Yang et al. [30].

#### 4.3.2. Phenolics

Phenolic compounds are secondary metabolites of plants that are natural antioxidants. In their structure, they have at least one aromatic ring with one or more -OH groups. Phenolics can be divided into flavonoids (anthocyanins, flavanols, flavanones, flavonols, flavones, isoflavones and chalcones) and nonflavonoid molecules (phenolic acids, stilbenes, lignans, tannins and curcuminoids). They are most often combined with other molecules, such as sugars or organic acids, making glycosides [45]. Phenolics have health-promoting properties, such as antioxidant, anti-diabetic, antihypertensive, anti-inflammatory, anti-allergenic, antiatherogenic, antimicrobial, antithrombic, cardioprotective, vasodilatory, neuroprotective, anti-platelet-aggregation, gene-encoding antioxidant enzyme-stimulating, epigenetic regulation and cytotoxic [45,79,80]. Phenolics are reactive and thermally sensitive [81]. The literature reports that quantitative and qualitative changes in phenolics of food matrices are associated with numerous processes, such as plant cell disruption, degradation, hydroxylation, isomerization, oxidative polymerization, methylation, isoprenylation, glycosylation and dimerization. This can lead to changes in biological activity, destruction, modification or formation of new phenolic compounds [35,82–84]. Zhang et al. [84], in their study, pointed out the following four possible reasons for the increase in the contents of selected phenolic groups following heating at 110 °C for 8.6 s of red raspberry juice: (1) release of phenols due to partial hydrolysis; (2) release of phenolic compounds from cells that ruptured because of heating; (3) thermal inactivation of oxidases; and (4) change in the numbers and positions of hydroxyl groups in phenolics due to thermal treatment. Confirmations of the presented possible transformations of phenolic compounds are provided in tables in this manuscript, where it can be seen that with the same preservation parameters, the degree of change in the content of individual phenolic compounds can vary significantly.

One of the most important groups of phenolic compounds are anthocyanins. They are glycosides or acylglycosides of anthocyanidins (or aglycones). They have a C6-C3-C6 skeleton. The main anthocyanins in food are cyanidin, delphinidin, pelargonidin, petunidin, peonidin and malvidin. Their color and stability depend on the pH of the environment, ranging from red (methylation, the most stable), through blue (hydroxylation, the least stable) to colorless [85]. During pasteurization, significant degradation usually occurs, resulting in changes in their colors and nutritional values (Table 2). Their stability and degradation depend on the preservation parameters used (temperature, time, heat dose, presence of oxygen and light) and on the properties of the heated product (pH, presence of enzymes, copigments, proteins, sugars and metal ions, chemical structures and concentrations of anthocyanin compounds). Copigmentation also plays a significant role in stabilizing anthocyanins [81,85]. The literature reports that anthocyanin retention can be positively affected by a mild heating step (~50 °C) [85].

**Table 2.** Effect of UHT processing on the contents of selected bioactive compounds.

| Matrix | Parameters | Phenolic Compounds | Vitamins | Other Bioactive Ingredients | Reference |
|---|---|---|---|---|---|
| Acerola juice, acerola juice + inulin, acerola juice + gluco-oligosaccharides | 136 °C/4.1 s | - | All—vitamin C ↑ * | - | [76] |
| Mango nectar | 110 °C/8.6 s | TPC—no statistical difference | Vitamin C—no statistical difference | TCC—no statistical difference | [44] |
| Watermelon juice | 110 °C/2 s<br>120 °C/2 s<br>135 °C/2 s | TPC ↓ * at 110 and no statistical difference at 120 and 135 °C | - | - | [24] |
| Pomegranate juice | 110 °C/8.6 s | Total monomeric anthocyanins 29.3% ↓ | Vitamin C 40.8% ↓ * | - | [70] |
| Orange juice | 130 °C/2 s | TPC—7.2% ↓ | Vitamin C 7.2% ↓ * | TCC 13.7% ↓ * | [22] |
| Açai juice | 138 °C/6 s | TPC ↓ *<br>Anthocyanins ↓ * | Vitamin C ↓ * | | [51] |

**Table 2.** *Cont.*

| Matrix | Parameters | Phenolic Compounds | Vitamins | Other Bioactive Ingredients | Reference |
|---|---|---|---|---|---|
| Carrot juice | 110 °C/8.6 s | TPC 11.5% ↓ * | - | Lutein 5.1% ↓<br>α-Carotene 6.2% ↓ *<br>β-Carotene 43.0% ↓ *<br>Falcarindiol 29.7% ↑ *<br>Falcarindiol-3-acetate 37.4% ↑ *<br>Falcarinol 29.1% ↑ * | [52] |
| Pepper and orange juice | 110 °C/8.6 s | TPC—no statistical difference | Vitamin C ↓ * | | [42] |
| Tomato juice | 110 °C/8.6 s | Quercetin ↓ *<br>Caffeic acid ↓<br>Chlorogenic acid—no statistical difference | Vitamin C ↓ * | TCC 7.3% ↓, total lycopene 8.1% ↓, 15-*cis*-phytoene 5.3% ↓, all-*trans* -phytoene 5.5% ↓, all-*trans*-lutein 5.6% ↓, 13-*cis*-lutein 12.8% ↓ *, 13-*cis*-β-carotene 8.0%↓, 15-*cis*-β-carotene from 0 to 0.208 µg/g, all-trans-β-carotene 17.8% ↓ *, *cis*-β-carotene 7.9% ↓, 9-*cis*-β-carotene 12.3% ↓, 15-*cis*-lycopene 57.6% ↑ *, 13-*cis*-lycopene from 0 to 0.432 µg/g, 9,13-*cis*,*cis*-lycopene 9.3% ↓, 9-*cis*-lycopene 22.0% ↓ *, all-*trans*-lycopene 9.2% ↓ | [18] |
| Red grapefruit juice | 110 °C/8.6 s | TPC 7.7% ↓ * | Ascorbic acid 27.9% ↓ * | - | [53] |
| Korla pear juice | 110 °C/8.6 s | TPC 4.7% ↓ * | Ascorbic acid 13.4% ↓ * | - | [54] |
| Carambola juice | 110 °C/8.6 s | TPC ↓ *<br>Total flavonols—no statistical difference | Ascorbic acid ↓ | - | [36] |
| Mulberry juice | 110 °C/8.6 s | TPC 4.0% ↑ * | - | - | [27] |
| Cloudy ginger juice | 110 °C/8.6 s | TPC 14.7% ↓ * | - | Gingerols 14.2% ↓ * | [38] |
| Papaya beverage | 110 °C/8.6 s | TPC 12.7% ↓ * | - | TCC 1.2% ↓ | [39] |
| Litchi juice | 134 °C/4 s | TPC 19.8% ↓ *<br>TFC 40.1% ↓ *<br>Rutin 19.3% ↓ *, (−)-epicatechin 36.5% ↓ *, chlorogenic acid 18.7% ↓ * | - | - | [86] |
| Cloudy pomegranate juice | 110 °C/8.6 s | TPC 7.5% ↓ *<br>Anthocyanins 13.2% ↓ * | - | - | [37] |
| Clear and cloudy Se-enriched kiwifruit juices | 110 °C/8.6 s | TPC 5.2% ↑ * and 2.5% ↓ | Ascorbic acid 38.4% ↓ * and 26.6% ↓ * | Total selenium 4.9% ↓ * and 27.3% ↓ *<br>Chlorophyll 80.9% ↓ * and 48.5% ↓ * | [25] |
| Apricot nectar | 110 °C/8.6 s | TPC 96.9% ↑ *<br>(+)-catechin 4.7% ↑ *, chlorogenic acid 12.2% ↑ *, neochlorogenic acid 14.6% ↑ *, (−)-epicatechin 5.0% ↓, ferulic acid 5.7% ↑, caffeic acid 12.0% ↑ *, *p*-coumaric acid 14.3% ↓ * | - | TCC 1.5% ↓, β-carotene 2.6% ↑, α-carotene 44.2% ↑ *, β-cryptoxanthin 13.5% ↓, zeaxanthin 2.8% ↑, lutein 2.7% ↑ | [20] |
| Red prickly pear juice | 130 °C/3 s | TPC 2.5% ↑ | - | Betacyanins 63.1% ↓ *<br>Betaxanthins 45.0% ↓ * | [40] |
| Red raspberry juice | 110 °C/8.6 s | TPC 31.4% ↓ *<br>TFC 25.5% ↑<br>Total proanthocyanidins content 5.6% ↑<br>Total monomer anthocyanins content ↓* | - | - | [84] |
| Black carrot juice | 130 °C/5 s | TFC 14.2% ↓ *<br>Total anthocyanins content 8.6% ↓ *<br>TPC ↓ * | - | TCC ↓ | [17] |
| Cloudy pomegranate juice | 110 °C/8.6 s | Total monomeric anthocyanins content 29.3% ↓ *<br>Cyanidin-3-*O*-glucoside ↓ *<br>Cyanidin-3,5-*O*-diglucoside ↓ *<br>Delphinidin-3-*O*-glucoside ↓ *<br>Delphinidin-3,5-*O*-diglucoside ↓ *<br>Afzelechin-delphinidin-3-*O*-hexosid ↓<br>Pelargonidin-3-*O*-glucoside ↓ *<br>Pelargonidin-3,5-*O*-diglucoside ↓ * | Vitamin C 40.8% ↓ * | - | [70] |

**Table 2.** *Cont.*

| Matrix | Parameters | Phenolic Compounds | Vitamins | Other Bioactive Ingredients | Reference |
|---|---|---|---|---|---|
| Freshly squeezed lettuce juice | 115 °C/5 s | - | Vitamin C 85.1% ↓ *<br>Vitamin E 13.3% ↓ *<br>Vitamin K1 44.2% ↓ *<br>Vitamin B1 25% ↓<br>Vitamin B2 4.7% ↑<br>Vitamin B3 2.2% ↓<br>Vitamin B6 29.7% ↓ *<br>Vitamin B9 31.7% ↓ *<br>Vitamin B12 12.3% ↓ * | Total chlorophyll 14.1% ↓ *<br>Chlorophyll a 13.7% ↓ *<br>Chlorophyll b 14.9% ↓ *<br>Total β-carotene 35.7% ↓ *<br>(all E)-β-carotene 44.2% ↓ *<br>(9Z)-β-carotene 182.7% ↑ * | [12] |
| Mulberry juice | 110 °C/8.6 s | Total anthocyanin content 3.6% ↓ * | - | - | [55] |

↑—Increase; ↓—decrease. * Statistically significant change. TPC—total phenolic content; TFC—total flavonoid content; TCC—total carotenoid content.

As a result of the UHT preservation, a significant decrease of 31.4% in the total phenolic content (TPC) of red raspberry juice was observed by Zhang et al. [84]. Chmiel et al. [87] reported a reduction in the TPC of 10–28% in cloudy apple juices, Chen et al. [38] in cloudy ginger juice of 14.7%, Chen et al. [39] in papaya beverage of 12.7%, Gao et al. [53] in red grapefruit juice of 7.7%, and Zhao et al. [54] in korla pear juice of 4.7%. However insignificant changes were noted by Huang et al. [36] in carambola juice. Bao et al. [17], because of UHT preservation of black carrot juice, observed significant decreases in total flavonoid and total anthocyanins contents of 14.2% and 8.6%, respectively. On the other hand, after UHT processing, Zhang et al. [12] observed increases in flavonoid and proanthocyanidins contents of 25.5% and 5.6%, respectively. Increases in TPC contents were noted by Wang et al. [35] and Zou et al. [27] after UHT in purple sweet potato nectar and mulberry juice. No significant changes in TPC were observed by Xu et al. [42] in a pepper and orange juice blend and by Liu et al. [44] in mango nectars. The type of food matrix is of great importance. Xu et al. [25] showed that the effect of beverage clarity is relevant. These authors found that in preserved clear, Se-enriched kiwifruit juice, there was a significant decrease in TPC of 5.2%, while in the cloudy version, there was a minimal increase of 0.5%. UHT preservation caused a significant decrease in quercetin content in tomato juice, while there were no significant changes in chlorogenic or caffeic acid contents [18]. Researchers confirm that the phenolics contents of preserved samples decreased during storage, which may be related to oxidation degradation [35]. Following preservation at 110 °C/8.6 s and refrigerated storage for 25 days, the TPC decreased by 90.7% [42]. However, less degradation of phenolics was reported by Liu et al. [44], at about 17.0% and 25.2% at 4 °C and 25 °C, respectively. It is emphasized that thermal inactivation of enzymes can also lead to the retention of a significant portion of anthocyanins in a sample immediately after preservation, as well as during storage. Also, further condensation of breakdown products with organic acids or carbohydrates can cause a decrease in the contents of these bioactive components [27,37,39,83,84]. In pomegranate juice there was a 29.3% decrease in TPC [70], and in mulberry juice there was a decrease of less than 4% [27]. Some researchers state that the anthocyanin content decreased during storage by 75.3% after 25 days of storage at 4 °C in pepper and orange juice blend [42] and by 7.8% and 34.1% during 12 weeks of storage of purple sweet potato nectar at 4 °C and 25 °C, respectively [35].

The HTST technique can perform significant extraction of phenolics from a raw food matrix. In sea blackthorn juice, there was a 6.6% increase in TPC following preservation at 100 °C for 15 s due to cell disruption [56]. In peach juice, preservation by HTST resulted in an increase from 617.2 mg (raw juice) to 1000.3 mg GAE/L (72 °C/15 s). An upward trend was also shown during storage (by 11.4% over 28 days) but to a significantly lesser extent than in raw juice (by 140.3% over 28 days) [31]. In addition, an increase in TPC content was also shown by Mena et al. [57]. Also, studies have been described in which phenolics are largely degraded [5] or only slightly, like in lemonade and citrus juice by 3.6% and 7.6%, respectively [64]. Morales-de la Peña et al. [82] observed, after the HTST process, a decrease in the phenolic acids content of less than 3%, and, on the other hand, an increase in flavonoids of 61.5%. Nonsignificant changes in TPC and total ellagitannin content in an HTST-preserved (71.1 °C/3 s) blackberry–soy–flaxseed beverage were observed by

Bonilla et al. [67]. A significant decrease of 27.5% in total monomeric anthocyanins content was shown by Yuan et al. [70]. Despite the differences in the preservations of phenolics, their degradation occurred during prolonged storage as a result of heating in different food matrices and using different parameters [88]. A different conclusion was reached by Morales-de la Peña et al. [82], who noted increases in phenolic acids and decreases in the flavonoid contents in an HTST-preserved fruit-juice–soymilk beverage during 56 days of storage. Deng et al. [5] showed a relatively stable TPC during storage (Table 3). No significant effect of HTST on anthocyanins was observed by Mena et al. [57] in pomegranate–lemon juice and pomegranate juice because of the short processing time (90 °C/5 s). In addition, a decrease in anthocyanin content of 27.6% was reported by Yuan et al. [70].

#### 4.3.3. Other Bioactive Substances

Xu et al. [25] showed a small decrease of 4.9% in the selenium content of clear Se-enriched kiwifruit juice preserved in parameters of 110 °C/8.6 s, while there was a significant decrease in this element on level of 25% in cloudy juice. The content of chlorophyll, the compound responsible for the attractive and green color of kiwifruit and the juices derived from them, also dropped significantly, by 80.9% and 47.8%, respectively. The researchers found that such a change is related to the thermal denaturation of a protein in chlorophylls. Bioactive compounds that also determine the color of products, and have antioxidant properties, are carotenoids. Chen et al. [39], Wang et al. [18] and Liu et al. [44] showed that heating under conditions of 110 °C/8.6 s did not change the content of carotenoids. It can lead to isomerization of various compounds and the formation of new compounds that were absent before the preservation process, such as the *cis*-13-lycopene in tomato juice [18]. In lettuce juice preserved by UHT (115 °C/5 s), there were significant decreases in the total chlorophyll content by 14.1% and total β-carotene content by 35.5% [12]. The same researchers observed also a significant decrease in vitamins E, K1, B6, B9 and B12, and negligible decreases in B1 and B3, and a negligible increase in B2. Mesta-Vicuña et al. [40] indicate degradation of betacyanins and betaxanthins by 63.1% and 45%, respectively, as a result of heating at 80 °C for 30 s. These thermal conditions result in a 14.2% increase in gingerol content due to disruption of plant cells and higher extraction of the inner compounds [38].

Thermal processing causes isomerization of carotenoids and can result in a reduction in their content, for example, of 20.5% in sea buckthorn juice preserved with HTST [56]. On the other hand, heating can cause their extraction through cell wall degradation or the formation of new carotenoid compounds, as well as concentrating the sample and thus the bioactive components [88]. Mesta-Vicuña et al. [40] indicated degradation of betacyanins and betaxanthins by 14.8% and 14.4%, respectively, following heating at 80 °C for 30 s (Table 3).

**Table 3.** Effect of HTST processing on the content of selected bioactive compounds.

| Matrix | Parameters | Phenolic Compounds | Vitamins | Other Bioactive Ingredients | References |
|---|---|---|---|---|---|
| Fruit juice–soymilk beverage | 90 °C/60 s | Total phenolic acids 2.8% ↓ *, caffeic acid 11.6% ↑ *, chlorogenic acid 19.4% ↓ *, coumaric acid 2.7% ↓ *, ferulic acid 29.1% ↑ *, sinapic acid 13.0% ↓ *, TFC 61.5% ↑ *, hesperidin 590.1% ↑ *, rutin 43.0% ↓ *, narirutin 22.5% ↑ *, quercetin 16.6% ↑ *, apigenin 31.3% ↓ | - | *Cis*-violacanthin + antheraxanthin 8.3% ↓, *cis*-antheraxanthin 17.4% ↓, lutein 38.9% ↓ *, zeaxanthin 26.3% ↓, α-cryptoxanthin 25.0% ↓, β-cryptoxanthin 23.8% ↓ *, α-carotene 0% and β-carotene 16.3% ↓ | [82] |
| Orange juice | 90 °C/20 s | - | Vitamin A 16.9% ↑ | TCC 12.6% ↓, neoxanthin+9-*cis*-violaxanthin 21.1% ↓, antheraxanthin 14.8% ↑, lutein 12.1% ↓, zeaxanthin 17.9% ↓, isolutein 6.8% ↓, β-cryptoxanthin 14.6% ↓, α-carotene 34.1% ↓, 9-*cis*-α-carotene 24.1% ↓, phytoene + phytofluene 7.7% ↓ and 7,8,7′,8′-tetrahidrolycopene 9.9% ↓, 9.7% ↓ and 17.5% ↓ | [89] |

**Table 3.** *Cont.*

| Matrix | Parameters | Phenolic Compounds | Vitamins | Other Bioactive Ingredients | References |
|---|---|---|---|---|---|
| Tomato juice | 90 °C/30 s<br>90 °C/60 s | TPC 1.1% ↑ and 0%, chlorogenic acid 0.5% ↑ and 0.7% ↑, ferulic acid 1.1% ↑ and 1.1% ↓, *p*-coumaric acid 0% and 3.1% ↓, caffeic acid 0% and 4.7% ↑, quercetin 3.9% ↓ and 3.4% ↓, kaempferol 1.8% ↓ and 0% | Vitamin A 2.0% ↑ and 5.1% ↑ | TCC 2.1% ↑ * and 2.1% ↑ *, lycopene 4.6% ↑ * and 7.2% ↑ *, neurosporene 2.3% ↓ * and 5.4% ↓ *, γ-carotene 5.6% ↓ * and 3.4% ↓ *, ζ-carotene 5.0% ↓ and 5.0% ↓, phytofluene 6.2% ↑ * and 5.4% ↑ * and phytoene 2.4% ↓ * and 11.6% ↓ * | [88] |
| Clarified and cloudy pomegranate juices | 90 °C/5 s | Total monomeric anthocyanin content 1.2% ↓ and 40.8% ↓ *<br>TPC—no difference | - | - | [81] |
| Sea buckthorn juice | 100 °C/15 s | TPC 6.5% ↑ * | Vitamin C 14.3% ↓ * | TCC 20.5% ↓ * | [56] |
| Acerola juice, acerola juice + inulin, acerola juice + gluco-oligosaccharides | 90 °C/2 s | - | Vitamin C ↑ * | - | [76] |
| Nonconcentrated and concentrated sea buckthorn juice | 100 °C/15 s | TPC—no significant difference | Vitamin C—no significant difference | - | [30] |
| Cloudy apple juice | 98 °C/50 s | TPC 22.7% ↓ * | - | - | [5] |
| Strawberry juice | 72 °C/15 s | TPC 3.4% ↑ | Vitamin C 0.6% ↑ | - | [26] |
| Açai juice | 90 °C/6 s | TPC—↓ *<br>Anthocyanins—no significant difference | Vitamin C—no significant difference | - | [51] |
| Lemonade<br>Citrus juice<br>Green juice | 75 °C/90 s | TPC 3.6% ↓, 7.6% ↓ and 0.5% ↑ | Vitamin C 91.7% ↓ *<br>Vitamin C 12.1% ↓ *<br>Vitamin C—nondetected in raw and processed | - | [64] |
| Orange juice | 90 °C/60 s | TPC 19.0% ↓ *<br>TFC 14.7% ↑, narirutin 10.8% ↓, hesperidin 28.8% ↑ *, eriocitrin 6.3% ↓, eriodictyol 12.5% ↓, naringenin 9.6% ↓, hesperetin 14.3% ↓ and kaempferol 39.6% ↑ * | L-ascorbic acid 20.1% ↓ *<br>Vitamin A 39.6% ↓ * | TCC 40.8% ↓ *, lutein 21.7% ↓, zeaxanthin 24.0% ↓ *, β-cryptoxanthin 32.3% ↓ *, α-carotene 42.5% ↓ * and β-carotene 39.0% ↓ * | [90] |
| Orange juice | 92 °C/30 s<br>85 °C/15 s | Total flavones 0.3% ↓ and 0.2% ↑, vicenin-2 2.0% ↑ and 1.8% ↑, apigenin-d 2.3% ↓ and 0.9% ↑, total flavanones 7.4% ↑ and 4.6% ↑, naringin-d 5.7% ↑ and 4.3% ↑, narirutin 0.4% ↑ and 4.3% ↑, hesperidin 7.9% ↑ and 4.5% ↑, didymin 11.2% ↑ and 9.2% ↑ and TFC 6.9% ↓ and 4.3% ↓ | - | TCC 16.6% ↓ * and 10.1% ↓ *, (Z)-antheraxanthin isomers 20.0% ↓ and 13.3% ↓, all-(E)-violaxanthin + (Z)- violaxanthin isomers 31.4% ↓ * and 20.0% ↓, (Z)-luteoxanthin isomer 0% and 2.6% ↑, (9Z)-violaxanthin + (Z)-antheraxanthin isomer 17.0% ↓ and 13.1% ↓, (Z)-luteoxanthin isomer 23.7% ↓ * and 15.3% ↓, lutein 2.0% ↓ and 3.9% ↑, zeaxanthin 24.0% ↓ * and 18.0% ↓ *, (9Z)- or (9 Z)-antheraxanthin 20.6% ↓ and 12.2% ↓, zeinoxanthin 21.6% ↓ and 16.2% ↓, β-cryptoxanthin 14.9% ↓ and 7.9% ↓, α-carotene 16.7% ↓ and 8.3% ↓, β-carotene 13.0% ↓ and 4.3% ↓ and phytoene 17.5% ↓ and 10.5% ↓ | [91] |
| Cloudy apple juice | 72 °C/15 s<br>85 °C/30 s | - | Vitamin C ↓ * and ↓ * | - | [74] |
| Orange juice | 72 °C/20 s | - | Dehydroascorbic acid ↓ *<br>Ascorbic acid—no significant difference | TCC<br>–no significant difference | [4] |
| Pomegranate juice (MW—Mollar de Elche varietal juice + Wonderful varietal juice, ML—Mollar de Elche + lemon juice, M100—Molar de Elche) | 90 °C/5 s | Anthocyanins 6.6% ↓, 2.5% ↓ and 3.7% ↑<br>Punicalagins 80% ↑ *, 866.7% ↑ * and 975% ↑ *<br>Punicalagin-like 18.2% ↓, 1.6% ↑ and 27.5% ↓ *<br>Punicalin 4.5% ↑, 27.7% ↓ and 3.0% ↓<br>Ellagic acid 2.7% ↓, 39.4% ↓ * and 12.1% ↓ | Vitamin C 59.6% ↓ *, 31.5% ↓ * and 12.9% ↑ | - | [57] |

**Table 3.** *Cont.*

| Matrix | Parameters | Phenolic Compounds | Vitamins | Other Bioactive Ingredients | References |
|---|---|---|---|---|---|
| Red prickly pear juice | 80 °C/30 s | TPC 2.0% ↓ | - | Betacyanins 14.8% ↓ * Betaxanthins 14.4% ↓ * | [40] |
| Peach juice | 72 °C/15 s | TPC 62.1% ↑ * | Ascorbic acid 22.0% ↓ * | - | [31] |
| 3 Beverages varied in terms of ingredients (blackberry juice, soy beverage, ground flaxseed, water, stabilizer and sweetener) | 71.1 °C/3 s | TPC 3.2% ↑, 8.0% ↓ and 0.2% ↓ Total ellagitannin 2.9% ↓, 15.4% ↓ and 7.4% ↑ Cyanidin-3-*O*-glucoside 11.4% ↑, 8.3% ↓ and 15.6% ↑ Cyanidin-3-*O*-malonyl-glucoside 2.4% ↓, 17.8% ↓ and 3.2% ↓ Daidzein 3.4% ↑, 17.5% ↓ and 5.5% ↓ Genistein 11.9% ↑, 9.9% ↓ and 8.5% ↑ Secoisolariciresinol 14.9% ↑, 1.5% ↓ and 9.5% ↓ | - | - | [67] |
| Cloudy pomegranate juice | 85 °C/30 s | Total monomeric anthocyanins content 27.5% ↓ * Cyanidin-3-*O*-glucoside ↑ Cyanidin-3,5-*O*-diglucoside ↓ * Delphinidin-3-*O*-glucoside ↓ Delphinidin-3,5-*O*-diglucoside ↓ Afzelechin-delphinidin-3-*O*-hexoside ↑ Pelargonidin-3-*O*-glucoside ↓ Pelargonidin-3,5-*O*-diglucoside ↓ * | Vitamin C 19.8% ↓ * | - | [70] |
| Pomegranate fermented beverage | 72 °C/15 s | TPC 14.1% ↓ TFC 3.4% ↓ Total anthocyanins content 17.5% ↓ * | - | - | [58] |
| Orange juice | 76.8 °C/15 s | - | Vitamin C—no significant difference | - | [78] |
| Raspberry juice | 85 °C/60 s 85 °C/20 s | Lamberirianin C 3.9% ↓ and 7.3% ↓ Sanguiin H-6 3.7% ↓ and 7.5% ↓ Ellagic acid conjugate 1 1.4% ↓ and 4.2% ↓ Ellagic acid conjugate 2 53.0% ↑ * and 47.1% ↑ * Ellagic acid conjugate 3 21.1% ↓ * and 22.8% ↓ * Ellagic acid 4.3% ↓ and 4.3% ↓ Cyanidin-3-*O*-sophoroside 7.0% ↓ and 9.0% ↓ Cyanidin-3-*O*-rutinoside 7.7% ↓ and 10.1% ↓ | - | - | [92] |
| Blueberry juice | 95 °C/15 s | TPC 1.1% ↓ Anthocyanin 9.2% ↓ | Vitamin C 18.5% ↓ * | - | [49] |
| Orange juice | 70, 80, 90 and 100 °C, time 2, 5, 10, 15 and 30 s | - | Vitamin C—for all temperatures ↓; for 70 °C/10 s 8% ↓, 80 °C/10 s 13% ↓, 90 °C/10 s 15% ↓ and 100 °C/10 s 18% ↓ | - | [14] |
| Apple juice with raspberry | 85 °C/6 s | TPC ↓, total flavonoids ↓ | - | - | [13] |
| Juice from "Baya Marisa" or "Golden Delicious" apples | 85 °C/30 s | Total hydroxycinnamic acids 42.0% ↓ * and 64.3% ↓ * Total hydroxybenzoic acids 33.6% ↓ * and nondetected in "Golden Delicious" apples Total dihydrochalcones 32.9% ↓ and 2.0% ↓ Total flavanols 79.5% ↓ * and 72.2% ↓ * Total anthocyanins 66.5% ↓ * and nondetected in "Golden Delicious" apples TPC 52.0% ↓ * and 63.7% ↓ * | - | - | [93] |
| Orange juice | 90 °C/20 s | TPC 5.3% ↓ | Vitamin C 0.1% ↓ | TCC 12.6% ↓ | [94] |
| Strawberry puree–kale juice mix | 72 °C/60 s | Total anthocyanins content ↓ | - | - | [61] |
| Whey–grape juice drink | 72 °C/15 s | Total anthocyanins content 8.5% ↓, monomeric anthocyanins 25.0% ↓ and polymeric anthocyanins 2.0% ↓ | - | - | [95] |

Table 3. *Cont.*

| Matrix | Parameters | Phenolic Compounds | Vitamins | Other Bioactive Ingredients | References |
|---|---|---|---|---|---|
| Valencia orange juice | 73.9 °C/30 s<br>92.2 °C/31 s | - | Ascorbic acid 3.8% ↓<br>and 2.6% ↓ | Total carotenes 5.5% ↑<br>and 14.0% ↓<br>TCC 8.5% ↑ and 13.6% ↓<br>Total carotenoid fatty acids<br>esters 7.7% ↑ and 15.1% ↓<br>α-carotene 12.5% ↑ and 7.1% ↓<br>β-carotene 20.6% ↑ and 4.1% ↑ | [96] |
| Orange juice | 90 °C/60 s | - | Vitamin C 17.6% ↓ | - | [97] |
| Apple and cranberry juice blend | 72 °C/26 s | TPC 3.0% ↓<br>Cyanidin-3-*O*-glucoside 14.6% ↓ | - | - | [98] |

↑—Increase; ↓—decrease. * Statistically significant change. TPC—total phenolic content; TFC—total flavonoid content; TCC—total carotenoid content.

### 4.4. Antioxidant Activity

Antioxidant activity (AA) is closely related to the contents of bioactive components found in food, such as phenolics, as well as their bioaccessibility and bioavailability. The literature points to numerous methods for determining total antioxidant activity in foods based on single-electron transfer reaction or SET assays or based on a hydrogen atom transfer reaction or HAT assays [79].

Heating alters the content and profile of bioactive compounds found in plant raw materials and, consequently, as well as antioxidant activities. A study by Chen et al. [39] conducted on a papaya beverage showed that the AA measured by FRAP and DPPH radical methods decreased significantly immediately after preservation (110 °C/8.6 s), while there was no change in carotenoids content and a decrease in total phenols content, which are strong antioxidants. Decreases in antioxidant activities were shown by Chen et al. [38] (8.2% for DPPH radical and 6.8% for FRAP) and Chen et al. [37] (10.2% for DPPH radical and 6.0% for FRAP), with a significant decrease in TPC, and by Gao et al. [53] in red grapefruit juice (1.4% for DPPH radical and 8.5% for FRAP), as well as other researchers [17,42]. On the other hand, Zou et al. [27] observed no significant change in AA using the DPPH radical method but noted a significant increase using the FRAP method immediately after preservation. In the second case, the phenomena was associated with an increase in TPC and the tendency of phenolics to undergo polymerization reactions. The opposite relationship was shown by Liu et al. [44]. Wang et al. [35] indicated neither an increase nor a decrease in AA after preservation of purple sweet potato nectar by the UHT method, which was associated with no change in phenolics content, while over 12 weeks of storage, AA decreased by more than 20%. Zhao et al. [54] observed no significant change in AA despite significant decreases in TPC and ascorbic acid content. A decrease in AA during storage was also demonstrated by Liu et al. [44].

HTST under conditions of 72 °C/15 s resulted in a significant (77.5%) increase in antioxidant activity immediately after treatment. Over the course of preservation, the AA in the preserved sample increased from 1554.64 (0 day) to 2489.16 µmol TE/L (28 day). However, from 14 to 28 days of preservation, the AA in heated peach juice was lower than in a nonprocessed one [31]. Yildiz and Aadil [26] came to similar conclusions, as follows: in strawberry juice immediately after the process, AA increased by 27% and was significantly higher by the 7th day of storage compared to fresh juice. In contrast, the AA values for ultrasound or high-pressure homogenized juices were significantly higher throughout the working principles of these processes [26,31]. On the other hand, the AA was not affected by thermal processing [30,57]. In contrast, some researchers point to a significant reduction in AA due to decreases in antioxidants, such as vitamin C and carotenoids, in sea buckthorn juice as a result of HTST [56].

### 4.5. Enzymes Activity

Preservation using high temperatures results in a significant reduction in the activity or complete inactivation of enzymes responsible for changes in the quality characteristics and sensory attributes of food. The higher the temperature, the lower the enzymatic activity remains [37]. The application of appropriate heat treatment can lead to complete inactivation of enzymes (Table 4). The main enzymes found in raw fruits and vegetables are polyphenyl oxidase (PPO), peroxidase (POD), pectinomethylesterase (PME) and phenylalanine ammonia lyase (PAL). They are responsible for numerous reactions, most often leading to changes in quality characteristics [20]. PPO and POD are enzymes largely responsible for enzymatic browning reactions in fruits and vegetables (undesirable brown color). Their action also leads to an unpleasant taste and a loss of nutritional properties [99]. PAL is responsible for the biosynthesis of phenolic compounds in plants, converting phenylalanine into flavonoids, phenolic acids or anthocyanins [84]. PME causes cloud loss by allowing for calcium pectates to precipitate and clarify the juice [4,31]. Enzyme inactivation usually involves the destruction of secondary and tertiary structures by physical (heat, ultrasound, and high pressure) and chemical factors [2].

In the storage process, changes in enzyme activity can occur. For example, changes in phenolic enzyme (PPO and POD) activity can arise because of reactions between phenolic compounds and proteins or phenolics oxidation. The first reaction leads to the formation of an inactive enzyme–substrate complex or a change in the enzyme's catalytic site. The second leads to a reduction in the amount of substrate and the formation of products that inhibit enzymes [74].

The combination of temperature and exposure time can affect enzyme inactivation differently. A detailed summary of the effects of the applied UHT and HTST preservation temperatures on enzymatic activity in juices, nectars and beverages is included in Table 4. Wang et al. [24] showed that UHT conditions of 120 °C/2 s and 135 °C/2 s resulted in more than a 10% PPO residual rate in watermelon juice, while 110 °C/2 s resulted in about a two times higher PPO residual rate. Because of the residual PPO activity in the juice, an inverse relationship occurred in the phenolics content. PPO is responsible for catalyzing the *o*-hydroxylation of *o*-phenols and *o*-diphenols. Juices preserved at 120 °C and 135 °C for two seconds did not differ in phenolics content from nonpreserved juices, while juice treated at 110 °C had significantly lower levels of phenolics. Prolonged exposure to high temperature can effectively increase the inactivation of tissue enzymes. Huang et al. [20] demonstrated the complete inactivation of PPO and POD in apricot nectar, Krapfenbauer et al. [100] of PPO in cloudy apple juice, and Gao et al. [53] of POD and PME in red grapefruit juice using 110 °C/8.6 s. In contrast, the 110 °C/8.6 s treatment caused the complete inactivation of PPO and the half inactivation of PAL in red raspberry juice, while POD activity remained at the same level [84].

The HTST preservation process caused a significant reduction in the activity or complete inactivation of tissue enzymes [20]. Yildiz [31] showed that under conditions of 72 °C/15 s, decreases of 13.6% and 13.0% in PPO and PME activities occurred, respectively, compared to raw juice. During 28 days of storage, PPO and PME activities in the HTST juices increased by 16.7% and 19.4%. Similar changes were observed for PPO in raw juice; however, PME increased by 99.7%. With the HTST process, we can obtain a product with less turbidity and change in color during storage, which is important in consumer evaluations. In addition, Wibowo et al. [74] showed that conditions of 72 °C/15 s caused inactivation of PPO, POD and PME by more than 90% in apple juice. Also, significant inactivation of PPO and POD was demonstrated by the treatment described in [101]. Even better results were achieved by Wibowo et al. [74]; they obtained complete inactivation of PPO, POD and PME using conditions of 85 °C/15 s, while Deng et al. [5] deactivated PPO and POD using the same parameters.

Another important enzyme found in fruits and vegetables is lipooxygenase (LOX). Polyunsaturated fatty acids undergo oxygenation catalyzed by LOX. This enzyme also plays an important role in shaping plant flavors. Heating 85 °C/15 s allowed for the residual activity of LOX to reach 58.2% in cucumber juice drink, while the enzyme activity decreased during storage [66]. Aguiló-Aguayo et al. [100] indicated that the LOX activities of strawberry juices preserved at 90 °C for 60 and 30 s were 37.7% and 45.3%, respectively [102]. Aguiar et al. [103] developed enzymic time–temperature integrators (TTIs) with rapid detection for the evaluation of continuous HTST pasteurization processes, using temperatures of 70–85 °C and times of 10–60 s.

**Table 4.** Effects of UHT and HTST processing on enzymatic activity in different beverages.

| Matrix | Parameters | Enzymes—Residual Activities | Reference |
|---|---|---|---|
| | | UHT | |
| Mango nectar | 110 °C/8.6 s | Acid invertase 8.6% * | [45] |
| Watermelon juice | 110 °C/2 s, 120 °C/2 s, 135 °C/2 s | For 120 °C and 135 °C, 2 times lower residual activity of PPO than at 110 °C | [24] |
| Red grapefruit juice | 110 °C/8.6 s | PPO, POD and PME—completely inactivated | [53] |
| Apricot nectar | 110 °C/8.6 s | PPO, POD and PME—completely inactivated | [20] |
| red raspberry juice | 110 °C/8.6 s | PAL ↓ *; PPO—completely inactivated; PPO—equal to that of fresh juice | [84] |
| | | HTST | |
| Strawberry juice | 90 °C/60 s<br>90 °C/30 s | 22.2% * for PME, 76.2% * for PG<br>48% * for PME and 96.8% * for PG | [104] |
| Strawberry juice | 90 °C/60 s<br>90 °C/30 s | Lipoxygenase 37.7% * and 45.3% *, b-glucosidase—slight increase in activity 7.9% and 4.1% | [102] |
| Sea buckthorn juice | 100 °C/15 s | SOD 51.3% * | [56] |
| Orange and carrot juice | 98 °C/21 s | PME—2% * | [46] |
| Cloudy apple juice | 98 °C/50 s | PPO and POD—completely inactivated | [5] |
| Cloudy apple juice | 72 °C/15 s<br>85 °C/30 s | PPO, POD and PME—up to 10% *<br>PPO, POD and PME—completely inactivated | [74] |
| Orange juice | 72 °C/20 s | PME 15% *; POD—completely inactivated | [4] |
| Peach juice | 72 °C/15 s | PPO 86.4%, PME 87.0% | [31] |
| Orange juice | 70, 80, 90 and 100 °C and time 2, 5, 10, 15 and 30 s | PME—inactivation of 99% was only achieved at 90 °C and 100 °C | [14] |

↓—decrease. * Statistically significant change. PPO—polyphenol oxidase, POD—peroxidase, PAL—phenylalanine ammonia-lyase; PME—pectinmethylesterase; SOD—superoxide dismutase.

### 4.6. *Hydroxymethylfurfural*

At high temperatures, Maillard reactions take place, leading to the formation of compounds such as hydroxymethylfurfural (HMF) and furfural. The reactions occur between amino acids and reducing sugars, in addition to ascorbic acid transformations. HMF and others are compounds characterized by genotoxic, cytotoxic and mutagenic risks [4,19,45,47]. The allowed maximum level of 10 mg for HMF per liter of juice are recommended by the AIJN (European Fruit Juice Association) [74].

Liu et al. [45] showed that heating at 110 °C/8.6 s increased the HMF content in mango nectar by 70.6% compared to nonpreserved juice. They also observed an increase in its content in UHT-preserved samples after 16 weeks of 55.2% at 4 °C and 177.8% at 25 °C. In apricot nectar subjected to UHT, HMF was not detected [20].

Varied changes are observed in the literature for the HMF content in samples preserved by the HTST method. Wibowo et al. [74] indicated that in orange juice pasteurized by the HTST method, HMF was not immediately detected after the process; however, it started to accumulate after 20 weeks of storage at 20 °C, 8 weeks at 28 °C, 3 weeks at 35 °C and 4 days at 42 °C. The researchers reported that the appearance of HMF may have been related to a change in the composition of the sugars present in the product. Tests on orange juice were also performed by Ağçam et al. [19] using the parameters of 90 °C/10 s and 90 °C/20 s. The HMF contents immediately after preservation were 1.7 and 4.7 ppb, respectively; at the same time, this compound was not observed in the raw sample. During 180 days of storage, increases of 1315.6% and 826.7% were observed. Aguiló-Aguayo et al. [104] noted significant increases of 50.6% and 77.1% in the HMF content of strawberry juice after preservation at 90 °C/30 s and 90 °C/60 s, while no significant changes occurred during storage for 63 days at 4 °C. A statistically insignificant increase in the amount of HMF, as a result of HTST, was demonstrated by Cortés et al. [69]. Rivas et al. [46] observed no effect of HTST on HMF content compared to raw and pulsed electric field preserved samples. In addition, the same researchers showed no significant changes in HMF content during storage at 2 °C for 10 weeks. In a study by Vervoort et al. [4], no measurable quantities of HMF were found under the HTST-processing conditions applied.

### 4.7. Volatile Compounds

Volatile compounds determine the aroma and flavor of foods. Their transformations can lead to organoleptic changes that are unacceptable to consumers.

As a result of UHT preservation, the neral content increased, while the geranial content decreased in cloudy ginger juice [37]. Also, Wang et al. [24] showed the degradation of typical volatile contents of watermelon juice (C9 alcohol and C9 aldehyde) by 15.4%, 10.9% and 10.1% after applying the conditions of 110 °C, 120 °C and 135 °C for 2 s, respectively. However, higher temperatures (120 °C and 135 °C) reduced the ADH activity and had a greater degrading effect on the C9 alcohol content rather than C9 aldehyde content. Wang et al. [43] investigated different preservation methods in watermelon juice. The typical volatile contents of UHT juice (135 °C/2 s) were significantly lower than those in unprocessed juice. The heated samples showed the presence of two compounds (acetic acid, 2-methyl butyric acid), which were absent in the raw sample. Liu et al. [105], in watermelon juice processed at 126 °C/15 s, noticed an increase in the relative contents of esters (58.4%) and aldehydes but a decrease in alcohols (80.1%), ketones (59.5%) and alkanes compared to raw juice. Nonanal and 3,7-dimethyl-2,6-octadienal (characteristic compounds for watermelon flavor) decreased by 59.2% and 92.1% in UHT watermelon juice. In the UHT juice, the content of amyl butyrate increased (80.1%), accounting for the "cooking flavor". Also, Wang et al. [18] pointed to the formation of the compound dimethyl sulfide, responsible for the "cooked flavor" of tomato juice, following heating at 110 °C/8.6 s. In addition, in raw tomato juice, the contents of alcohols, aldehydes, and ketones accounted for 46%, 18% and 23% of all flavor compounds, respectively, whereas in juice after UHT they were 36%, 34% and 18%, respectively. Liu et al. (2022) [28] indicated that UHT has the effect of raising the contents of the esters and ketones responsible for the specific flavor of melon juice, while it causes a decrease in the contents of alcohols, aldehydes and total volatile components. Zhao et al. [54] indicated that 20 volatile compounds were present in the control sample, but following UHT (110 °C/8.6 s), only 13 remained in korla pear juice. In comparison, there were 17 compounds in the HPP-preserved sample. Liu et al. [41] compared the UHT and HPP processes in an analysis of the key odorants of clear cucumber juice. In the case of (Z)-6-nonenal and (E,Z)-3,6-nonandien-1-ol, there was no significant difference between the processed and raw juices. (E,Z)-2,6-nonadienal and (E)-2-nonenal showed significant increases in their contents after preservation but to a much lesser extent in the UHT sample than in the HPP one. Other observations made by Bao et al. [17] included that the heating of black carrot juice at 130 °C/5 s did not change the overall odor profile.

Zhu et al. [49] compared blueberry juice after HTST and PEF preservation, indicating that the decreases in the compounds were considerably lower for PEF than with the HTST method, particularly in esters. HTST causes the aroma of blueberry juice to be less diverse. Passion fruit juice also showed HTST-preserved volatile compounds to a lesser extent than when using the HHP preservation technique. The concentrations of esters, alcohols, ketones and hydrocarbons decreased in the HTST sample by 23.9, 44.2, 12.5 and 39.7%, respectively [11]. The volatile acidities of a pomegranate fermented beverage, as a result of heating at 72 °C for 15 s, did not change significantly compared to the nonprocessed sample, while PEF and traditional pasteurization significantly lowered them [58]. A study by Atuonwu et al. [78] indicated that it was not possible to distinguish between the microwave- and HTST-treated samples of orange juice. Tian et al. [32] compared HPP, HTST (85 °C/30 s) and UHT (110 °C/8.6 s) in terms of the volatiles compounds of cloudy pomegranate juice. Alcohols were more sensitive to thermal sterilization than other kinds of volatile compounds. The UHT and HTST methods preserved volatile compounds to a lesser extent because of the direct effect of the thermal treatments on the small-molecule flavor compounds. For the same reason, linalool was observed in the HTST and UHT samples in contrast to the pressurized and raw samples. In addition, a significant increase in the content of a pentadecanoic acid (131.1% for HTST and 228.8% for UHT), a compound responsible for the waxy aroma and off-odor of cloudy pomegranate juices, was observed in the high-temperature-treated samples.

### *4.8. Sensory Quality*

Highly significant is the impact of preservation methods on sensory qualities. The high temperatures lead to changes in the contents of various ingredients, transformations of substances and occurrences of chemical processes that change the taste, smell and color of products.

Xu et al. [25] evaluated the quality of selenium-enriched kiwi juice treated at 110 °C for 8.6 s. They showed that clear and cloudy preserved juices received significantly lower scores for color and appearance, flavor and overall acceptability compared to their raw counterparts. Only for taste was there no significant difference among trained panelists. In addition, in all distinguishing factors, thermally preserved juices were rated worse than HPP-treated juices. Chen et al. [39] came to different conclusions; the UHT-preserved papaya beverage did not differ in color and appearance, while it had worse mouth feel, flavor and overall acceptability compared to raw juice. In contrast, raw clear cucumber juice did not differ in color and appearance from the UHT sample [41]. When stored, cucumber juices received lower scores after 20 days than immediately after preservation. Liu et al. [41] and Xu et al. [25] showed that thermally preserved juices scored lower than HPP juices in all distinguishing factors.

de Souza et al. [64] conducted sensory evaluations of lemonade, citrus juice and green juice. The HTST-preserved lemonade scored better in flavor and overall liking than the raw sample, and the preserved green juice scored worse in overall liking and more than 30% worse in color than raw juice. In contrast, the preserved citrus juice was not significantly different from its untreated version. The researchers also sensory evaluated those products preserved by other methods—mostly, the HPP method and UV-C light changed the ratings to a lesser extent compared to HTST. Rivas et al. [46] found that juice subjected to HTST preservation showed less acceptable odor and taste than samples subjected to PEF preservation. Walkling-Ribeiro et al. [50] compared orange juice preserved at 94 °C/26 s and a sample preserved with a combined thermosonification/PEF method. The panelists found no significant differences between these samples in terms of color, odor, sweetness, acidity, flavor and acceptability. However, the HTST-preserved juice was rated as being blander than the TS/PEF juice by twice as many evaluators. Deshaware et al. [16] found that an increase in the processing temperature for the same product can result in poorer flavor scores among evaluators. The HTST-preserved orange juice was rated worse overall and in terms of flavor and aroma than fresh juice; however, even lower scores were given to the UHT-preserved sample [22]. Polydera et al. [106], based on sensory evaluations, determined the shelf-life of orange juice, which they indicated as being 60 days at 0 °C and 47 days at 5 °C. This is about one-third to one-half shorter times than when HPP treatment was used, indicating a lower level of preservation of the product's relevant sensory attributes.

## 5. Conclusions

The HTST and UHT methods are promising methods in the fruit and vegetable industry. The ability to modify the processing parameters can allow for the selection of adequate preservation parameters for individual products and better results than other unconventional methods, such as HPP or PEF. HTST and UHT are methods characterized by the ability to inactivate unwanted microorganisms, allowing for product shelf-lives that last for many weeks.

In the case of basic physicochemical characteristics, such as total soluble solids, pH or acidity, the HTST and UHT methods most often do not determine significant changes that negatively affect the quality of the final product. In the case of color, contents of bioactive components and the antioxidant and enzymatic activities it is significantly important to choose the right temperature and duration. Excessively high temperatures or exposure times can lead to the degradation of key bioactive components, such as vitamins, carotenoids or phenolic compounds, while the right choice of parameters can allow these compounds to be extracted from the food matrix, transformed or, at least, to remain at similar levels as in raw samples. Accordingly, the transformation of natural colorants due to heat leads to changes in the absolute color, which can be important in consumer evaluations. The HTST and UHT methods show highly effective inactivation of enzymes such as POD, PPO or PME, allowing for better quality of products to be preserved including during storage. In the case of volatile compounds, these thermal methods most often lead to changes in their profiles, leading to flavor and aroma degradation to some extent.

It is presumed that in the near future the HTST and UHT methods will be dynamically improved upon on an industrial scale in the fruit and vegetable industry because of their numerous benefits compared to the traditional pasteurization process commonly used in the industry. Better retention of bioactive components, less sensory degradation and equally or even more effective inactivation of microorganisms, as well as lower financial costs of these processes are the most important advantages in the rapid heating of food. A challenge for the described preservation methods is to investigate their application in terms of adapting process parameters to products with varying raw material particle sizes, such as beverages enriched with different raw materials or multivegetable soups with solid pieces. A little-developed area of research is the application of HTST and UHT in a combined method, for example with HPP.

**Author Contributions:** Conceptualization, N.P., S.K. and B.K.; methodology, N.P.; software, N.P.; validation, N.P.; formal analysis, N.P.; investigation, N.P.; resources, N.P. and B.K.; data curation, N.P.; writing—original draft preparation, N.P.; writing—review and editing, N.P., S.K. and B.K.; visualization, B.K.; supervision, S.K. and B.K.; project administration, N.P.; funding acquisition, S.K. and B.K. All authors have read and agreed to the published version of the manuscript.

**Funding:** This research received no external funding.

**Institutional Review Board Statement:** Not applicable.

**Informed Consent Statement:** Not applicable.

**Data Availability Statement:** Not applicable.

**Conflicts of Interest:** The authors declare no conflicts of interest.

## Abbreviations

| | |
|---|---|
| HPP | high-pressure processing; |
| PEF | pulsed electric field; |
| HTST | high temperature, short time; |
| UHT | ultra-high temperature; |
| TSS | total soluble solids; |
| HPH | high-pressure homogenization; |

| | |
|---|---|
| US | ultrasounds; |
| TAB | total aerobic bacteria; |
| Y&M | yeasts and molds; |
| TA | titratable acidity; |
| TPC | total phenolic content; |
| TFC | total flavonoid content; |
| TCC | total carotenoid content; |
| GAE | gallic acid; |
| AA | antioxidant activity; |
| FRAP | ferric-reducing antioxidant power assay; |
| DPPH | 2,2-diphenyl-1-picrylhydrazyl; |
| TE | Trolox equivalents; |
| PPO | polyphenyl oxidase; |
| POD | peroxidase; |
| PME | pectinomethylesterase; |
| PAL | phenylalanine ammonia lyase; |
| LOX | lipooxygenase; |
| SOD | superoxide dismutase; |
| HMF | hydroxymethylfurfural. |

## References


1. Saikumar, A.; Singh, A.; Dobhal, A.; Arora, S.; Junaid, P.M.; Badwaik, L.S.; Kumar, S. A review on the impact of physical, chemical, and novel treatments on the quality and microbial safety of fruits and vegetables. *Syst. Microbiol. Biomanufacturing* **2024**, *4*, 575–597. [CrossRef]
2. Liu, Y.; Deng, J.; Zhao, T.; Yang, X.; Zhang, J.; Yang, H. Bioavailability and mechanisms of dietary polyphenols affected by non-thermal processing technology in fruits and vegetables. *Curr. Res. Food Sci.* **2024**, *8*, 100715. [CrossRef] [PubMed]
3. Kruszewski, B.; Domian, E.; Nowacka, M. Influence of High-Pressure Homogenization on the Physicochemical Properties and Betalain Pigments of Red Beetroot (*Beta vulgaris* L.) Juice. *Molecules* **2023**, *28*, 2018. [CrossRef] [PubMed]
4. Vervoort, L.; Van Der Plancken, I.; Grauwet, T.; Timmermans, R.A.H.; Mastwijk, H.C.; Matser, A.M.; Hendrickx, M.E.; Van Loey, A. Comparing equivalent thermal, high pressure and pulsed electric field processes for mild pasteurization of orange juice: Part II: Impact on specific chemical and biochemical quality parameters. *Innov. Food Sci. Emerg. Technol.* **2011**, *12*, 466–477. [CrossRef]
5. Deng, H.; Zhao, P.T.; Yang, T.G.; Meng, Y.H. A comparative study of the cloudy apple juice sterilized by high-temperature short-time or high hydrostatic pressure processing: Shelf-life, phytochemical and microbial view. *Food Sci. Technol.* **2022**, *42*, e63620. [CrossRef]
6. Duhan, S.; Kar, A. Optimization of process parameter combinations for pasteurization of sugarcane (*Saccharum officinarum*) juice using continuous flow microwave system. *Indian J. Agric. Sci.* **2018**, *88*, 1253–1257. [CrossRef]
7. Sattar, S.; Imran, M.; Mushtaq, Z.; Ahmad, M.H.; Arshad, M.S.; Holmes, M.; Maycock, J.; Nisar, M.F.; Khan, M.K. Retention and stability of bioactive compounds in functional peach beverage using pasteurization, microwave and ultrasound technologies. *Food Sci. Biotechnol.* **2020**, *29*, 1381–1388. [CrossRef]
8. Askin, B.; Türkyılmaz, M.; Özkan, M.; Küçüköner, E. Changes in anthocyanins and colour of black mulberry (*Morus nigra*) juice during clarification and pasteurization. *J. Food Meas. Charact.* **2022**, *16*, 784–792. [CrossRef]
9. Bhagat, B.; Chakraborty, S. Potential of pulsed light treatment to pasteurize pomegranate juice: Microbial safety, enzyme inactivation, and phytochemical retention. *LWT* **2022**, *159*, 113215. [CrossRef]
10. Kruszewski, B.; Zawada, K.; Karpiński, P. Impact of High-Pressure Homogenization Parameters on Physicochemical Characteristics, Bioactive Compounds Content, and Antioxidant Capacity of Blackcurrant Juice. *Molecules* **2021**, *26*, 1802. [CrossRef]
11. Zou, W.; Niu, H.; Yi, J.; Zhou, L. Passion fruit juicing with or without seeds treated by high-pressure processing and thermal pasteurization: Effects on the storage stability of enzymes and quality properties. *Innov. Food Sci. Emerg. Technol.* **2024**, *91*, 103554. [CrossRef]
12. Zhang, J.; Cheng, J.; Li, Z.; Weng, M.; Zhang, X.; Tang, X.; Pan, Y. Effects of ultra-high pressure, thermal pasteurization, and ultra-high temperature sterilization on color and nutritional components of freshly-squeezed lettuce juice. *Food Chem.* **2024**, *435*, 137524. [CrossRef] [PubMed]
13. Estrada-Beltrán, A.E.; Salas-Salazar, N.A.; Quintero-Ramos, A.; Parra-Quezada, R.A.; Soto-Caballero, M.C.; Rodríguez-Roque, M.J.; Chávez-Martínez, A.; Flores-Cordova, M.A. Effect of UV-C Radiation and Thermal Treatment on Volatile Compounds, Physicochemical, Microbiological and Phytochemical Parameters on Apple Juice (*Malus domestica*) with Raspberry (*Rubus idaleus* L.). *Beverages* **2024**, *10*, 7. [CrossRef]
14. Amaro, K.C.; Russo, G.; Fan, D.L.; Gut, J.A.W.; Tadini, C.C. Modeling and experimental validation of the time-temperature profile, pectin methylesterase inactivation, and ascorbic acid degradation during the continuous flow microwave-assisted pasteurization of orange juice. *Food Bioprod. Process.* **2024**, *144*, 191–202. [CrossRef]

15. Ren, G.; Wan, K.; Kong, H.; Guo, L.; Wang, Y.; Liu, X.; Wei, G. Recent advance in biomass membranes: Fabrication, functional regulation, and antimicrobial applications. *Carbohydr. Polym.* **2023**, *305*, 120537. [CrossRef]
16. Deshaware, S.; Gupta, S. Influence of different pasteurization techniques on antidiabetic, antioxidant and sensory quality of debittered bitter gourd juice during storage. *Food Chem.* **2019**, *285*, 156–162. [CrossRef]
17. Bao, S.; Yin, D.; Zhao, Q.; Zhou, Y.; Hu, Y.; Sun, X.; Liu, X.; Ma, T. Comprehensive evaluation of the effect of five sterilization methods on the quality of black carrot juice based on PCA, TOPSIS and GRA models. *Food Chem. X* **2023**, *17*, 100604. [CrossRef]
18. Wang, X.; Chen, F.; Ma, L.; Liao, X.; Hu, X. Non-volatile and volatile metabolic profiling of tomato juice processed by high-hydrostatic-pressure and high-temperature short-time. *Food Chem.* **2022**, *371*, 131161. [CrossRef]
19. Agcam, E.; Akyildiz, A.; Akdemir Evrendilek, G. A comparative assessment of long-term storage stability and quality attributes of orange juice in response to pulsed electric fields and heat treatments. *Food Bioprod. Process.* **2016**, *99*, 90–98. [CrossRef]
20. Huang, W.; Bi, X.; Zhang, X.; Liao, X.; Hu, X.; Wu, J. Comparative study of enzymes, phenolics, carotenoids and color of apricot nectars treated by high hydrostatic pressure and high temperature short time. *Innov. Food Sci. Emerg. Technol.* **2013**, *18*, 74–82. [CrossRef]
21. Aguiló-Aguayo, I.; Plaza, L.; Bobo, G.; Abadias, M.; Viñas, I. Pome Fruit Juices. In *Innovative Technologies in Beverage Processing*; Aguiló-Aguayo, I., Plaza, L., Eds.; John Wiley & Sons, Ltd.: Chichester, UK, 2017; pp. 1–25, ISBN 9781118929346.
22. Zvaigzne, G.; Karklina, D.; Moersel, J.T.; Kuehn, S.; Krasnova, I.; Seglina, D. Ultra-high temperature effect on bioactive compounds and sensory attributes of orange juice compared with traditional processing. *Proc. Latv. Acad. Sci. Sect. B Nat. Exact Appl. Sci.* **2017**, *71*, 486–491. [CrossRef]
23. Dash, K.K.; Fayaz, U.; Dar, A.H.; Shams, R.; Manzoor, S.; Sundarsingh, A.; Deka, P.; Khan, S.A. A comprehensive review on heat treatments and related impact on the quality and microbial safety of milk and milk-based products. *Food Chem. Adv.* **2022**, *1*, 100041. [CrossRef]
24. Wang, Y.; Guo, X.; Ma, Y.; Zhao, X.; Zhang, C. Effect of ultrahigh temperature treatment on qualities of watermelon juice. *Food Sci. Nutr.* **2018**, *6*, 594–601. [CrossRef] [PubMed]
25. Xu, X.; Deng, J.; Luo, D.; Bao, Y.; Liao, X.; Gao, H.; Wu, J. Comparative study of high hydrostatic pressure and high temperature short time processing on quality of clear and cloudy Se-enriched kiwifruit juices. *Innov. Food Sci. Emerg. Technol.* **2018**, *49*, 1–12. [CrossRef]
26. Yildiz, G.; Aadil, R.M. Comparison of high temperature-short time and sonication on selected parameters of strawberry juice during room temperature storage. *J. Food Sci. Technol.* **2020**, *57*, 1462–1468. [CrossRef]
27. Zou, H.; Lin, T.; Bi, X.; Zhao, L.; Wang, Y.; Liao, X. Comparison of High Hydrostatic Pressure, High-PressureCarbon Dioxide and High-Temperature Short-Time Processing on Quality of Mulberry Juice. *Food Bioprocess Technol.* **2016**, *9*, 217–231. [CrossRef]
28. Liu, X.; Wang, R.; Liu, H.; Wang, Y.; Shi, Y.; Zhang, C. High-pressure treatment enhanced aromatic compound concentrations of melon juice and its mechanism. *Front. Nutr.* **2022**, *9*, 1052820. [CrossRef]
29. Pandiselvam, R.; Mitharwal, S.; Rani, P.; Shanker, M.A.; Kumar, A.; Aslam, R.; Barut, Y.T.; Kothakota, A.; Rustagi, S.; Bhati, D.; et al. The influence of non-thermal technologies on color pigments of food materials: An updated review. *Curr. Res. Food Sci.* **2023**, *6*, 100529. [CrossRef]
30. Yang, P.; Wang, Y.; Wu, X.; Geng, H.; Liao, X.; Zhao, L. Effect of High Pressure Processing and High-Temperature Short-Time Sterilization on the Quality of Sea Buckthorn Juice. *Shipin Kexue/Food Sci.* **2022**, *43*, 23–32. [CrossRef]
31. Yildiz, G. Application of ultrasound and high-pressure homogenization against high temperature-short time in peach juice. *J. Food Process Eng.* **2019**, *42*, e12997. [CrossRef]
32. Tian, J.; Cheng, F.; Yun, Y.; Yi, J.; Cai, S.; Zhou, L. Characterization of the flavor, sensory quality and in vitro bioaccessibility in cloudy pomegranate juice treated by high pressure and thermal processing. *J. Sci. Food Agric.* **2023**, *103*, 666–679. [CrossRef] [PubMed]
33. Soni, A.; Brightwell, G. Effect of Hurdle Approaches Using Conventional and Moderate Thermal Processing Technologies for Microbial Inactivation in Fruit and Vegetable Products. *Foods* **2022**, *11*, 1811. [CrossRef] [PubMed]
34. Cebrián, G.; Condón, S.; Mañas, P. Physiology of the inactivation of vegetative bacteria by thermal treatments: Mode of action, influence of environmental factors and inactivation kinetics. *Foods* **2017**, *6*, 107. [CrossRef]
35. Wang, Y.; Liu, F.; Cao, X.; Chen, F.; Hu, X.; Liao, X. Comparison of high hydrostatic pressure and high temperature short time processing on quality of purple sweet potato nectar. *Innov. Food Sci. Emerg. Technol.* **2012**, *16*, 326–334. [CrossRef]
36. Huang, H.W.; Chen, B.Y.; Wang, C.Y. Comparison of high pressure and high temperature short time processing on quality of carambola juice during cold storage. *J. Food Sci. Technol.* **2018**, *55*, 1716–1725. [CrossRef] [PubMed]
37. Chen, D.; Xi, H.; Guo, X.; Qin, Z.; Pang, X.; Hu, X.; Liao, X.; Wu, J. Comparative study of quality of cloudy pomegranate juice treated by high hydrostatic pressure and high temperature short time. *Innov. Food Sci. Emerg. Technol.* **2013**, *19*, 85–94. [CrossRef]
38. Chen, D.; Pan, S.; Chen, J.; Pang, X.; Guo, X.; Gao, L.; Liao, X.; Wu, J. Comparing the Effects of High Hydrostatic Pressure and Ultrahigh Temperature on Quality and Shelf Life of Cloudy Ginger Juice. *Food Bioprocess Technol.* **2016**, *9*, 1779–1793. [CrossRef]
39. Chen, D.; Pang, X.; Zhao, J.; Gao, L.; Liao, X.; Wu, J.; Li, Q. Comparing the effects of high hydrostatic pressure and high temperature short time on papaya beverage. *Innov. Food Sci. Emerg. Technol.* **2015**, *32*, 16–28. [CrossRef]
40. Mesta-Vicuña, G.; Quintero-Ramos, A.; Meléndez-Pizarro, C.O.; Galicia-García, T.; Sánchez-Madrigal, M.Á.; Delgado, E.; Ruiz-Gutiérrez, M.G. Physical, Chemical and Microbiological Properties during Storage of Red Prickly Pear Juice Processed by a Continuous Flow UV-C System. *Appl. Sci.* **2022**, *12*, 3488. [CrossRef]

41. Liu, F.; Zhang, X.; Zhao, L.; Wang, Y.; Liao, X. Potential of high-pressure processing and high-temperature/short-time thermal processing on microbial, physicochemical and sensory assurance of clear cucumber juice. *Innov. Food Sci. Emerg. Technol.* **2016**, *34*, 51–58. [CrossRef]
42. Xu, Z.; Lin, T.; Wang, Y.; Liao, X. Quality assurance in pepper and orange juice blend treated by high pressure processing and high temperature short time. *Innov. Food Sci. Emerg. Technol.* **2015**, *31*, 28–36. [CrossRef]
43. Wang, Y.; Li, W.; Ma, Y.; Zhao, X.; Zhang, C. Effect of Thermal Treatments on Quality and Aroma of Watermelon Juice. *J. Food Qual.* **2018**, *2018*, 9242675. [CrossRef]
44. Liu, F.; Wang, Y.; Li, R.; Bi, X.; Liao, X. Effects of high hydrostatic pressure and high temperature short time on antioxidant activity, antioxidant compounds and color of mango nectars. *Innov. Food Sci. Emerg. Technol.* **2014**, *21*, 35–43. [CrossRef]
45. Liu, F.; Li, R.; Wang, Y.; Bi, X.; Liao, X. Effects of high hydrostatic pressure and high-temperature short-time on mango nectars: Changes in microorganisms, acid invertase, 5- hydroxymethylfurfural, sugars, viscosity, and cloud. *Innov. Food Sci. Emerg. Technol.* **2014**, *22*, 22–30. [CrossRef]
46. Rivas, A.; Rodrigo, D.; Martínez, A.; Barbosa-Cánovas, G.V.; Rodrigo, M. Effect of PEF and heat pasteurization on the physical-chemical characteristics of blended orange and carrot juice. *LWT* **2006**, *39*, 1163–1170. [CrossRef]
47. Santhirasegaram, V.; Razali, Z.; George, D.S.; Somasundram, C. Comparison of UV-C treatment and thermal pasteurization on quality of Chokanan mango (*Mangifera indica* L.) juice. *Food Bioprod. Process.* **2015**, *94*, 313–321. [CrossRef]
48. Timmermans, R.A.H.; Mastwijk, H.C.; Knol, J.J.; Quataert, M.C.J.; Vervoort, L.; Van der Plancken, I.; Hendrickx, M.E.; Matser, A.M. Comparing equivalent thermal, high pressure and pulsed electric field processes for mild pasteurization of orange juice. Part I: Impact on overall quality attributes. *Innov. Food Sci. Emerg. Technol.* **2011**, *12*, 235–243. [CrossRef]
49. Zhu, N.; Zhu, Y.; Yu, N.; Wei, Y.; Zhang, J.; Hou, Y.; Sun, A.D. Evaluation of microbial, physicochemical parameters and flavor of blueberry juice after microchip-pulsed electric field. *Food Chem.* **2019**, *274*, 146–155. [CrossRef]
50. Walkling-Ribeiro, M.; Noci, F.; Cronin, D.A.; Lyng, J.G.; Morgan, D.J. Shelf life and sensory evaluation of orange juice after exposure to thermosonication and pulsed electric fields. *Food Bioprod. Process.* **2009**, *87*, 102–107. [CrossRef]
51. Linhares, M.d.F.D.; Alves Filho, E.G.; Silva, L.M.A.; Fonteles, T.V.; Wurlitzer, N.J.; de Brito, E.S.; Fernandes, F.A.N.; Rodrigues, S. Thermal and non-thermal processing effect on açai juice composition. *Food Res. Int.* **2020**, *136*, 109506. [CrossRef]
52. Zhang, Y.; Liu, X.C.; Wang, Y.; Zhao, F.; Sun, Z.; Liao, X. Quality comparison of carrot juices processed by high-pressure processing and high-temperature short-time processing. *Innov. Food Sci. Emerg. Technol.* **2016**, *33*, 135–144. [CrossRef]
53. Gao, G.; Zhao, L.; Ma, Y.; Wang, Y.; Sun, Z.; Liao, X. Microorganisms and Some Quality of Red Grapefruit Juice Affected by High Pressure Processing and High Temperature Short Time. *Food Bioprocess Technol.* **2015**, *8*, 2096–2108. [CrossRef]
54. Zhao, L.; Wang, Y.; Hu, X.; Sun, Z.; Liao, X. Korla pear juice treated by ultrafiltration followed by high pressure processing or high temperature short time. *LWT* **2016**, *65*, 283–289. [CrossRef]
55. You, Y.; Li, N.; Han, X.; Guo, J.; Zhao, Y.; Liu, G.; Huang, W.; Zhan, J. Influence of different sterilization treatments on the color and anthocyanin contents of mulberry juice during refrigerated storage. *Innov. Food Sci. Emerg. Technol.* **2018**, *48*, 1–10. [CrossRef]
56. Hou, Z.; Zhang, Y.; Qin, X.; Zhao, L.; Wang, Y.; Liao, X. High pressure processing for sea buckthorn juice with higher superoxide dismutase activity. *J. Food Nutr. Res.* **2018**, *57*, 252–263.
57. Mena, P.; Martí, N.; Saura, D.; Valero, M.; García-Viguera, C. Combinatory Effect of Thermal Treatment and Blending on the Quality of Pomegranate Juices. *Food Bioprocess Technol.* **2013**, *6*, 3186–3199. [CrossRef]
58. Rios-Corripio, G.; de la Peña, M.M.; Welti-Chanes, J.; Guerrero-Beltrán, J.Á. Pulsed electric field processing of a pomegranate (*Punica granatum* L.) fermented beverage. *Innov. Food Sci. Emerg. Technol.* **2022**, *79*, 103045. [CrossRef]
59. Noci, F.; Riener, J.; Walkling-Ribeiro, M.; Cronin, D.A.; Morgan, D.J.; Lyng, J.G. Ultraviolet irradiation and pulsed electric fields (PEF) in a hurdle strategy for the preservation of fresh apple Juice. *J. Food Eng.* **2008**, *85*, 141–146. [CrossRef]
60. Walkling-Ribeiro, M.; Noci, F.; Cronin, D.A.; Lyng, J.G.; Morgan, D.J. Shelf life and sensory attributes of a fruit smoothie-type beverage processed with moderate heat and pulsed electric fields. *LWT* **2010**, *43*, 1067–1073. [CrossRef]
61. Stübler, A.S.; Lesmes, U.; Juadjur, A.; Heinz, V.; Rauh, C.; Shpigelman, A.; Aganovic, K. Impact of pilot-scale processing (thermal, PEF, HPP) on the stability and bioaccessibility of polyphenols and proteins in mixed protein- and polyphenol-rich juice systems. *Innov. Food Sci. Emerg. Technol.* **2020**, *64*, 102426. [CrossRef]
62. Amaral, G.V.; Silva, E.K.; Costa, A.L.R.; Alvarenga, V.O.; Cavalcanti, R.N.; Esmerino, E.A.; Guimarães, J.T.; Freitas, M.Q.; Sant'Ana, A.S.; Cunha, R.L.; et al. Whey-grape juice drink processed by supercritical carbon dioxide technology: Physical properties and sensory acceptance. *LWT* **2018**, *92*, 80–86. [CrossRef]
63. Jittanit, W.; Suriyapornchaikul, N.; Nithisopha, S. The Comparison between the Quality of Lime Juices Produced by Different Preservation Techniques. *Procedia—Soc. Behav. Sci.* **2013**, *91*, 691–696. [CrossRef]
64. de Souza, V.R.; Popović, V.; Bissonnette, S.; Ros, I.; Mats, L.; Duizer, L.; Warriner, K.; Koutchma, T. Quality changes in cold pressed juices after processing by high hydrostatic pressure, ultraviolet-c light and thermal treatment at commercial regimes. *Innov. Food Sci. Emerg. Technol.* **2020**, *64*, 102398. [CrossRef]
65. Torregrosa, F.; Esteve, M.J.; Frígola, A.; Cortés, C. Ascorbic acid stability during refrigerated storage of orange-carrot juice treated by high pulsed electric field and comparison with pasteurized juice. *J. Food Eng.* **2006**, *73*, 339–345. [CrossRef]
66. Zhao, L.; Wang, S.; Liu, F.; Dong, P.; Huang, W.; Xiong, L.; Liao, X. Comparing the effects of high hydrostatic pressure and thermal pasteurization combined with nisin on the quality of cucumber juice drinks. *Innov. Food Sci. Emerg. Technol.* **2013**, *17*, 27–36. [CrossRef]

67. Bonilla, A.I.; Usaga, J.; Cortés, C.; Pérez, A.M. Effect of thermal treatment on selected bioactive compounds and physicochemical properties of a blackberry-soy-flaxseed beverage. *NFS J.* **2024**, *35*, 12–15. [CrossRef]
68. Charles-Rodríguez, A.V.; Nevárez-Moorillón, G.V.; Zhang, Q.H.; Ortega-Rivas, E. Comparison of thermal processing and pulsed electric fields treatment in pasteurization of apple juice. *Food Bioprod. Process.* **2007**, *85*, 93–97. [CrossRef]
69. Cortés, C.; Esteve, M.J.; Frígola, A. Color of orange juice treated by High Intensity Pulsed Electric Fields during refrigerated storage and comparison with pasteurized juice. *Food Control* **2008**, *19*, 151–158. [CrossRef]
70. Yuan, L.; Cheng, F.; Yi, J.; Cai, S.; Liao, X.; Lao, F.; Zhou, L. Effect of high-pressure processing and thermal treatments on color and in vitro bioaccessibility of anthocyanin and antioxidants in cloudy pomegranate juice. *Food Chem.* **2022**, *373*, 131397. [CrossRef]
71. Lee, P.Y.; Kebede, B.T.; Lusk, K.; Mirosa, M.; Oey, I. Investigating consumers' perception of apple juice as affected by novel and conventional processing technologies. *Int. J. Food Sci. Technol.* **2017**, *52*, 2564–2571. [CrossRef]
72. Feszterová, M.; Kowalska, M.; Mišiaková, M. Stability of Vitamin C Content in Plant and Vegetable Juices under Different Storing Conditions. *Appl. Sci.* **2023**, *13*, 10640. [CrossRef]
73. Wibowo, S.; Essel, E.A.; De Man, S.; Bernaert, N.; Van Droogenbroeck, B.; Grauwet, T.; Van Loey, A.; Hendrickx, M. Comparing the impact of high pressure, pulsed electric field and thermal pasteurization on quality attributes of cloudy apple juice using targeted and untargeted analyses. *Innov. Food Sci. Emerg. Technol.* **2019**, *54*, 64–77. [CrossRef]
74. Verbeyst, L.; Bogaerts, R.; Van der Plancken, I.; Hendrickx, M.; Van Loey, A. Modelling of Vitamin C Degradation during Thermal and High-Pressure Treatments of Red Fruit. *Food Bioprocess Technol.* **2013**, *6*, 1015–1023. [CrossRef]
75. Feszterová, M.; Mišiaková, M.; Kowalska, M. Bioactive Vitamin C Content from Natural Selected Fruit Juices. *Appl. Sci.* **2023**, *13*, 3624. [CrossRef]
76. Alves Filho, E.G.; Silva, L.M.A.; de Brito, E.S.; Wurlitzer, N.J.; Fernandes, F.A.N.; Rabelo, M.C.; Fonteles, T.V.; Rodrigues, S. Evaluation of thermal and non-thermal processing effect on non-prebiotic and prebiotic acerola juices using 1H qNMR and GC–MS coupled to chemometrics. *Food Chem.* **2018**, *265*, 23–31. [CrossRef]
77. Fonteles, T.V.; Alves Filho, E.d.G.; Karolina de Araújo Barroso, M.; de Fátima Dantas Linhares, M.; Rabelo, M.C.; e Silva, L.M.A.; Sousa de Brito, E.; Wurlitzer, N.J.; Rodrigues Pereira, E.P.; Ferreira, B.M.; et al. Protective effect of inulin on thermally treated acerola juice: In vitro bioaccessibility of bioactive compounds. *Food Biosci.* **2021**, *41*, 101018. [CrossRef]
78. Atuonwu, J.C.; Leadley, C.; Bosman, A.; Tassou, S.A. High-pressure processing, microwave, ohmic, and conventional thermal pasteurization: Quality aspects and energy economics. *J. Food Process Eng.* **2020**, *43*, e13328. [CrossRef]
79. Matute, A.; Tabart, J.; Cheramy-Bien, J.P.; Kevers, C.; Dommes, J.; Defraigne, J.O.; Pincemail, J. Ex vivo antioxidant capacities of fruit and vegetable juices. Potential in vivo extrapolation. *Antioxidants* **2021**, *10*, 770. [CrossRef]
80. Prestes, A.A.; Canella, M.H.; Helm, C.V.; Gomes da Cruz, A.; Prudencio, E.S. The use of cold pressing technique associated with emerging nonthermal technologies in the preservation of bioactive compounds in tropical fruit juices: An overview. *Curr. Opin. Food Sci.* **2023**, *51*, 101005. [CrossRef]
81. Vegara, S.; Mena, P.; Martí, N.; Saura, D.; Valero, M. Approaches to understanding the contribution of anthocyanins to the antioxidant capacity of pasteurized pomegranate juices. *Food Chem.* **2013**, *141*, 1630–1636. [CrossRef]
82. Morales-De La Peña, M.; Salvia-Trujillo, L.; Rojas-Graü, M.A.; Martín-Belloso, O. Changes on phenolic and carotenoid composition of high intensity pulsed electric field and thermally treated fruit juice-soymilk beverages during refrigerated storage. *Food Chem.* **2011**, *129*, 982–990. [CrossRef] [PubMed]
83. Siguemoto, É.S.; Purgatto, E.; Hassimotto, N.M.A.; Gut, J.A.W. Comparative evaluation of flavour and nutritional quality after conventional and microwave-assisted pasteurization of cloudy apple juice. *LWT* **2019**, *111*, 853–860. [CrossRef]
84. Zhang, W.; Liang, L.; Pan, X.; Lao, F.; Liao, X.; Wu, J. Alterations of phenolic compounds in red raspberry juice induced by high-hydrostatic-pressure and high-temperature short-time processing. *Innov. Food Sci. Emerg. Technol.* **2021**, *67*, 102569. [CrossRef]
85. Luiza Koop, B.; Nascimento da Silva, M.; Diniz da Silva, F.; Thayres dos Santos Lima, K.; Santos Soares, L.; José de Andrade, C.; Ayala Valencia, G.; Rodrigues Monteiro, A. Flavonoids, anthocyanins, betalains, curcumin, and carotenoids: Sources, classification and enhanced stabilization by encapsulation and adsorption. *Food Res. Int.* **2022**, *153*, 110929. [CrossRef] [PubMed]
86. Liu, L.; Zeng, Q.; Zhang, R.; Wei, Z.; Deng, Y.; Zhang, Y.; Tang, X.; Zhang, M. Comparative study on phenolic profiles and antioxidant activity of litchi juice treated by high pressure carbon dioxide and thermal processing. *Food Sci. Technol. Res.* **2015**, *21*, 41–49. [CrossRef]
87. Chmiel, T.; Van Droogenbroeck, B.; Kusznierewicz, B. The influence of pasteurization methods on the phenolic profiles of cloudy apple juices. *J. Int. Soc. Antioxid. Nutr. Health* **2016**, *3*, 3–4. [CrossRef]
88. Odriozola-Serrano, I.; Hernández-Jover, T.; Martín-Belloso, O. Comparative evaluation of UV-HPLC methods and reducing agents to determine vitamin C in fruits. *Food Chem.* **2007**, *105*, 1151–1158. [CrossRef]
89. Cortés, C.; Esteve, M.J.; Rodrigo, D.; Torregrosa, F.; Frígola, A. Changes of colour and carotenoids contents during high intensity pulsed electric field treatment in orange juices. *Food Chem. Toxicol.* **2006**, *44*, 1932–1939. [CrossRef]
90. Silva, V.M.; Sato, A.C.K.; Barbosa, G.; Dacanal, G.; Ciro-Velásquez, H.J.; Cunha, R.L. The effect of homogenisation on the stability of pineapple pulp. *Int. J. Food Sci. Technol.* **2010**, *45*, 2127–2133. [CrossRef]
91. Stinco, C.M.; Sentandreu, E.; Mapelli-Brahm, P.; Navarro, J.L.; Vicario, I.M.; Meléndez-Martínez, A.J. Influence of high pressure homogenization and pasteurization on the in vitro bioaccessibility of carotenoids and flavonoids in orange juice. *Food Chem.* **2020**, *331*, 127259. [CrossRef]

92. Sójka, M.; Nowakowska, A.; Hejduk, A. Influence of enzymatic clarification, filtration, and pasteurization on ellagitannin and anthocyanin content in raspberry juices. *Eur. Food Res. Technol.* **2024**, *250*, 351–359. [CrossRef]
93. Juhart, J.; Medic, A.; Jakopic, J.; Veberic, R.; Hudina, M.; Stampar, F. Use of HPLC-MS to Determine the Loss of Metabolites in Apple Juices under Different Storage Conditions. *Foods* **2023**, *12*, 2822. [CrossRef] [PubMed]
94. Esteve, M.J.; Frigola, A. The Effects of Thermal and Non-thermal Processing on Vitamin C, Carotenoids, Phenolic Compounds and Total Antioxidant Capacity in Orange Juice. *Tree For. Sci. Biotechnol.* **2008**, *2*, 128–134.
95. Amaral, G.V.; Silva, E.K.; Cavalcanti, R.N.; Martins, C.P.C.; Andrade, L.G.Z.S.; Moraes, J.; Alvarenga, V.O.; Guimarães, J.T.; Esmerino, E.A.; Freitas, M.Q.; et al. Whey-grape juice drink processed by supercritical carbon dioxide technology: Physicochemical characteristics, bioactive compounds and volatile profile. *Food Chem.* **2018**, *239*, 697–703. [CrossRef] [PubMed]
96. Etzbach, L.; Stolle, R.; Anheuser, K.; Herdegen, V.; Schieber, A.; Weber, F. Impact of different pasteurization techniques and subsequent ultrasonication on the in vitro bioaccessibility of carotenoids in valencia orange (*Citrus sinensis* (L.) Osbeck) juice. *Antioxidants* **2020**, *9*, 534. [CrossRef]
97. Elez-Martínez, P.; Martín-Belloso, O. Effects of high intensity pulsed electric field processing conditions on vitamin C and antioxidant capacity of orange juice and gazpacho, a cold vegetable soup. *Food Chem.* **2007**, *102*, 201–209. [CrossRef]
98. Caminiti, I.M.; Noci, F.; Muñoz, A.; Whyte, P.; Morgan, D.J.; Cronin, D.A.; Lyng, J.G. Impact of selected combinations of non-thermal processing technologies on the quality of an apple and cranberry juice blend. *Food Chem.* **2011**, *124*, 1387–1392. [CrossRef]
99. Soltani Firouz, M.; Farahmandi, A.; Hosseinpour, S. Recent advances in ultrasound application as a novel technique in analysis, processing and quality control of fruits, juices and dairy products industries: A review. *Ultrason. Sonochem.* **2019**, *57*, 73–88. [CrossRef]
100. Krapfenbauer, G.; Kinner, M.; Gössinger, M.; Schönlechner, R.; Berghofer, E. Effect of thermal treatment on the quality of cloudy apple juice. *J. Agric. Food Chem.* **2006**, *54*, 5453–5460. [CrossRef]
101. Aguilar-Rosas, S.; Ballinas-Casarrubias, M.; Elias-Ogaz, L.; Martin-Belloso, O.; Ortega-Rivas, E. Enzyme activity and colour changes in apple juice pasteurised thermally and by pulsed electric fields. *Acta Aliment.* **2013**, *42*, 45–54. [CrossRef]
102. Aguiar, H.d.F.; Yamashita, A.S.; Gut, J.A.W. Development of enzymic time-temperature integrators with rapid detection for evaluation of continuous HTST pasteurization processes. *LWT* **2012**, *47*, 110–116. [CrossRef]
103. Aguiló-Aguayo, I.; Oms-Oliu, G.; Soliva-Fortuny, R.; Martín-Belloso, O. Changes in quality attributes throughout storage of strawberry juice processed by high-intensity pulsed electric fields or heat treatments. *LWT* **2009**, *42*, 813–818. [CrossRef]
104. Aguiló-Aguayo, I.; Oms-Oliu, G.; Soliva-Fortuny, R.; Martín-Belloso, O. Flavour retention and related enzyme activities during storage of strawberry juices processed by high-intensity pulsed electric fields or heat. *Food Chem.* **2009**, *116*, 59–65. [CrossRef]
105. Liu, Q.; Huang, G.; Ma, C.; Li, G.; Wang, R. Effect of ultra-high pressure and ultra-high temperature treatments on the quality of watermelon juice during storage. *J. Food Process. Preserv.* **2021**, *45*, e15723. [CrossRef]
106. Polydera, A.C.; Stoforos, N.G.; Taoukis, P.S. Comparative shelf life study and vitamin C loss kinetics in pasteurised and high pressure processed reconstituted orange juice. *J. Food Eng.* **2003**, *60*, 21–29. [CrossRef]

applied sciences

MDPI

*Review*

# Valorization of Date Fruit (*Phoenix dactylifera* L.) Processing Waste and By-Products: A Review

**Linghong Shi [1], Thaiza Serrano Pinheiro de Souza [1], Farhad Ahmadi [1], Ali Imran [1], Frank R. Dunshea [1], Colin Barrow [2] and Hafiz Ansar Rasul Suleria [1,*]**

[1] School of Agriculture, Food and Ecosystem Sciences, Faculty of Science, The University of Melbourne, Parkville, VIC 3010, Australia; linghongs@student.unimelb.edu.au (L.S.); thaiza.serranopinheirodesou@unimelb.edu.au (T.S.P.d.S.); farhad.ahmadi@unimelb.edu.au (F.A.); imran.a@unimelb.edu.au (A.I.); fdunshea@unimelb.edu.au (F.R.D.)

[2] Center for Sustainable Bioproducts, School of Life and Environmental Sciences, Deakin University, Waurn Ponds, Geelong, VIC 3217, Australia; colin.barrow@deakin.edu.au

* Correspondence: hafiz.suleria@unimelb.edu.au; Tel.: +61-470-439-670

**Abstract:** The date is a well-established and important crop that holds economic significance. However, a substantial amount of waste in the form of low-grade dates and date pits is generated and accounts for 10 to 15% of the total production. Given the substantial amount of nutrients in these by-products and the large volume of waste generated, there is a promising opportunity to utilize them to create valuable commodities like fiber and phenolic compounds, which hold a high market value. This review presents a summary of the chemical and nutritional composition of dates and their by-products and aims to investigate the possibility of utilizing date processing by-products and waste as an eco-friendly resource for various chemical and biological processes like composting and extraction of value-added compounds, as well as providing insight into the date processing industry and typical methods employed for the beneficial use of date waste. In addition, this review also addresses the current challenges and future perspectives in date waste valorization expectations.

**Keywords:** date palm; processing waste; valorization; nutritional and biochemical composition; high-value compounds

## 1. Introduction

Date palm (*Phoenix dactylifera* L.) is an important agricultural crop that has been widely cultivated across the globe. Currently, over a thousand varieties of date palms have been recognized worldwide including highly acclaimed varieties like Ajwa, Medjool, Khalas, Deglet Noor, and others [1]. Dates have primarily been cultivated in dry and semi-dry regions for more than five millennia [2]. The global production of date palms has steadily increased over the past decade, indicating a growing demand for this fruit (Figure 1). Currently, the majority of dates are produced in the Middle East and North Africa, accounting for 90% of worldwide output [3].

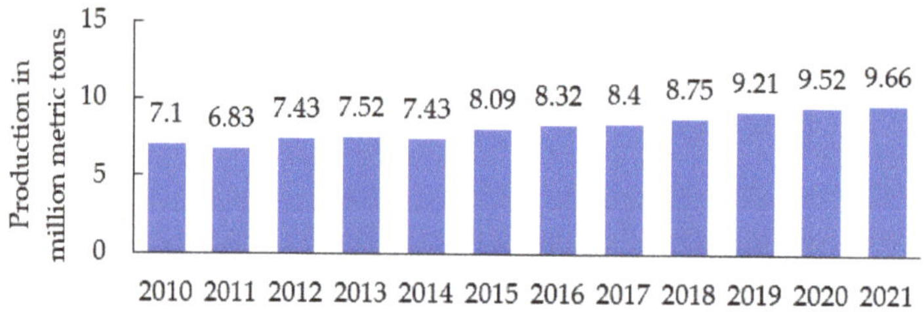


**Figure 1.** Global production of date palm from 2010 to 2021 [4].

The date fruit processing industry plays an increasingly important role in the agro-industrial sector, as these fruits can be transformed into a variety of products, such as date syrup, alcohol, date powder, and paste [5]. Apart from the fruit pulp, other date palm components can also substantially contribute to the agricultural economy. For instance, the date palm stem can be utilized for various purposes such as, for example, to produce boats, roofing, paper, and fiber [6]. Furthermore, local handicrafts such as nets and mats can be created from date foliage [7].

In addition, date fruits provide an economical and abundant supply of vital nutrients, including carbohydrates like soluble sugars and fiber, minerals, and vitamins [2,8]. The date fruit also has minimal quantities of fats and proteins. Furthermore, date fruits, including their pulp and seeds, encompass significant nutraceutical elements (such as phenolic compounds, phenolic acids, cinnamic acid derivatives, flavones, anthocyanidins, isoflavones, and volatile compounds) that deliver practical benefits [9]. These benefits encompass antioxidant, antimicrobial, anti-inflammatory, antimutagenic, hepato-protective, gastro-protective, anticancer, and immune-stimulating properties [2,10,11]. Habib et al. [12] explored the health benefits of date palm fruit seed extract and its mechanism of action. Their study reveals that the extract inhibits free radicals, labile iron activities, and DNA/protein damage, suggesting its potential to protect against oxidative damage and programmed cell death due to iron-catalyzed ferroptosis. Additionally, date palm fruit seed extract inhibits enzymes linked to various diseases. These findings support the use of date palm fruit seed extract in functional foods and nutraceuticals, encouraging further in vivo and clinical trials.

A considerable quantity of waste is produced during the processing of date fruit, including the date pits and the date press cake (produced during date juice extraction). Date pits represent an average reduction of 10% in the weight of the whole fruit [13]. Even though date palm agricultural wastes contain valuable components such as dietary fibers, phenolic compounds, and other bioactive compounds, they are often discarded or used inefficiently [14,15].

While many researchers have explored date palm cultivation, its potential uses, and applications in therapy, there is limited availability of comprehensive reviews in the literature that extends beyond the chemistry and pharmacology of date fruits. Therefore, in this review, we explore the potential of utilizing bioprocessing technologies to maximize the value of date fruit processing by-products and waste materials, aiming for the complete utilization of these waste products that are typically disposed of into the environment. This comprehensive review covers various aspects, including the nutritional value and biochemical characteristics of date fruits, their medicinal and pharmacological properties, date fruit processing and the generation of products and by-products, current waste management practices, and challenges and opportunities for enhancing value through bioprocessing, the production of fermented products from date palm fruits, and a brief discussion of future trends in the valorization of date palm fruit processing by-products and waste materials.

## 2. Types of Date Palm Waste

The process of harvesting date palm fruit typically results in substantial losses due to fruit falling from the tree prematurely and issues during storage and conditioning. These lost dates are commonly referred to as "date by-products" and are considered of low grade due to their inadequate texture, being either too hard or too soft and being contaminated with fungi or infested by insects. Owing to these quality problems and safety risks, these lost dates were generally discarded in the past [16]. Currently, these by-products can be used as animal feed compost, among other purposes, thereby enhancing their value to a certain extent [17].

The processing of pitted dates generates two major types of waste: the whole date fruits that do not meet quality requirements, and the seeds that are removed during the production process. The seed waste generated from the pitting process of date fruit makes up about 18% of the fruit's total weight [18] (Figure 2). Considering that the world production of dates reached 9.66 million tons in 2021, this would potentially generate over 900,000 tons of date seeds [4]. Indeed, date seeds, previously considered waste, have now gained recognition as a valuable source of fiber and phenolic compounds. Furthermore, they can be converted into products that have added value. Moreover, the oil extracted from date seeds can be converted into value-added products.

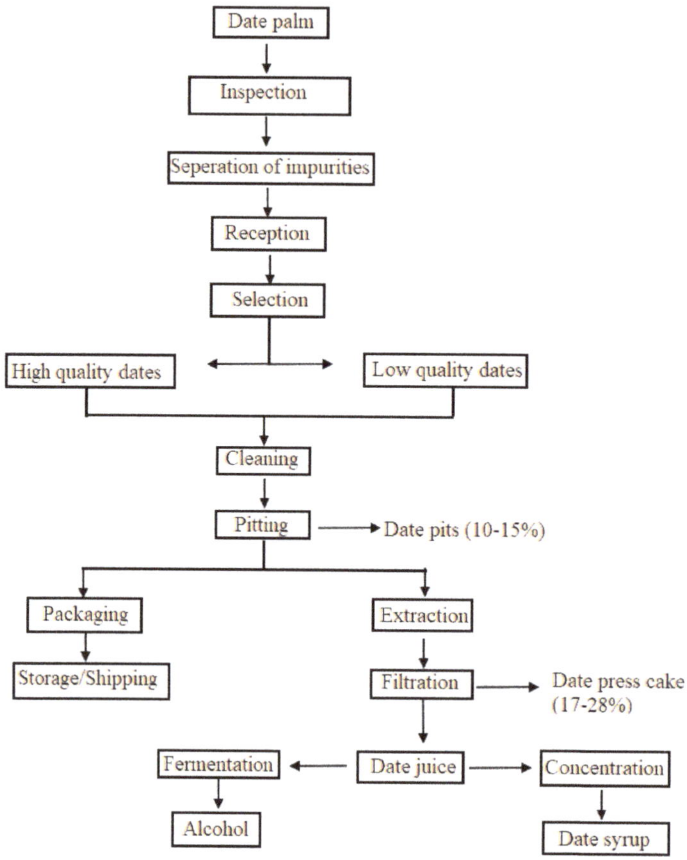


**Figure 2.** Flowchart of pitted date processing and date juice processing.

Date seed oil (DSO) can be extracted from date seeds using different methods, including Soxhlet extraction, hydrothermal and ultrasonic-assisted methods, and supercritical and subcritical $CO_2$ techniques [10,19]. DSO has numerous applications in the food industry, including its use in cooking, frying, and seasoning. DSO can be used as a replacement for palm olein, a common ingredient in many food products [20]. Previous reports have also mentioned that DSO has been used in margarine production and as an alternative to vegetable oils like corn and sunflower oil in mayonnaise production [19]. DSO can also be used in cosmetic formulations such as body creams, shaving soap, and shampoo [21]. Research has revealed that DSO has properties that help reduce cellular and oxidative rancidity, and it can protect human skin from UV irradiation, making it beneficial for skincare and cosmetic products [21]. In addition, DSO can be utilized in other non-food applications, such as biogas production of deoxygenated hydrocarbons, where DSO can serve as a substrate [22]. DSO's potential applications in both food and non-food industries highlight its versatility and potential value as a by-product of date processing.

Low-grade dates, unsuitable for direct consumption, can be used to produce other products such as date paste and date juice. Date juice can be further processed to produce date spread, syrup, and liquid sugar or fermented to produce wine, alcohol, and vinegar. Date juice production typically produces date press cake (DPC), a fibrous by-product that remains after the filtration of date juice (Figure 2). This by-product is often bulky in nature and has a high content of moisture and carbohydrates [23]. Date juicing can produce about 17–28% of DPC with high nutritional value, although the conventional use of DPC is still for animal feed or directly discarded [17].

## 3. Nutritional and Biochemical Composition of Date Fruit and Date By-Products

Date palms contain abundant essential macro- and micro-nutrients, including carbohydrates, fatty acids, protein, amino acids, minerals and salts, vitamins, and dietary fiber. These compounds can be powerful substances found in dates and date wastes that have the potential to positively influence health. Functional ingredients and dietary supplements could be generated by employing various methods to identify and separate these compounds from date waste. This approach could reduce waste and promote sustainability in the date palm industry (Figure 3).

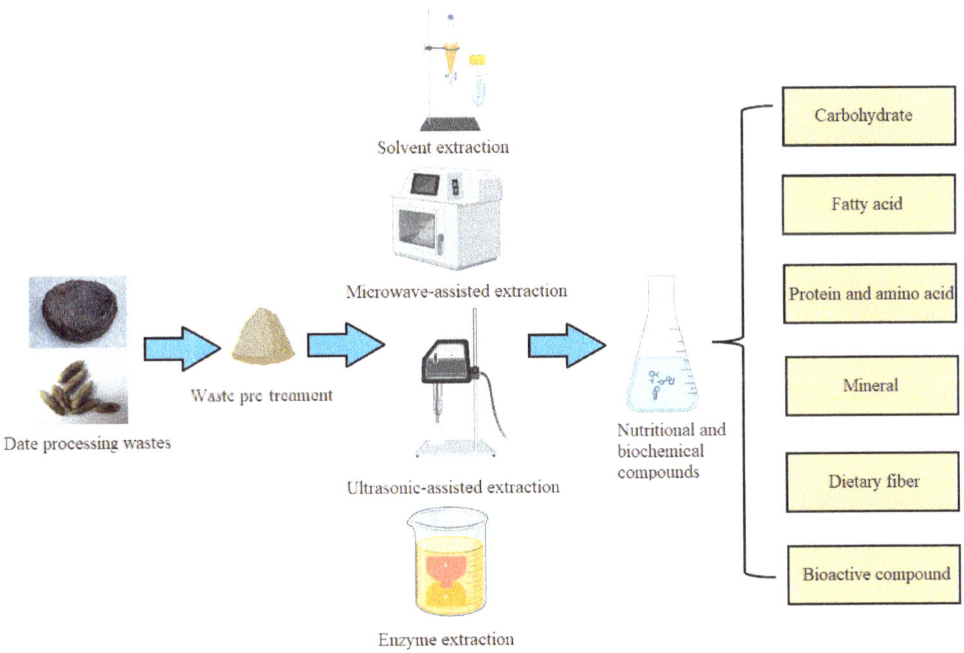


**Figure 3.** Extraction of nutritional and biochemical substances from date waste using various extraction methods.

The proximate nutritional composition of date fruit and date pit is shown in Table 1. Dates at fresh edible stages may have a moisture content of around 40%, and the high nutritional value of dates is due to their high content of sugar, potassium, calcium, magnesium, and iron as well as vitamins (B1 and B2) and niacin [2,24,25]. Date seed contains 8–12% moisture, 60–80% fiber, 4–14% fat, and 5–7% protein. The date seed also contains various phytochemicals, including tannins, flavonoids, terpenoids, saponin, anthraquinone, and alkaloids, as well as essential minerals like potassium and calcium [26]. Date seed fiber contains both soluble and insoluble dietary fibers, and the content of dietary fiber content in seeds is proven to be higher than date pulp [27]. The main components of insoluble fiber are hemicellulose, cellulose, and lignin [21]. Attia et al. [27] reported that the high fiber content in date seeds can substitute corn and barley in the diet. Date seeds contain different fiber fractions, including acid detergent fiber, neutral detergent fiber, hemicellulose, cellulose, and lignin. Acid detergent fiber is a combination of cellulose and lignin, ranging from 39.6% to 57.5% [26]. Neutral detergent fiber is a combination of hemicellulose, cellulose, and lignin and ranges from 51.6% to 75.0% [26]. Hemicellulose content in date seeds ranges from 12.0% to 17.5%, while cellulose ranges from 26.1% to 42.5%. Additionally, lignin content in date seeds falls within the range of 7.21% to 11.0% [26].

The proximate nutritional composition of DPC is also shown in Table 1. Majzoobi et al. [23] also found that magnesium is the mineral with the highest content in DPC, which has an average of 959.5 mg/kg, followed by phosphorus (853.4 mg/kg) and calcium (460.5 mg/kg). Surprisingly, DPC has an average phenolic content value of 17.8 mg/g, which is very close to fresh date fruit (18 mg/g), which showed that high antioxidant compounds remain in DPC after date juice extraction [23]. The composition of DPC was found to vary, with crude protein ranging from 4.1% to 10.5%, neutral detergent fiber at 48%, acid detergent fiber at 25.6%, and nitrogen-free extract at 40.1% [28]. Due to its high fiber and antioxidant content, the DPC left over from processing dates can be useful as an ingredient in healthy foods. The natural brown color of DPC makes it a good option for darker foods like baked goods, and its ability to hold onto the water is an advantage in products like salad dressings and instant soups [23]. The particle size of DPC is also important to consider, as it can affect its chemical composition, nutritional value, and usefulness in different foods [23].

**Table 1.** The proximate nutritional composition of date fruit, pit, and date press cake.

| Component | Date Fruit (per 100 g) | Date Pit (per 100 g) | Date Press Cake (per 100 g) | References |
|---|---|---|---|---|
| Moisture (g) | 12.60–50.40 | 1.42–4.14 | 6.11–13.4 | [18,23,29–31] |
| Carbohydrate (g) | 76.69–90.18 | 83.39–85.55 | 66.2–79.1 | [18,23,31] |
| Protein (g) | 1.60–3.53 | 3.20–5.00 | 4.10–10.5 | [18,23,31] |
| Fat (g) | 0.32–1.09 | 6.30–7.40 | 4.92–5.12 | [18,23,31] |
| Ash (g) | 2.08–2.50 | 1.14–1.50 | 2.78 | [18,31] |
| Dietary fiber (g) | 8.10–12.70 | 64.50–79.84 | 11.7–12.4 | [23,29,31,32] |
| Calcium (mg) | 15.46–53.82 | 28.9–38.8 | 41.74–50.35 | [21–23] |
| Phosphorus (mg) | 52.19–77.94 | 83.6–68.3 | 54.17–116.51 | [21–23] |
| Sodium (mg) | 6.25–17.52 | 10.25–10.4 | 0.16–0.25 | [21–23] |
| Potassium (mg) | 281.74–478.29 | 229–293 | 54.17–116.51 | [21–23] |
| Magnesium (mg) | 42.17–70.38 | 51.7–58.4 | 82.57–109.33 | [21–23] |
| Iron (mg) | 0.84–1.51 | 2.30–2.21 | 7.14–9.01 | [21–23] |

### 3.1. Carbohydrate

The sugar composition of dates varies depending on the species and the maturity stage of the fruit. Typically, the date palm reaches its highest sugar concentration at its mature (Tamar) stage [29]. Most date cultivars contain glucose, fructose, and sucrose, which are easily absorbed by the body to provide energy. In fresh dates, the concentration of inverted sugars is higher than in semi-dried dates. In semi-dried dates, the ratio of inverted sugars to sucrose is similar, while in dried dates, the concentration of sucrose is higher than that of inverted sugars [33]. The overall sugar content in dates consists of both reducing and non-reducing sugars, and these proportions vary noticeably across different types of dates. The total sugar levels in various date cultivars ranged from 59.0% to 73.9% (dry weight basis). Among these cultivars, the reducing sugars concentration ranged from 52.8% to 69.0%, while the quantity of non-reducing sugars fell between 4.65% and 7.66% [34].

### 3.2. Fatty Acids

Fatty acids play a crucial role in various physiological functions in the human body, including energy storage, cell membrane structure, and signaling [35]. The fatty acid composition of the date palm varies depending on the cultivar, geographic location, and maturity stage of the fruit [36]. The major fatty acids present in the date palm include palmitic acid (C16:0), oleic acid (C18:1), linoleic acid (C18:2), and stearic acid (C18:0) [37].

Palmitic acid is a saturated fatty acid accounting for around 10–15% of the total fatty acid content in date palm fruit [37]. Palmitic acid has been associated with an increased risk of cardiovascular disease and insulin resistance when consumed in excess. However, moderate consumption of palmitic acid from natural sources such as date palm fruit may not have negative health effects [38]. Stearic acid is another saturated fatty acid that accounts for around 5–10% of the total fatty acid content in date palm fruit [39]. The consumption of high amounts of saturated fatty acids is related to high LDL cholesterol levels in the blood, raising the risk of heart disease and stroke [40].

Oleic acid is a monounsaturated fatty acid that accounts for around 40–50% of the total fatty acid content in date palm fruit, which is the highest among all of the fatty acids, and linoleic acid is a polyunsaturated fatty acid that accounts for around 8–19% [37]. Unsaturated acids can improve heart health by lowering LDL cholesterol levels and enhancing insulin sensitivity [41]. Oleic acid is also involved in synthesizing various hormones and cell membrane structures [42]. Furthermore, date seed oil, rich in linoleic acid, has skin-enhancing and protective properties due to its essential fatty acids that our body cannot produce naturally. Insufficient levels of these essential fatty acids can lead to skin issues such as dryness, flakiness, fragile nails, and hair thinning. Numerous skincare products utilize linoleic acid to address conditions like acne vulgaris, skin disorders, and sunburn [29].

### 3.3. Protein and Amino Acids

Proteins are large organic molecules with one or multiple lengthy sequences of amino acids. Traditionally, individuals have obtained protein by consuming animal-based sources such as meat, eggs, and milk. However, the research on alternative protein sources to replace expensive animal proteins has led researchers to consider materials previously considered wastes but rich in protein. While the protein content of date palm may not be as high as its carbohydrate content, its essential amino acids provide significant benefits to human health [29]. Recovering functional proteins from date waste can increase the value of discarded low-grade fruits and date peels, skins, and seeds.

Date fruits have a protein range of 2.13% to 3.12%, while protein content in date seed is 4.81% to 5.84 [21,43]. Rambabu et al. [43] reported the presence of essential amino acids in dates, with significant proportions of glutamine, aspartic acid, glycine, proline, histidine, and valine. In their research, it was observed that all tested date varieties predominantly contained glutamine and aspartic amino acids, with varying percentages ranging from 24.23% to 38.82% for glutamine and 20.98% to 32.11% for aspartic acid. Additionally, glycine, proline, histidine, and valine were also present in notable amounts, contributing to the overall amino acid composition of dates, with concentrations ranging from 11.60% to 18.03%, 9.83% to 14.19%, 3.47% to 6.20%, and 8.23% to 12.02%, respectively. Other essential amino acids, such as glycine, proline, and valine, were also at significant levels. These amino acids play a crucial role in the formation and synthesis of proteins in the body, involving the production of neurotransmitters or the formation of collagen, as the protein is the structure that forms the connective tissues in the body, including cartilage, tendons, and ligaments [44]. Also, the isoleucine content in dates is over 800 times greater than that found in apples [45]. According to the research conducted by El-Dreny and Shaheen [46], the quantity of lysine found in dates is remarkably high. Specifically, the lysine content in date seeds, at 4.05 g per 100 g of protein, is higher than the lysine in wheat flour by approximately 1.8 times, which is 2.46 g per 100 g of protein.

### 3.4. Minerals

Minerals are significant constituents of the date palm, and their content plays a vital role in evaluating the nutritional quality of dates. Tripler et al. [47] reported that there are at least 15 essential minerals in date palms, including phosphorus, potassium, sodium, zinc, manganese, magnesium, copper, and iron. Minerals play an essential role in the human body, including building strong bones, transmitting nerve impulses, regulating the stability of human hormones, and regulating the standard heartbeat.

Rambabu et al. [43] evaluated 11 date fruit cultivars and found that potassium was the major mineral element present in all date varieties, ranging from 281.7 to 478.3 mg/100 g. Iron was present in good levels in all date varieties, with a maximum of 1.51 mg/100 g in Raziz date flesh. The mineral content may vary among different cultivars of dates due to other factors such as variety, soil type, and amount of fertilizer [48]. However, for individuals diagnosed with hypertension, a dietary regimen supplying lower sodium intake while increasing potassium intake in their diet may assist in lowering blood pressure level. The use of date palm by-products with high potassium and low sodium content presents an opportunity in the field of medicine for the treatment of high blood pressure [49].

### *3.5. Dietary Fibers*

The date palm is a valuable source of dietary fiber, and a significant portion of that fiber can be found in the flesh of the fruit [37]. Date palm dietary fiber contains soluble and insoluble fibers, whereas both have different physiological effects on the body. Soluble fibers such as pectin and hemicellulose can be dissolved in water and form a gel-like substance in the digestive tract [50]. Besides that, soluble fibers can regulate blood sugar levels by inhibiting the rapid digestion and absorption of carbohydrates and lipids, lowering cholesterol levels, and promoting the growth of healthy gut bacteria [51]. Insoluble fibers such as cellulose and lignin can add bulk to fecal matter and maintain and promote bowel movement, which can help prevent constipation and other digestive disorders [51].

The main dietary fiber compounds in date palms include cellulose, hemicellulose, pectin, and lignin. Cellulose is a linear polymer of glucose units that forms the structural component of plant cell walls, and its content in date palm can be over 30 g/100 g [52]. Hemicellulose is a branched polymer composed of several sugars, including xylose, arabinose, and galactose. Dates contain a significant amount of hemicellulose, ranging from approximately 24.0% to 45.1% of their overall composition [52]. The date also has a high value of pectin, which is a non-cellulosic fiber. However, the pectin amount varies widely with ripening stages of the date fruit because the process involves increased activity of fiber-hydrolyzing enzymes (pectinase). High-quality dates generally have elevated pectin and low lignin, whereas the opposite pattern results in less palatable to inedible fruit [24]. The pectin content in date palm can range from 0.78 to 1.12% [53]. Lignin is a complex phenolic polymer that gives plant cell walls rigidity and resistance to degradation, ranging from 22 to 23% in dates [54].

Recent studies indicate that fibers from date palms have prebiotic properties, which can aid in the growth of beneficial gut bacteria. Bamigbade et al. [55] reported that date palm fibers increased the abundance of beneficial bacteria, such as *Bifidobacterium* and *Lactobacillus*. Another study showed that date palm fibers can reduce inflammation in the human body, which may benefit those with inflammatory bowel diseases [26]. The considerable amount of dietary fiber in dates may be beneficial for individuals with diabetes by the regulation of blood sugar levels. Alkaabi et al. [56] studied the impact of date consumption on blood sugar levels in healthy individuals and those with type 2 diabetes mellitus. The findings indicated that the five types of dates examined in the study have a low glycemic index and did not cause significant increases in postprandial glucose levels when consumed by diabetic individuals. Therefore, including dates in a balanced and healthy diet may offer benefits for individuals with diabetes. This explanation was also supported by Goff et al. [57], who suggested that a diet rich in dietary fiber may aid in balancing blood sugar levels, likely owing to the presence of fiber and bioactive compounds in dates, which can slow down the absorption of glucose from a meal.

### 3.6. Bioactive Compounds—Phenolic Compounds and Antioxidant Activity

The date fruit is a rich source of phenolic compounds that possess biological activity due to the presence of a benzene hydroxyl ring with a carboxyl group [58]. The date fruit contains a range of phenolic compounds, including phenolic acids, flavonoids, and high-molecular-weight compact polymers [3,36]. There are four primary categories of phenolic compounds found in plants, namely phenolic acids, flavonoids, stilbene, and lignin. They can be categorized based on their dissimilarities in structure, the number of phenol rings, and the type of molecular linkage [59].

The polyphenols exhibit their antioxidant properties by eliminating free radicals and reactive oxygen species, thereby inhibiting the oxidative process, which may help prevent diseases associated with oxidative stress, and improving glycemic control, lowering blood pressure, and improving lipid profiles [27]. Phenolic acids and flavonoids are the main polyphenols present in the fruit pulp and seeds, and due to the higher concentration of phenolic compounds in the seeds, the antioxidant capacity of the seeds is often greater than that of the pulp. Hilary et al. [60] analyzed polyphenols in Khalas date seeds in three human-consumable forms: date seed powder, date seed pita bread, and date seed extract and identified 27, 29, and 15 compounds via HPLC-ESI-UV/MS/MS and confirmed by UPLC-QTOF MS/MS, respectively. These compounds included hydroxycinnamic acids, hydroxybenzoic acids, hydroxyphenyl acetic acid, flavanols, flavonols, and flavones. Flavan-3-ols were the most abundant polyphenols. The study revealed that polymeric proanthocyanidins are present in date seeds and can potentially undergo gut microbial metabolism. During the process of in vitro digestion, there was an increase in the number of polyphenols present in the intestinal phase, indicating a favorable level of bioaccessibility. Several key phenolic acids and flavones were actively transported across Caco-2 monolayers, suggesting significant bioavailability from date seeds, while some procyanidins remained undetected in transport, possibly due to their binding to carbohydrates and protein precipitation properties. Several studies have evaluated the antioxidant activity of date palm, and some of these studies are listed in Table 2. However, it is worth noting that the content of phenolic compounds in dates can differ based on the variety of dates.

**Table 2.** Studies on the phenolic compounds and antioxidant properties of date fruits.

| Date Varieties | Method(s) | Findings | Reference |
|---|---|---|---|
| Ajwah, Safawy, and Sukkari (Saudi Arabia) | Antioxidant activity by DPPH, hydroxyl radical scavenging and total antioxidant capacity assay. The qualitative phenolic composition was determined using UPLC-QTOF-MS. | The extracts showed DPPH radical scavenging ($IC_{50}$ 103–177 μg/mL) and hydroxyl radical scavenging ($IC_{50}$ 1.1–1.55 mg/mL) activity and total antioxidant capacity ($IC_{50}$ 87–192 μg/mL). UPLC-QTOF-MS revealed a total of 22 compounds in these date cultivars classified into common phenolics, flavonoids, sterols, and phytoestrogens. | [61] |
| 13 varieties from UAE; 10 varieties from Pakistan; 1 variety from Tunisia | Total phenolic content and antioxidant activity using the Folin–Ciocalteu, $ABTS^+$, FRAP, and DPPH assays. | The total phenolic content varied 46–397 mg GAE/100 g fresh weight (FW) using the Folin–Ciocalteu method, 0.9–4.3 μmol TE/100 g FW using $ABTS^+$, 355–2421 μmol TE/100 g FW using FRAP, and 0.0–1.8 mg/mL using DPPH. | [62] |
| Medjool (Mexico) | Total phenolic content by the Folin–Ciocalteu method, antioxidant activity by DPPH, ABTS, and β-carotene assay. | Total contents of phenolic compounds (pulp: 1.16 mg of GAE/100 g; seeds: 13.73 mg of GAE/100 g) and antioxidant activity (β-carotene, 65.50% and 47.75%; DPPH, 0.079 $IC_{50}$ g/L and 0.0046 $IC_{50}$ g/L; and ABTS, 13.72 $IC_{50}$ g/L and 0.238 $IC_{50}$ g/L). | [13] |
| 19 varieties from Pakistan | Total phenolic content by the Folin–Ciocalteu method. The qualitative phenolic composition was determined using HPLC. | The range of the total phenolic contents ranged from 142.52 ± 0.64 to 298.02 ± 0.95 mg GAE/100 g on fresh fruit weight basis. Seven phenols of chlorogenic acid, caffeic acid, vanillic acid, gallic acid, cinnamic acid, 3,5-DHB, and 2,5-DHB were identified and quantified in four varieties of date seeds. | [34] |
| Kabkab, Rabbi, Zahedi, and Mazafat (Iran) | Total phenolic content by the Folin–Ciocalteu method and phenolic compound identification and quantification by HPLC analysis. | The phenolic contents for the date pit extracts ranged 1483–3377 mg GAE/100 g dw. | [63] |

GAE, gallic acid equivalents; TE, Trolox equivalents.

Compared with date fruit, date seed has a relatively higher content of antioxidant compounds, especially phenolic compounds. In addition, date seeds have a high level of total polyphenol content compared to other fruits such as grapes and other types of seeds like nut seeds. The high polyphenol content in date seed has been proven to possess a variety of pharmacological effects, including anti-inflammatory, anticancer, and antimutagenic activities [26,64]. Current studies show that date seeds contain phenolic acids such as caffeic, chlorogenic, *p*-coumaric, ferulic, gallic, syringic, and vanillic acid [65]. Some of the important phenolic compounds that can be isolated from date pits are listed in Table 3.

**Table 3.** Important phenolic compounds isolated from date pits.

| Compound Name | Classification | Structure | Molecular Formula | Extraction and Identification Method | References |
|---|---|---|---|---|---|
| *p*-coumaric acid | Hydroxycinnamic acid | 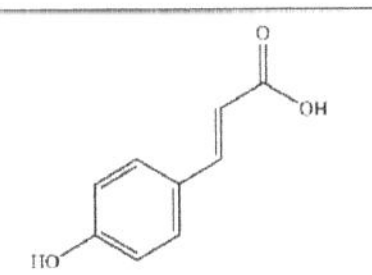 | $C_6H_4(OH)COOH$ | Solvent extraction; Ultrasonic bath extraction; HPLC | [65,66] |
| Caffeic acid | Hydroxycinnamic acid | 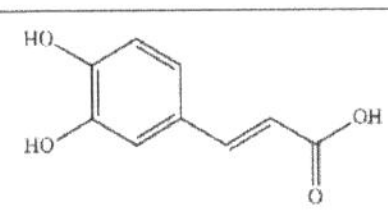 | $C_9H_8O_4$ | Solvent extraction; HPLC | [65] |
| Rutin | Flavonoid glycoside | 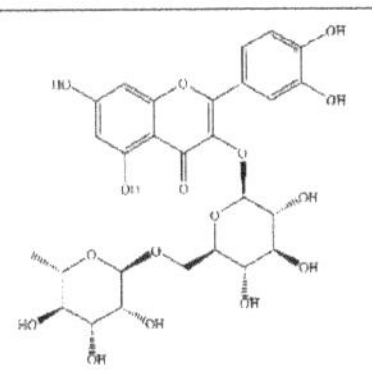 | $C_{27}H_{30}O_{16}$ | Solvent extraction; HPLC; LC-MS | [65,67] |
| Gallic acid | Phenolic acid | 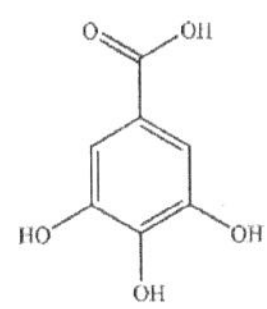 | $C_6H_2(OH)_3COOH$ | Solvent extraction; HPLC | [65] |
| Catechin/epiccatechin | Flavan-3-ol | 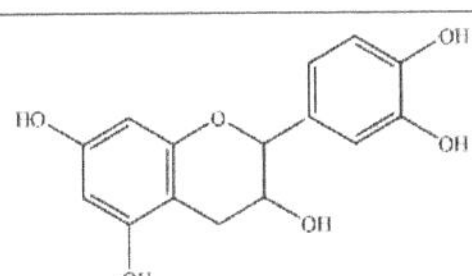 | $C_{15}H_{14}O_6$ | Ultrasonic-assisted extraction; HPLC | [68] |
| Ferulic acid | Hydroxycinnamic acid | 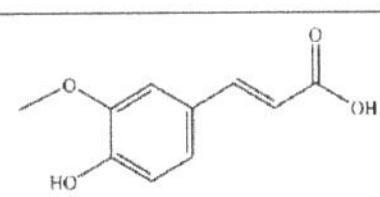 | $C_6H_8O_4$ | Solvent extraction; HPLC; LC-MS | [65,67] |
| Vanillic acid | Benzoic acid derivative | 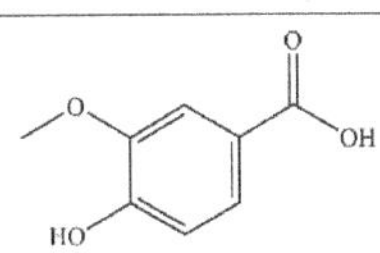 | $C_8H_8O_4$ | Solvent extraction; HPLC | [65] |

Iranian date seeds were previously reported to have high antioxidant activity, with 37.42 mmol $Fe^{II}$/100 g, showing a potential to be applied in medicinal and commercial areas. Also, the phenolic content of Iranian date seeds reached a high content of 3541 mg gallic acid/100 g dry plant [69]. The Mabseeli variety of date seeds has a substantial concentration of phenolic compounds, including gallic acid, protocatechuic acid, *p*-hydroxybenzoic acid, vanillic acid, caffeic acid, *p*-coumaric acid, ferulic acid, *m*-coumaric acid, and *o*-coumaric acid. These compounds have been identified in this particular type of date seed [70]. Alshwyeh [67] analyzed three varieties of dates, Ajwa, Khalas Alkharj, and Al-Qasim, using LC-MS and LC-MS/MS, and a large group of unique phenolic compounds was identified, including caffeic acid and ferulic acid. John and Shahidi [68] reported that the total soluble phenolic content of the date seeds ranged from 68.7 to 82.6 mg GAE/g, and several phenolic compounds were also identified in their date seed extracts, including proanthocyanidin dimers, catechin, epicatechin, and others. The extract of soluble phenolic from date seeds is effective in preventing DNA strand scission by 74.2% and can efficiently inhibit the mRNA level of COX-2 at concentrations as low as 5 μg/mL, which represents the potency of date seed samples in the inhibition of radical-induced DNA scission.

## 4. Valorization Approaches of Date Processing Waste

The valorization approaches of date processing waste are essential for several reasons. First and foremost, the disposal of date processing waste can have significant environmental impacts. The process of organic waste decomposition in landfills produces methane, a powerful greenhouse gas that contributes to climate change [71]. By utilizing valorization approaches for date processing waste, it will be possible to redirect waste away from landfills and decrease the emission of greenhouse gases. In addition to the environmental benefits, valorization approaches for date processing waste can also lead to economic opportunities. Furthermore, the extraction of high-value compounds from date processing waste can create new markets for the food, cosmetic, and pharmaceutical industries [72]. Moreover, valorization approaches for date processing waste can also become a source of renewable energy. The production of biofuels and organic waste can be subjected to anaerobic digestion to produce electricity and heat, which can be used to power homes, buildings, and industrial processes [73]. Another reason to utilize date processing waste is to improve the sustainability of the date industry and reduce costs. Disposing of date processing waste can be expensive, so finding ways to convert it into valuable products may reduce costs and make the industry more sustainable [17]. The valorization of date processing waste, by transforming it into value-added products, represents an innovative solution that addresses both waste reduction and the creation of new economic opportunities (Figure 4).

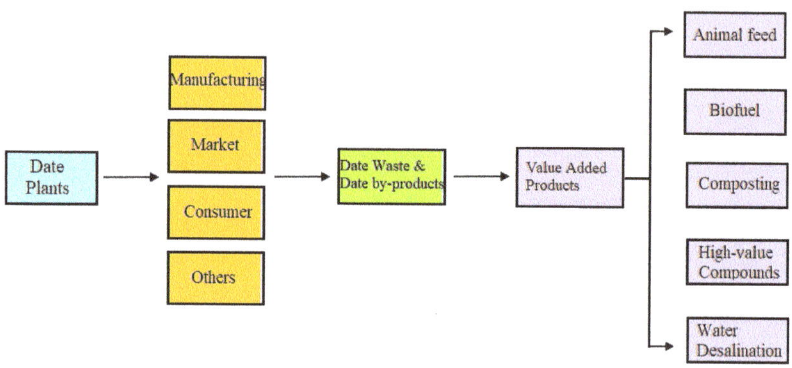


**Figure 4.** Opportunities of date waste valorization.

Overall, the valorization approaches of date processing waste are crucial for sustainable development. These approaches may create a source of renewable energy, animal feed, and materials and help to reduce the environmental impact of waste disposal and create economic opportunities. The possible valorization methods of date by-products are listed in Table 4.

**Table 4.** Possible applications of date by-products for their valorization.

| By-Products | Possible Use | Product | References |
|---|---|---|---|
| Low-grade dates | Juice and following product production | Date juice; concentration: date spread, syrup and liquid sugar; fermentation: wine, alcohol, and vinegar | [5,13] |
| | Composting | Nutrient-rich fertilizer | [74] |
| | Biofuel production | Ethanol | [75] |
| | Phenolic compound extraction | Used as raw materials or additives in food, health supplement, pharmaceutical, and cosmetic areas | [76] |
| Date processing wastes | Animal feed | Ostrich chick feed additive | [77] |
| | Fiber extraction | Functional fiber ingredients in food industry | [78] |
| | Soap production | Soap formula additives | [79] |
| | Food packaging | Improved film | [80,81] |
| Date pits | Oil production | Date pit oil | [10,19] |
| | Phenolic compound extraction | Used as raw materials or additives in food, health supplement, pharmaceutical, and cosmetic areas | [10,76] |
| | Water desalination | Coagulant and flocculant | [82] |
| Date leaves | Water desalination | Adsorbent | [83] |
| Date seed oil | Palm olein alternative | Cooking oil | [20] |
| | Biofuel production | Bio-jet fuel; green diesel | [22] |
| | Margarine production | Crucial ingredient | [19] |
| | Participate in cosmetic formulations | Body creams, shaving soap, and shampoo | [21] |

### 4.1. Conversion of Date Processing Waste to Animal Feed

The conversion of date processing waste into animal feed is an important valorization approach due to its nutritional benefits, abundance, and availability. Animal feed produced from date processing waste is high in fiber and carbohydrates, which are essential nutrients for livestock and poultry [16]. Boufennara et al. [84] found that date processing waste is particularly useful for animal feed production, as it is readily available in regions where the date industry is prevalent.

Various methods can convert date processing waste into animal feed, including ensiling, drying, and pelletization. Ensiling is a useful method for the safe and long-term preservation of moist biomass, while minimally impacting the nutritional quality of the biomass [85]. Drying is a practical technique that reduces the moisture content of a product, resulting in increased stability and facilitating its transportation. Pelletization can improve the handling and storage properties of the feed [86,87]. The use of date processing waste as animal feed has several advantages. Najafi et al. [77] observed that incorporating up to 30% whole date waste into the diets of ostrich chicks not only ensures uncompromised growth performance but also holds the potential to improve stress-related variables and antioxidant status in growing ostriches.

However, using date palm waste as a source of animal feed is also challenging. For example, economically, it raises questions about the cost-effectiveness of transforming waste into feed, as well as issues related to transportation. Nutritionally, questions arise about whether animal feed derived from date palm can meet the animals' dietary requirements and the need for potential formulation improvements. Furthermore, safety considerations encompass compliance with food laws and potential risks such as fungal and viral infections during the processing stages [88].

### 4.2. Production of Biofuels

Date processing waste can be converted into biofuels through various methods such as anaerobic digestion, fermentation, and pyrolysis [89]. The process of anaerobic digestion is characterized by the decomposition of organic matter in the absence of oxygen through the action of microorganisms, resulting in the production of biogas as a by-product [22]. Fermentation involves the conversion of sugars and carbohydrates into biofuels, such as ethanol and butanol [90]. Pyrolysis is a thermal decomposition process that converts organic matter into bio-oil, biochar, and syngas [91]. Biofuels produced from date processing waste have several advantages over conventional fossil fuels. For example, they are renewable, biodegradable, and emit fewer greenhouse gases [92]. Biofuels can also provide economic opportunities for farmers and feed producers by creating a new market for date processing waste [93].

Research has shown that date processing waste can be a valuable biofuel feedstock. The substantial lignocellulosic composition and fatty acid characteristics found in date palm seeds offer a compelling opportunity for cost-effective and sustainable biofuel production as a readily available secondary biomass resource [94]. The obtained date-seed-derived pyrolysis oil used as a potential feedstock for producing alternative fuels is chemically and thermally stable and contains some value-added chemicals [95]. Rambabu et al. [22] investigated the use of tantalum phosphate (TaPa) as a hydroprocessing catalyst for the drop-in biofuel production from date palm seed oil under mild experimental conditions. The optimized one-step reaction yielded high quantities of deoxygenated hydrocarbons, including 53.6% bio-jet fuel ($C_9$–$C_{15}$) and 35.9% green diesel ($C_{14}$–$C_{20}$). For instance, a study conducted by Taghizadeh-Alisaraei et al. [75] investigated the potential of date processing waste as a feedstock for biogas production through direct sugar fermentation, direct gasification, pretreatment hydrolysis fermentation, and pretreatment saccharification/co-fermentation or fermentation. According to their results, the date processing waste had a high bioethanol yield, making it a suitable feedstock for biogas production. Similarly, a study by Ahmad et al. [96] investigated the possible application of date processing waste for producing ethanol through fermentation. According to the study, the high concentration of fermentable sugars in date processing waste indicates that it could serve as a valuable raw material in bioethanol production.

Despite the potential of date processing waste for biofuel production, some challenges must be addressed. For example, the high lignocellulosic content of date processing waste can make it difficult to convert it into biofuels using certain methods, such as fermentation [97]. In addition, the variability of date processing waste composition can also affect the quality and quantity of the biofuels produced [75]. So, producing biofuels using date processing waste is a promising valorization approach that can provide a renewable and sustainable alternative to fossil fuels. However, there are still challenges to be overcome for an efficient conversion of this waste into fuel.

### 4.3. Composting

The decline in soil fertility causes a serious environmental challenge, leading to various harmful consequences [98]. Firstly, it reduces crop yields and quality, impacting the sustainability of agricultural production. Secondly, declining soil fertility can trigger soil erosion, leading to land infertility and disrupting the ecological balance. Additionally, fertility decline may result in soil acidification, alkalization, or salinization, threatening plant growth. Most importantly, reduced fertility puts additional pressure on agriculture, prompting farmers to use excessive chemical fertilizers and pesticides, causing soil and water pollution [99].

Therefore, composting as an effective soil improvement method was introduced to solve this problem. Composting transforms organic waste into organic fertilizers, enhancing soil organic matter content, which also improves soil structure, enhances soil aeration, water retention, and provides essential nutrients for plant growth. Moreover, the organic matter generated during composting enhances soil resilience, reducing reliance on chemical fertilizers and promoting soil fertility restoration, thus increasing crop yields [100]. And date processing waste is an excellent source of organic matter, as it is rich in carbohydrates, fiber, and other nutrients, being a suitable candidate to be composed. The composting of date processing waste can be achieved through various methods, such as windrow composting, vermicomposting, and in-vessel composting. Windrow composting involves piling the date processing waste into long rows and periodically turning it to promote aeration and decomposition [101]. Vermicomposting involves using earthworms to break down waste, producing a high-quality compost rich in beneficial microorganisms and nutrients [102]. In-vessel composting involves using enclosed containers to control the temperature, moisture, and aeration of the composting process, producing faster and more efficient decomposition of the waste [103].

The compost produced from date processing waste can be used as a soil amendment or fertilizer, providing numerous benefits for plant growth and soil health. For example, a study by Abid et al. [104] showed that the application of date processing waste compost significantly increased the growth of tomato plants and improved the soil structure and fertility. Similarly, Benabderrahim et al. [74] found that the application of date processing waste compost improved the growth and yield of forage alfalfa and increased the soil's nutrient content.

Overall, composting is a valuable valorization approach for date processing waste, providing a sustainable and nutrient-rich organic fertilizer for plant growth and soil health. With further research and development, this approach could become a cost-effective and environmentally friendly solution for the date industry.

### *4.4. Extraction of High-Value Compounds*

Date processing waste is a significant source of high-value compounds such as antioxidants, flavonoids, phenolic acids, and dietary fibers [17]. Different methods can be used to extract high-value compounds, especially phenolic compounds, from date processing waste, including solvent extraction, supercritical fluid extraction, and microwave-assisted extraction. Solvent extraction is one of the most commonly used methods, which involves the use of solvents such as methanol, ethanol, and acetone to extract target compounds from the waste [105]. For instance, Habchi et al. [76] conducted a study investigating the extraction of phenolic compounds from date pits using reflux solvent extraction. In their results, the extraction process was efficient, and the phenolic extracts had high antioxidant activity. Similarly, a study by Ghafoor et al. [10] explored the use of supercritical fluid extraction to extract phenolic compounds such as phenolic acids and flavonoids from date palm pits. This method utilizes supercritical carbon dioxide to extract the compounds. The study found that the extraction process had higher efficiency than conventional extraction and Soxhlet extraction, and the extracts had potential applications in the food and pharmaceutical industries. This method has the advantage of being environmentally friendly and producing high-quality extracts [10]. Extraction with microwave assistance is a relatively newer method that involves using microwave radiation to extract the compounds from the waste. This method is known to be efficient, fast, and environmentally friendly and was previously applied by Pourshoaib et al. [106] to recover phenolic compounds from date palm seeds, and the extracts were obtained within 5 min as a fast extraction.

#### 4.4.1. Antibacterial Compound and Antioxidant Compound

The extracted high-value compounds from date processing waste can be applied in the food, pharmaceutical, and cosmetic industries. For instance, natural preservatives derived from high-value compounds extracted from date processing waste can be used to increase the shelf life of food products without relying on synthetic preservatives and the same for use as colorants [107,108]. Additionally, these compounds have potential health benefits, such as reducing the risk of chronic diseases and improving gut health [26]. Furthermore, phenolic compounds extracted from date processing waste have potential as natural remedies for various skin conditions such as acne and wrinkles [21]. High antioxidant phenolic extract from date fruit syrup waste is an effective additive in soap formulations. Rambabu et al. [79] investigated a more environmentally friendly approach to soap manufacturing by utilizing natural ingredients as a potential substitute for synthetic chemicals commonly employed in soap production. As a result, the soaps produced from the waste extract of date fruit syrup exhibited physicochemical properties that were similar to those of commercially available turmeric soap. Furthermore, antibacterial assays demonstrated the enhanced bactericidal properties of date fruit syrup waste extract soaps against Gram-positive *Streptococcus pyogenes* and Gram-negative *Pseudomonas aeruginosa* bacteria.

A series of gelatin blend films with date fruit waste extract were developed by Rangaraj et al. [80]. The hydrophilic properties of the extract improved water solubility but reduced tensile strength and increased flexibility in the films. Films with 25% extract had higher water vapor permeability due to plasticizing effects. These films released active phenolic compounds and had better antioxidant capacity than fatty food simulants in aqueous food. In canola oil storage tests, films with extract showed lower peroxide and Totox values than pure gelatin films. Date fruit waste extract is a promising natural antioxidant for enhancing food packaging film functionality and performance. Farousha et al. [81] also developed date-seed-extract-encapsulated mesoporous MCM-41 material, to extend the shelf life of food in food packaging applications, and the in vitro release performance and antioxidant and antimicrobial activities were investigated. As a result, date seed extract was successfully encapsulated into MCM-41 with a 1:1 weight ratio using vacuum adsorption, achieving a high encapsulation efficiency of 91%. DSE@MCM-41 exhibited sustained release in both acidic (pH 5.2) and alkaline (pH 7.4) environments without a burst effect, following a first-order release model regulated by the carrier structure. It showed a more sustained release of bioactive compounds and antioxidants than free date seed extract when stored at 4 °C for 90 days. Additionally, DSE@MCM-41 demonstrated stable antibacterial activity even after 72 h against *E. coli* and *S. aureus*, attributed to sustained DSE release from the encapsulated MCM-41 matrix. These findings suggest the potential of DSE-encapsulated MCM-41 as a nanocarrier for sustained-release food packaging systems to extend food shelf life.

#### 4.4.2. Dietary Fiber

Date processing waste has high dietary fiber content, including cellulose, hemicellulose, and lignin. Studies have indicated that these types of fibers found in diets offer several health advantages, such as enhancing the digestive system and minimizing the probability of developing long-term illnesses [109,110]. Several studies have investigated the extraction of dietary fibers from date processing waste using different methods, including enzymatic hydrolysis, acid and alkali treatments, and microwave-assisted extraction [78,111]. The extracted dietary fibers can be used as functional ingredients in the food industry, such as in the production of high-fiber bread, pasta, and snacks, as well as in the development of dietary supplements [112].

#### 4.4.3. Minerals

Similarly, the date processing waste is a rich source of essential minerals such as K, Mg, Ca, and Fe, and the recovery of these minerals from date processing waste can offer a sustainable and cost-effective solution for the food and supplement industry. The extraction of minerals from date processing waste can be achieved through different methods, such as acid extraction, enzymatic hydrolysis, and microwave-assisted extraction [27]. The extracted minerals can be used in various applications, such as in the production of dietary supplements or as food fortifiers to increase the nutritional value of processed foods.

Extracting valuable compounds from date processing waste is a promising method to produce value-added products using this abundant waste stream. Using green extraction methods and identifying novel bioactive compounds offer opportunities for economic and environmental sustainability in the date industry. Further research and development in this field are needed to optimize the extraction processes and to identify new applications for the extracted compounds.

### *4.5. Water Desalination and Purification*

Water scarcity is a growing concern worldwide, and many regions are turning to desalination as a means of addressing the issue. Desalination refers to the technique of separating salts and other minerals from seawater or brackish water to make it potable and suitable for consumption and irrigation purposes [113]. However, desalination is an energy-intensive process and can have environmental impacts, such as the discharge of brine into the ocean [113]. One potential solution to these challenges is the use of date processing waste in desalination processes.

Date processing waste, such as date pits and leaves, contain high amounts of tannins and other organic compounds that can be used in the desalination process. The potential uses of date processing waste include its application in seawater pretreatment as a preliminary step before desalination [114]. Pretreatment is a critical step in desalination that involves removing suspended solids, colloids, and other impurities from the feed water [114]. Date processing waste can be used as a natural coagulant and flocculant to enhance the removal of impurities during pretreatment [114].

Rambabu et al. [115] investigated the utilization of date palm empty fruit bunches (DPEFBs) as a sustainable biosorbent for hexavalent chromium ($Cr^{6+}$) removal from synthetic wastewater. The pretreated DPEFB was analyzed for its morphology and surface chemistry using scanning electron microscopy, energy dispersive elemental analysis, and Fourier-transform infrared spectroscopy. Various biosorption parameters, including pH, biosorbent dosage, contact time, temperature, initial feed concentration, and agitation speed, were examined for their impact on $Cr^{6+}$ removal by DPEFB, and they found that the DPEFB sorbent exhibited an isoelectric point at pH 2, becoming dehydronated above this value to capture positively charged $Cr^{6+}$ ions. Optimal conditions for chromium removal were identified as pH 2, 0.3 g dosage, 100 rpm agitation speed, 120 min contact time, 50 mg/L initial feed concentration, and 30 °C operational temperature. Furthermore, DPEFB demonstrated reusability with $NaNO_3$ as an effective regenerant, allowing for up to three successive biosorption–desorption cycles without significant loss in efficiency. This study establishes DPEFB as an efficient, cost-effective, and eco-friendly biosorbent for removing toxic $Cr^{6+}$ ions from wastewater, offering a sustainable approach to address this environmental concern. Date-palm-coir-waste-derived nano-activated carbon (DPC-AC) as a cost-effective solution for removing 2,4-DPA herbicide from the water was also investigated by Rambabu et al. [116]. DPC-AC, created through a one-step process, exhibited strong adsorption properties with a 98.6% removal rate under optimal conditions. It followed Langmuir isotherm and pseudo-second-order kinetics, and thermodynamic analysis confirmed its efficiency. The study also highlighted the nanosorbent's electrostatic and chemical interactions with 2,4-DPA, showcasing its potential for water treatment. Additionally, cost analysis and regeneration tests emphasized its economic feasibility and reusability, turning date palm coir waste into a valuable resource for pollution control.

In a recent study conducted by Rambabu et al. [83], they successfully demonstrated the effective treatment of aqueous effluents containing the 2,4-DPA herbicide using modified date palm leaf waste with sulfuric acid. The resulting product exhibited an amorphous structure characterized by numerous cavities, pore connectivity, and oxygen-bearing functionalities, facilitating the efficient removal of 2,4-DPA from water sources. The adsorption process was favorable, spontaneous, and exothermic from a thermodynamic standpoint. Additionally, reusability tests confirmed that the product could be regenerated using a NaOH regenerant and employed for up to five cycles to remediate 2,4-DPA-contaminated wastewater. Crucially, the adsorbent product demonstrated a remarkable remediation efficiency of 69.4% when applied to the treatment of actual agricultural runoff contaminated with 2,4-DPA herbicide, underscoring its readiness and potential for real-world applications. This study effectively valorized date palm leaf waste as a valuable adsorbent through a simple and cost-effective method for addressing the issue of agricultural wastewater pollution caused by harmful herbicides.

The process of pyrolysis was employed to create biochar from date seed biomass, and the current study reveals that the porous activated carbon known as DSAC-1:1.5 is a practical and high-performing electrode material suitable for capacitive deionization in desalination applications [82]. According to the study by Hai et al. [82], an optimal weight ratio of 1:1.5 (biochar: KOH) resulted in mesoporous date seed biochar-activated carbon with a uniform structure. It displayed amorphous and graphitic carbon formation due to KOH activation, offering a mixed micro- and mesoporous nature. Electrochemical tests revealed a specific capacitance of 400 F $g^{-1}$ for DSAC-1:1.5, making it suitable for NaCl desalination, achieving 22.2 mg $gm^{-1}$ electrosorption capacity and 86.4% charge efficiency. The electrode proved reusable for six cycles without performance decline. This low-cost date seed biochar-activated carbon outperforms existing carbon-based electrodes in desalination applications.

While the use of date processing waste in water desalination shows promise, several challenges must be addressed. One such challenge is the variability of the chemical composition of date processing waste, which can affect the effectiveness of the pretreatment and the performance of the desalination membranes [117]. This challenge can be overcome through proper selection and processing of the raw materials to ensure consistency in the quality of the bioactive compounds [117]. However, most current research in this area has only been conducted on a laboratory scale, and more research is required to determine the feasibility of large-scale production and commercialization of these products [118,119]. Finally, the regulatory and environmental impacts of using date processing waste in desalination processes need to be further studied to ensure no unintended negative consequences.

## 5. Challenges and Future Perspectives

Date palm waste valorization presents opportunities and challenges critical to address for sustainable and environmentally friendly resource management. These waste materials can be transformed into valuable products through the above valorization processes. However, unlocking the full potential of date palm waste valorization requires overcoming several key challenges while considering future perspectives for sustainable practices.

One of the challenges in date palm waste valorization is the efficient collection and transportation of waste materials from date palm farms to processing facilities. Given the often remote and dispersed locations of date palm plantations, logistics and infrastructure for waste collection can be costly and complicated. Innovations in collection and transportation methods, such as mobile chipping and shredding units, can help mitigate this challenge by reducing transportation costs and increasing the feasibility of waste utilization.

The variability in the composition and quality of date palm waste is another significant obstacle. Date palm waste consists of various components, including fronds, pits, coir, and leaves, each with different properties and potential applications. This heterogeneity demands tailored valorization approaches for different waste components. Developing efficient separation and sorting techniques to segregate these components and optimize their utilization is essential for maximizing the value of date palm waste.

Future perspectives for date palm waste valorization involve exploring advanced technologies and innovations to enhance its efficiency and sustainability. One such innovation is the integration of biorefinery concepts, where multiple products and value streams are extracted from date palm waste. This approach can involve the simultaneous production of bioenergy, biofuels, bioplastics, and bioactive compounds from different waste components, leading to a more comprehensive and profitable resource utilization.

Biotechnological advancements, including the development of genetically modified date palm varieties with enhanced biomass properties or improved bioactive compound content, hold the potential for optimizing date palm waste valorization. Genetic engineering can increase yields and improve properties, making date palm waste a more attractive feedstock for various applications.

Furthermore, research into the utilization of date palm waste for producing bioenergy and biofuels, such as biogas, bioethanol, and biodiesel, is a popular and promising recent avenue. Advanced conversion technologies, such as anaerobic digestion, pyrolysis, and gasification, can efficiently convert date palm waste into renewable energy sources, reducing dependence on fossil fuels and mitigating greenhouse gas emissions.

## 6. Conclusions

Date palm fruit, known for its nutritional richness, holds great promise for producing value-added products in the food and nutraceutical industries through bioprocessing technologies. The substantial quantities of date seeds and discarded fruits offer untapped potential for bioprocessing, opening doors to new product possibilities. Additionally, the efficient utilization of date palm fruit, which is rich in carbohydrates, dietary fiber, and bioactive compounds like phenolic compounds, can be achieved through various extraction methods applied to date processing wastes, serving as essential ingredients for diverse value-added products across various industries. As stated in this review, date palm waste materials have versatile applications beyond traditional use, such as animal feed. Recent research has also shown that they can be utilized for biofuel production, composting, antimicrobial and antioxidant additives, and water desalination and purification. The abundance of date fruit by-products and waste, particularly the vast reserves of unused date seeds, holds immense potential for novel product development. Date-palm-growing countries can leverage this opportunity to fully exploit the date palm's potential, manage waste effectively, and create opportunities for rural and socioeconomic advancement.

**Author Contributions:** Conceptualization, formal analysis, validation, and investigation, L.S. and H.A.R.S.; resources, H.A.R.S.; writing—original draft preparation, L.S.; writing—review and editing, L.S., T.S.P.d.S., F.A., A.I., F.R.D., C.B., and H.A.R.S.; supervision, H.A.R.S., F.R.D., and C.B.; ideas sharing, H.A.R.S.; funding acquisition, H.A.R.S. All authors have read and agreed to the published version of the manuscript.

**Funding:** Hafiz Suleria is the recipient of an "Australian Research Council—Discovery Early Career Award" (ARC-DECRA—DE220100055) funded by the Australian Government. This research was funded by the University of Melbourne under the "McKenzie Fellowship Scheme" (grant No. UoM-18/21), the "Future Food Hallmark Research Initiative Funds (Grant No. UoM-21/23)", and the "Collaborative Research Development Grant (Grant No. UoM-21/23)" funded by the Faculty of Veterinary and Agricultural Sciences, the University of Melbourne, Australia.

**Institutional Review Board Statement:** Not applicable.

**Informed Consent Statement:** Not applicable.

**Data Availability Statement:** Not applicable.

**Acknowledgments:** We would like to thank NorVicFoods, a cofounded initiative between the University of Melbourne and the Victorian State Government—Department of Education and Training. We would also like to thank the Honours/Master/PhD and postdoctoral researchers of the Hafiz Suleria group from the School of Agriculture, Food and Ecosystem Sciences, Faculty of Science, University of Melbourne for their incredible support.

**Conflicts of Interest:** The authors declare no conflict of interest.

## References

1. Qadir, A.; Shakeel, F.; Ali, A.; Faiyazuddin, M. Phytotherapeutic potential and pharmaceutical impact of *Phoenix dactylifera* (date palm): Current research and future prospects. *J. Food Sci. Technol.* **2020**, *57*, 1191–1204. [CrossRef] [PubMed]
2. Fernández-López, J.; Viuda-Martos, M.; Sayas-Barberá, E.; Navarro-Rodríguez de Vera, C.; Pérez-Álvarez, J.Á. Biological, Nutritive, Functional and Healthy Potential of Date Palm Fruit (*Phoenix dactylifera* L.): Current Research and Future Prospects. *Agronomy* **2022**, *12*, 876. [CrossRef]

3. Al-Shwyeh, H.A. Date Palm (*Phoenix dactylifera* L.) Fruit as Potential Antioxidant and Antimicrobial Agents. *J. Pharm. Bioallied Sci.* **2019**, *11*, 1–11. [CrossRef] [PubMed]
4. Statista. Production of Dates Worldwide from 2010 to 2021 (in Million Metric Tons)*. Available online: https://www.statista.com/statistics/960247/dates-production-worldwide/ (accessed on 23 August 2023).
5. Salomón-Torres, R.; Valdez-Salas, B.; Norzagaray-Plasencia, S. Date palm: Source of foods, sweets and beverages. In *The Date Palm Genome, Vol. 2: Omics and Molecular Breeding*; Springer: Berlin/Heidelberg, Germany, 2021; pp. 3–26.
6. El-Mously, H.; Atef, E. Date Palm Byproducts: History of Utilization and Technical Heritage. In *Date Palm Fiber Composites*; Springer: Singapore, 2020; pp. 3–71. [CrossRef]
7. Darwish, E.A.F.; El-Mously, H.; Abdelrahman, A.A. The Technical Heritage of Date Palm Leaves Utilization in Traditional Handicrafts and Architecture in Egypt & the Middle East. In *By-Products of Palm Trees and Their Applications*; Materials Research Forum LLC: Millersville, PA, USA, 2019.
8. Ahmed, Z.F.R.; Al Shaibani, F.Y.Y.; Kaur, N.; Maqsood, S.; Schmeda-Hirschmann, G. Improving Fruit Quality, Bioactive Compounds, and Storage Life of Date Palm (*Phoenix dactylifera* L., cv. Barhi) Using Natural Elicitors. *Horticulturae* **2021**, *7*, 293. [CrossRef]
9. Echegaray, N.; Gullón, B.; Pateiro, M.; Amarowicz, R.; Misihairabgwi, J.M.; Lorenzo, J.M. Date fruit and its by-products as promising source of bioactive components: A review. *Food Rev. Int.* **2023**, *39*, 1411–1432. [CrossRef]
10. Ghafoor, K.; Sarker, M.Z.I.; Al-Juhaimi, F.Y.; Babiker, E.E.; Alkaltham, M.S.; Almubarak, A.K. Extraction and Evaluation of Bioactive Compounds from Date (*Phoenix dactylifera*) Seed Using Supercritical and Subcritical $CO_2$ Techniques. *Foods* **2022**, *11*, 1806. [CrossRef]
11. Bouaziz, F.; Ben Abdeddayem, A.; Koubaa, M.; Ellouz Ghorbel, R.; Ellouz Chaabouni, S. Date seeds as a natural source of dietary fibers to improve texture and sensory properties of wheat bread. *Foods* **2020**, *9*, 737. [CrossRef]
12. Habib, H.M.; El-Fakharany, E.M.; Souka, U.D.; Elsebaee, F.M.; El-Ziney, M.G.; Ibrahim, W.H. Polyphenol-rich date palm fruit seed (*Phoenix dactylifera* L.) extract inhibits labile iron, enzyme, and cancer cell activities, and DNA and protein damage. *Nutrients* **2022**, *14*, 3536. [CrossRef]
13. Salomón-Torres, R.; Ortiz-Uribe, N.; Valdez-Salas, B.; Rosas-González, N.; García-González, C.; Chávez, D.; Córdova-Guerrero, I.; Díaz-Rubio, L.; del Pilar Haro-Vázquez, M.; Mijangos-Montiel, J.L.; et al. Nutritional assessment, phytochemical composition and antioxidant analysis of the pulp and seed of medjool date grown in Mexico. *PeerJ* **2019**, *7*, e6821. [CrossRef]
14. Ogungbenro, A.E.; Quang, D.V.; Al-Ali, K.A.; Vega, L.F.; Abu-Zahra, M.R.M. Synthesis and characterization of activated carbon from biomass date seeds for carbon dioxide adsorption. *J. Environ. Chem. Eng.* **2020**, *8*, 104257. [CrossRef]
15. Mrabet, A.; Jiménez-Araujo, A.; Guillén-Bejarano, R.; Rodríguez-Arcos, R.; Sindic, M. Date Seeds: A Promising Source of Oil with Functional Properties. *Foods* **2020**, *9*, 787. [CrossRef] [PubMed]
16. Najjar, Z.; Stathopoulos, C.; Chockchaisawasdee, S. Utilization of date by-products in the food industry. *Emir. J. Food Agric.* **2020**, *32*, 808–815. [CrossRef]
17. Oladzad, S.; Fallah, N.; Mahboubi, A.; Afsham, N.; Taherzadeh, M.J. Date fruit processing waste and approaches to its valorization: A review. *Bioresour. Technol.* **2021**, *340*, 125625. [CrossRef] [PubMed]
18. Ibourki, M.; Azouguigh, F.; Jadouali, S.M.; Sakar, E.H.; Bijla, L.; Majourhat, K.; Gharby, S.; Laknifli, A. Physical fruit traits, nutritional composition, and seed oil fatty acids profiling in the main date palm (*Phoenix dactylifera* L.) varieties grown in Morocco. *J. Food Qual.* **2021**, *2021*, 5138043. [CrossRef]
19. Mrabet, A.; Rodríguez-Gutiérrez, G.; Guillén-Bejarano, R.; Rodríguez-Arcos, R.; Sindic, M.; Jiménez-Araujo, A. Optimization of date seed oil extraction using the assistance of hydrothermal and ultrasound technologies. *Grasas Aceites* **2022**, *73*, e457. [CrossRef]
20. Nehdi, I.A.; Sbihi, H.M.; Tan, C.P.; Rashid, U.; Al-Resayes, S.I. Chemical composition of date palm (*Phoenix dactylifera* L.) seed oil from six Saudi Arabian cultivars. *J. Food Sci.* **2018**, *83*, 624–630. [CrossRef]
21. Alharbi, K.L.; Raman, J.; Shin, H.-J. Date Fruit and Seed in Nutricosmetics. *Cosmetics* **2021**, *8*, 59. [CrossRef]
22. Rambabu, K.; Bharath, G.; Sivarajasekar, N.; Velu, S.; Sudha, P.N.; Wongsakulphasatch, S.; Banat, F. Sustainable production of bio-jet fuel and green gasoline from date palm seed oil via hydroprocessing over tantalum phosphate. *Fuel* **2023**, *331*, 125688. [CrossRef]
23. Majzoobi, M.; Karambakhsh, G.; Golmakani, M.; Mesbahi, G.; Farahnaki, A. Chemical composition and functional properties of date press cake, an agro-industrial waste. *J. Agric. Sci. Technol.* **2019**, *21*, 1807–1817.
24. Maqsood, S.; Adiamo, O.; Ahmad, M.; Mudgil, P. Bioactive compounds from date fruit and seed as potential nutraceutical and functional food ingredients. *Food Chem.* **2020**, *308*, 125522. [CrossRef]
25. Rybicka, I.; Kiewlicz, J.; Kowalczewski, P.Ł.; Gliszczyńska-Świgło, A. Selected dried fruits as a source of nutrients. *Eur. Food Res. Technol.* **2021**, *247*, 2409–2419. [CrossRef]
26. Alkhoori, M.A.; Kong, A.S.; Aljaafari, M.N.; Abushelaibi, A.; Erin Lim, S.H.; Cheng, W.H.; Chong, C.M.; Lai, K.S. Biochemical Composition and Biological Activities of Date Palm (*Phoenix dactylifera* L.) Seeds: A Review. *Biomolecules* **2022**, *12*, 1626. [CrossRef] [PubMed]
27. Attia, A.I.; Reda, F.M.; Patra, A.K.; Elnesr, S.S.; Attia, Y.A.; Alagawany, M. Date (*Phoenix dactylifera* L.) by-Products: Chemical Composition, Nutritive Value and Applications in Poultry Nutrition, an Updating Review. *Animals* **2021**, *11*, 1133. [CrossRef]
28. Fahmy, M.; Gado, H.; Matloup, O.; Kholif, S.; El-Bordiny, N. In vitro evaluation of ensiling and/or exogenous fibrolytic enzyme supplementation of date press cake. *Arab Univ. J. Agric. Sci.* **2019**, *27*, 347–355. [CrossRef]

29. Younas, A.; Naqvi, S.A.; Khan, M.R.; Shabbir, M.A.; Jatoi, M.A.; Anwar, F.; Inam-Ur-Raheem, M.; Saari, N.; Aadil, R.M. Functional food and nutra-pharmaceutical perspectives of date (*Phoenix dactylifera* L.) fruit. *J. Food Biochem.* **2020**, *44*, e13332. [CrossRef]
30. Sait, H.H.; Hussain, A.; Bassyouni, M.; Ali, I.; Kanthasamy, R.; Ayodele, B.V.; Elhenawy, Y. Hydrogen-Rich Syngas and Biochar Production by Non-Catalytic Valorization of Date Palm Seeds. *Energies* **2022**, *15*, 2727. [CrossRef]
31. Sheir, M.A. Innovative use of date (*Phoenix dactylifera* L.) press cake in the food industry. *Food Raw Mater.* **2022**, *10*, 2–9. [CrossRef]
32. Mrabet, A.; Hammadi, H.; Rodríguez-Gutiérrez, G.; Jiménez-Araujo, A.; Sindic, M. Date Palm Fruits as a Potential Source of Functional Dietary Fiber: A Review. *Food Sci. Technol. Res.* **2019**, *25*, 1–10. [CrossRef]
33. Alghamdi, A.A.; Awadelkarem, A.M.; Hossain, A.; Ibrahim, N.A.; Fawzi, M.; Ashraf, S.A. Nutritional assessment of different date fruits (*Phoenix dactylifera* L.) varieties cultivated in Hail province, Saudi Arabia. *Biosci. Biotechnol. Res. Commun.* **2018**, *11*, 263–269. [CrossRef]
34. Nadeem, M.; Qureshi, T.M.; Ugulu, I.; Riaz, M.N.; Ain, Q.U.; Khan, Z.I.; Ahmad, K.; Ashfaq, A.; Bashir, H.; Dogan, Y. Mineral, vitamin and phenolic contents and sugar profiles of some prominent date palm (*Phoenix dactylifera*) varieties of Pakistan. *Pak. J. Bot.* **2019**, *51*, 171–178. [CrossRef]
35. Batsale, M.; Bahammou, D.; Fouillen, L.; Mongrand, S.; Joubès, J.; Domergue, F. Biosynthesis and functions of very-long-chain fatty acids in the responses of plants to abiotic and biotic stresses. *Cells* **2021**, *10*, 1284. [CrossRef] [PubMed]
36. Al Juhaimi, F.; Özcan, M.M.; Uslu, N.; Ghafoor, K.; Babiker, E.E.; Mohamed Ahmed, I.A. Bioactive properties, fatty acid compositions, and phenolic compounds of some date palm (*Phoenix dactylifera* L.) cultivars. *J. Food Process. Preserv.* **2020**, *44*, e14432. [CrossRef]
37. Muñoz-Tebar, N.; Viuda-Martos, M.; Lorenzo, J.M.; Fernandez-Lopez, J.; Perez-Alvarez, J.A. Strategies for the Valorization of Date Fruit and Its Co-Products: A New Ingredient in the Development of Value-Added Foods. *Foods* **2023**, *12*, 1456. [CrossRef]
38. Schwingshackl, L.; Hoffmann, G. Monounsaturated fatty acids, olive oil and health status: A systematic review and meta-analysis of cohort studies. *Lipids Health Dis.* **2014**, *13*, 154. [CrossRef] [PubMed]
39. Alahyane, A.; ElQarnifa, S.; Ayour, J.; Elateri, I.; Ouamnina, A.; Ait-Oubahou, A.; Benichou, M.; Abderrazik, M. Date seeds (*Phoenix dactylifera* L.) valorization: Chemical composition of lipid fraction. *Braz. J. Biol.* **2022**, *84*, e260771. [CrossRef] [PubMed]
40. Islam, M.A.; Amin, M.N.; Siddiqui, S.A.; Hossain, M.P.; Sultana, F.; Kabir, M.R. Trans fatty acids and lipid profile: A serious risk factor to cardiovascular disease, cancer and diabetes. *Diabetes Metab. Syndr. Clin. Res. Rev.* **2019**, *13*, 1643–1647. [CrossRef]
41. Thomas, M.S.; Calle, M.; Fernandez, M.L. Healthy plant-based diets improve dyslipidemias, insulin resistance, and inflammation in metabolic syndrome. A narrative review. *Adv. Nutr.* **2022**, *14*, 44–54. [CrossRef]
42. Miro, C.; Nappi, A.; Sagliocchi, S.; Di Cicco, E.; Murolo, M.; Torabinejad, S.; Acampora, L.; Pastore, A.; Luciano, P.; La Civita, E.; et al. Thyroid Hormone Regulates the Lipid Content of Muscle Fibers, Thus Affecting Physical Exercise Performance. *Int. J. Mol. Sci.* **2023**, *24*, 12074. [CrossRef]
43. Rambabu, K.; Bharath, G.; Hai, A.; Banat, F.; Hasan, S.W.; Taher, H.; Mohd Zaid, H.F. Nutritional Quality and Physico-Chemical Characteristics of Selected Date Fruit Varieties of the United Arab Emirates. *Processes* **2020**, *8*, 256. [CrossRef]
44. Dalangin, R.; Kim, A.; Campbell, R.E. The Role of Amino Acids in Neurotransmission and Fluorescent Tools for Their Detection. *Int. J. Mol. Sci.* **2020**, *21*, 6197. [CrossRef]
45. Barakat, H.; Alfheeaid, H.A. Date Palm Fruit (*Phoenix dactylifera*) and Its Promising Potential in Developing Functional Energy Bars: Review of Chemical, Nutritional, Functional, and Sensory Attributes. *Nutrients* **2023**, *15*, 2134. [CrossRef] [PubMed]
46. El-Dreny, E.; Shaheen, M. Utilizing the date palm (*Phoenix canariensis*) seeds in the preparation of some special foods of high nutritional value. *Curr. Chem. Lett.* **2022**, *11*, 275–284. [CrossRef]
47. Tripler, E.; Shani, U.; Mualem, Y.; Ben-Gal, A. Long-term growth, water consumption and yield of date palm as a function of salinity. *Agric. Water Manag.* **2011**, *99*, 128–134. [CrossRef]
48. Borchani, C.; Besbes, S.; Blecker, C.; Masmoudi, M.; Baati, R.; Attia, H. Chemical properties of 11 date cultivars and their corresponding fiber extracts. *Afr. J. Biotechnol.* **2010**, *9*, 4096–4105.
49. Ali, Z.; Li, J.; Zhang, Y.; Naeem, N.; Younas, S.; Javeed, F. Dates (*Phoenix dactylifera*) and Date Vinegar: Preventive Role against Various Diseases and Related in Vivo Mechanisms. *Food Rev. Int.* **2022**, *38*, 480–507. [CrossRef]
50. Popoola-Akinola, O.O.; Raji, T.J.; Olawoye, B. Lignocellulose, dietary fibre, inulin and their potential application in food. *Heliyon* **2022**, *8*, e10459. [CrossRef] [PubMed]
51. Noorbakhsh, H.; Khorasgani, M.R. Date (*Phoenix dactylifera* L.) polysaccharides: A review on Chemical structure and nutritional properties. *J. Food Meas. Charact.* **2022**, *16*, 3240–3250. [CrossRef]
52. Kamal, H.; Habib, H.M.; Ali, A.; Show, P.L.; Koyande, A.K.; Kheadr, E.; Ibrahim, W.H. Food waste valorization potential: Fiber, sugar, and color profiles of 18 date seed varieties (*Phoenix dactylifera*, L.). *J. Saudi Soc. Agric. Sci.* **2023**, *22*, 133–138. [CrossRef]
53. Kamal-Eldin, A.; George, N.; Sobti, B.; AlRashidi, N.; Ghnimi, S.; Ali, A.A.; Andersson, A.A.M.; Andersson, R.; Antony, A.; Hamed, F. Dietary fiber components, microstructure, and texture of date fruits (*Phoenix dactylifera*, L.). *Sci. Rep.* **2020**, *10*, 21767. [CrossRef]
54. Zrelli, A.; Elfalleh, W.; Ghorbal, A.; Chaouachi, B. Valorization of date palm wastes by lignin extraction to be used for the improvement of polymeric membrane characteristics. *Period. Polytech. Chem. Eng.* **2022**, *66*, 70–81. [CrossRef]
55. Bamigbade, G.B.; Subhash, A.J.; Kamal-Eldin, A.; Nyström, L.; Ayyash, M. An Updated Review on Prebiotics: Insights on Potentials of Food Seeds Waste as Source of Potential Prebiotics. *Molecules* **2022**, *27*, 5947. [CrossRef] [PubMed]

56. Alkaabi, J.M.; Al-Dabbagh, B.; Ahmad, S.; Saadi, H.F.; Gariballa, S.; Ghazali, M.A. Glycemic indices of five varieties of dates in healthy and diabetic subjects. *Nutr. J.* **2011**, *10*, 59. [CrossRef] [PubMed]
57. Goff, H.D.; Repin, N.; Fabek, H.; El Khoury, D.; Gidley, M.J. Dietary fibre for glycaemia control: Towards a mechanistic understanding. *Bioact. Carbohydr. Diet. Fibre* **2018**, *14*, 39–53. [CrossRef]
58. Pinto, T.; Aires, A.; Cosme, F.; Bacelar, E.; Morais, M.C.; Oliveira, I.; Ferreira-Cardoso, J.; Anjos, R.; Vilela, A.; Gonçalves, B. Bioactive (Poly)phenols, Volatile Compounds from Vegetables, Medicinal and Aromatic Plants. *Foods* **2021**, *10*, 106. [CrossRef]
59. Hamad, H.A.M. Phenolic Compounds: Classification, Chemistry, and Updated Techniques of Analysis and Synthesis. In *Phenolic Compounds*; Farid, A.B., Ed.; IntechOpen: Rijeka, Croatia, 2021; Chapter 4. [CrossRef]
60. Hilary, S.; Tomás-Barberán, F.A.; Martinez-Blazquez, J.A.; Kizhakkayil, J.; Souka, U.; Al-Hammadi, S.; Habib, H.; Ibrahim, W.; Platat, C. Polyphenol characterisation of *Phoenix dactylifera* L. (date) seeds using HPLC-mass spectrometry and its bioaccessibility using simulated in-vitro digestion/Caco-2 culture model. *Food Chem.* **2020**, *311*, 125969. [CrossRef]
61. Zihad, S.M.N.K.; Uddin, S.J.; Sifat, N.; Lovely, F.; Rouf, R.; Shilpi, J.A.; Sheikh, B.Y.; Göransson, U. Antioxidant properties and phenolic profiling by UPLC-QTOF-MS of Ajwah, Safawy and Sukkari cultivars of date palm. *Biochem. Biophys. Rep.* **2021**, *25*, 100909. [CrossRef]
62. Alam, M.Z.; Alhebsi, M.S.R.; Ghnimi, S.; Kamal-Eldin, A. Inability of total antioxidant activity assays to accurately assess the phenolic compounds of date palm fruit (*Phoenix dactylifera* L.). *NFS J.* **2021**, *22*, 32–40. [CrossRef]
63. Radfar, R.; Farhoodi, M.; Ghasemi, I.; Khaneghah, A.M.; Shahraz, F.; Hosseini, H. Assessment of phenolic contents and antioxidant and antibacterial activities of extracts from four varieties of Iranian date Palm (*Phoenix dactylifera* L.) seeds. *Appl. Food Biotechnol.* **2019**, *6*, 173–184.
64. Selim, S.; Abdel-Mawgoud, M.; Al-Sharary, T.; Almuhayawi, M.S.; Alruhaili, M.H.; Al Jaouni, S.K.; Warrad, M.; Mohamed, H.S.; Akhtar, N.; AbdElgawad, H. Pits of date palm: Bioactive composition, antibacterial activity and antimutagenicity potentials. *Agronomy* **2021**, *12*, 54. [CrossRef]
65. Bouhlali, E.D.T.; Hmidani, A.; Bourkhis, B.; Khouya, T.; Harnafi, H.; Filali-Zegzouti, Y.; Alem, C. Effect of *Phoenix dactylifera* seeds (dates) extract in triton WR-1339 and high fat diet induced hyperlipidaemia in rats: A comparison with simvastatin. *J. Ethnopharmacol.* **2020**, *259*, 112961. [CrossRef]
66. Babiker, E.E.; Atasoy, G.; Özcan, M.M.; Juhaimi, F.A.; Ghafoor, K.; Ahmed, I.A.M.; Almusallam, I.A. Bioactive compounds, minerals, fatty acids, color, and sensory profile of roasted date (*Phoenix dactylifera* L.) seed. *J. Food Process. Preserv.* **2020**, *44*, e14495. [CrossRef]
67. Alshwyeh, H.A. Phenolic profiling and antibacterial potential of Saudi Arabian native date palm (*Phoenix dactylifera*) cultivars. *Int. J. Food Prop.* **2020**, *23*, 627–638. [CrossRef]
68. John, J.A.; Shahidi, F. Phenolic content, antioxidant and anti-inflammatory activities of seeds and leaves of date palm (*Phoenix dactylifera* L.). *J. Food Bioact.* **2019**, *5*, 120–130. [CrossRef]
69. Ardekani, M.R.S.; Khanavi, M.; Hajimahmoodi, M.; Jahangiri, M.; Hadjiakhoondi, A. Comparison of antioxidant activity and total phenol contents of some date seed varieties from Iran. *Iran. J. Pharm. Res. IJPR* **2010**, *9*, 141.
70. Al-Farsi, M.A.; Lee, C.Y. Optimization of phenolics and dietary fibre extraction from date seeds. *Food Chem.* **2008**, *108*, 977–985. [CrossRef]
71. Driga, A.M.; Drigas, A.S. Climate Change 101: How Everyday Activities Contribute to the Ever-Growing Issue. *Int. J. Recent Contrib. Eng. Sci. IT* **2019**, *7*, 22–31. [CrossRef]
72. Hai, A.; Bharath, G.; Rambabu, K.; Kannan, P.; Banat, F.; Taher, H.; Jayaraman, R.; Show, P.L. Pyrolysis of different date palm industrial wastes into high-quality bio-oils: A comparative study. *Clean Technol. Environ. Policy* **2021**, *23*, 55–64. [CrossRef]
73. Lattieff, F.A. A study of biogas production from date palm fruit wastes. *J. Clean. Prod.* **2016**, *139*, 1191–1195. [CrossRef]
74. Benabderrahim, M.A.; Elfalleh, W.; Belayadi, H.; Haddad, M. Effect of date palm waste compost on forage alfalfa growth, yield, seed yield and minerals uptake. *Int. J. Recycl. Org. Waste Agric.* **2018**, *7*, 1–9. [CrossRef]
75. Taghizadeh-Alisaraei, A.; Motevali, A.; Ghobadian, B. Ethanol production from date wastes: Adapted technologies, challenges, and global potential. *Renew. Energy* **2019**, *143*, 1094–1110. [CrossRef]
76. Habchi, A.; Dekmouche, M.; Hamia, C.; Saidi, M.; Yousfi, M.; Bouguerra, A. Extraction of Phenolic Compounds of six Algerian date (*Phoenix dactylifera* L.) cultivars from Ain-Saleh region, using Reflux method and Screening of Antioxidant Activity in vitro. *Asian J. Res. Chem.* **2021**, *14*, 161–167.
77. Najafi, S.; Ghasemi, H.A.; Hajkhodadadi, I.; Khodaei-Motlagh, M. Nutritional value of whole date waste and evaluating its application in ostrich diets. *Animal* **2021**, *15*, 100165. [CrossRef] [PubMed]
78. Chen, X.; Huang, C.; Jian, R.; Zhang, Y.; Wu, W.; Tang, X.; Tan, Q.; Sheng, Z.; Zhang, K. Optimization of microwave-assisted extraction of soluble dietary fiber from potato pulp and its properties. *J. Food Process. Preserv.* **2019**, *43*, e14204. [CrossRef]
79. Rambabu, K.; Edathil, A.A.; Nirmala, G.S.; Hasan, S.W.; Yousef, A.F.; Show, P.L.; Banat, F. Date-fruit syrup waste extract as a natural additive for soap production with enhanced antioxidant and antibacterial activity. *Environ. Technol. Innov.* **2020**, *20*, 101153. [CrossRef]
80. Rangaraj, V.M.; Rambabu, K.; Banat, F.; Mittal, V. Effect of date fruit waste extract as an antioxidant additive on the properties of active gelatin films. *Food Chem.* **2021**, *355*, 129631. [CrossRef]
81. Farousha, K.; Rangaraj, V.M.; Rambabu, K.; Haija, M.A.; Banat, F. Development of date seed extract encapsulated MCM-41: Characterization, release kinetics, antioxidant and antibacterial studies. *Food Biosci.* **2023**, *53*, 102563. [CrossRef]

82. Hai, A.; Bharath, G.; Babu, K.R.; Taher, H.; Naushad, M.; Banat, F. Date seeds biomass-derived activated carbon for efficient removal of NaCl from saline solution. *Process Saf. Environ. Prot.* **2019**, *129*, 103–111. [CrossRef]
83. Rambabu, K.; Bharath, G.; Avornyo, A.; Thanigaivelan, A.; Hai, A.; Banat, F. Valorization of date palm leaves for adsorptive remediation of 2,4-dichlorophenoxyacetic acid herbicide polluted agricultural runoff. *Environ. Pollut.* **2023**, *316*, 120612. [CrossRef]
84. Boufennara, S.; Bouazza, L.; de Vega, A.; Fondevila, M.; Amanzougarene, Z.; Lopez, S. In vitro assessment of nutritive value of date palm by-products as feed for ruminants. *Emir. J. Food Agric.* **2016**, *28*, 695–703.
85. McDonald, P.; Henderson, A.; Heron, S.J.E. *The Biochemistry of Silage*; Chalcombe Publications: Chalcombe, UK, 1991.
86. Gouda, G.A.; Kholif, A.E.; Hamdon, H.A.; Kassab, A.Y.; Patra, A.K. Utilization of Waste Date Palm Leaves Biomass Ensiled with Malic or Lactic Acids in Diets of Farafra Ewes under Tropical Conditions. *Animals* **2022**, *12*, 1432. [CrossRef]
87. Wang, H.; Li, L.; Zhang, N.; Zhang, T.; Ma, Y. Effects of Pelleting and Long-Term High-Temperature Stabilization on Vitamin Retention in Swine Feed. *Animals* **2022**, *12*, 1058. [CrossRef]
88. Ominski, K.; McAllister, T.; Stanford, K.; Mengistu, G.; Kebebe, E.G.; Omonijo, F.; Cordeiro, M.; Legesse, G.; Wittenberg, K. Utilization of by-products and food waste in livestock production systems: A Canadian perspective. *Anim. Front.* **2021**, *11*, 55–63. [CrossRef] [PubMed]
89. Lee, S.Y.; Sankaran, R.; Chew, K.W.; Tan, C.H.; Krishnamoorthy, R.; Chu, D.-T.; Show, P.-L. Waste to bioenergy: A review on the recent conversion technologies. *BMC Energy* **2019**, *1*, 4. [CrossRef]
90. Henstra, A.M.; Sipma, J.; Rinzema, A.; Stams, A.J.M. Microbiology of synthesis gas fermentation for biofuel production. *Curr. Opin. Biotechnol.* **2007**, *18*, 200–206. [CrossRef] [PubMed]
91. Igliński, B.; Kujawski, W.; Kiełkowska, U. Pyrolysis of Waste Biomass: Technical and Process Achievements, and Future Development—A Review. *Energies* **2023**, *16*, 1829.
92. Pandey, K.; Yadav, A.K.; Goel, C. Utilization of Food Waste for Biofuel Production. In *Food Waste to Green Fuel: Trend & Development*; Srivastava, N., Malik, M.A., Eds.; Springer: Singapore, 2022; pp. 1–23. [CrossRef]
93. Yahya, S.A.; Iqbal, T.; Omar, M.M.; Ahmad, M. Techno-Economic Analysis of Fast Pyrolysis of Date Palm Waste for Adoption in Saudi Arabia. *Energies* **2021**, *14*, 6048. [CrossRef]
94. Swathy, R.; Rambabu, K.; Banat, F.; Ho, S.-H.; Chu, D.-T.; Show, P.L. Production and optimization of high grade cellulase from waste date seeds by Cellulomonas uda NCIM 2353 for biohydrogen production. *Int. J. Hydrog. Energy* **2020**, *45*, 22260–22270. [CrossRef]
95. Bharath, G.; Hai, A.; Rambabu, K.; Banat, F.; Jayaraman, R.; Taher, H.; Bastidas-Oyanedel, J.-R.; Ashraf, M.T.; Schmidt, J.E. Systematic production and characterization of pyrolysis-oil from date tree wastes for bio-fuel applications. *Biomass Bioenergy* **2020**, *135*, 105523. [CrossRef]
96. Ahmad, A.; Naqvi, S.A.; Jaskani, M.J.; Waseem, M.; Ali, E.; Khan, I.A.; Faisal Manzoor, M.; Siddeeg, A.; Aadil, R.M. Efficient utilization of date palm waste for the bioethanol production through Saccharomyces cerevisiae strain. *Food Sci. Nutr.* **2021**, *9*, 2066–2074. [CrossRef]
97. Liu, Y.; Tang, Y.; Gao, H.; Zhang, W.; Jiang, Y.; Xin, F.; Jiang, M. Challenges and Future Perspectives of Promising Biotechnologies for Lignocellulosic Biorefinery. *Molecules* **2021**, *26*, 5411. [CrossRef]
98. Johnston, J.; Fellow, L.T.S. *Assessing Soil Fertility. The Importance of Soil Analysis and Its Interpretation*; Potash Development Association: York, UK, 2005.
99. Nikita, B.; Puneet Singh, C. Excessive and Disproportionate Use of Chemicals Cause Soil Contamination and Nutritional Stress. In *Soil Contamination*; Marcelo, L.L., Sonia, S., Eds.; IntechOpen: Rijeka, Croatia, 2020; Chapter 6. [CrossRef]
100. Ayilara, M.S.; Olanrewaju, O.S.; Babalola, O.O.; Odeyemi, O. Waste Management through Composting: Challenges and Potentials. *Sustainability* **2020**, *12*, 4456. [CrossRef]
101. Kuhlman, L.R. Windrow composting of agricultural and municipal wastes. *Resour. Conserv. Recycl.* **1990**, *4*, 151–160. [CrossRef]
102. Ali, U.; Sajid, N.; Khalid, A.; Riaz, L.; Rabbani, M.M.; Syed, J.H.; Malik, R.N. A review on vermicomposting of organic wastes. *Environ. Prog. Sustain. Energy* **2015**, *34*, 1050–1062. [CrossRef]
103. Iyengar, S.R.; Bhave, P.P. In-vessel composting of household wastes. *Waste Manag.* **2006**, *26*, 1070–1080. [CrossRef] [PubMed]
104. Abid, W.; Magdich, S.; Mahmoud, I.B.; Medhioub, K.; Ammar, E. Date Palm Wastes Co-composted Product: An Efficient Substrate for Tomato (*Solanum lycopercicum* L.) Seedling Production. *Waste Biomass Valorization* **2018**, *9*, 45–55. [CrossRef]
105. Vatai, T.; Škerget, M.; Knez, Ž. Extraction of phenolic compounds from elder berry and different grape marc varieties using organic solvents and/or supercritical carbon dioxide. *J. Food Eng.* **2009**, *90*, 246–254. [CrossRef]
106. Pourshoaib, S.J.; Rajabzadeh Ghatrami, E.; Shamekhi, M.A. Comparing ultrasonic- and microwave-assisted methods for extraction of phenolic compounds from Kabkab date seed (*Phoenix dactylifera* L.) and stepwise regression analysis of extracts antioxidant activity. *Sustain. Chem. Pharm.* **2022**, *30*, 100871. [CrossRef]
107. Benucci, I.; Lombardelli, C.; Mazzocchi, C.; Esti, M. Natural colorants from vegetable food waste: Recovery, regulatory aspects, and stability—A review. *Compr. Rev. Food Sci. Food Saf.* **2022**, *21*, 2715–2737. [CrossRef]
108. Al-Khalili, M.; Al-Habsi, N.; Rahman, M.S. Applications of date pits in foods to enhance their functionality and quality: A review. *Front. Sustain. Food Syst.* **2023**, *6*, 1101043. [CrossRef]
109. Gill, S.K.; Rossi, M.; Bajka, B.; Whelan, K. Dietary fibre in gastrointestinal health and disease. *Nat. Rev. Gastroenterol. Hepatol.* **2021**, *18*, 101–116. [CrossRef]

110. Hojsak, I.; Benninga, M.A.; Hauser, B.; Kansu, A.; Kelly, V.B.; Stephen, A.M.; Lopez, A.M.; Slavin, J.; Tuohy, K. Benefits of dietary fibre for children in health and disease. *Arch. Dis. Child.* **2022**, *107*, 973–979. [CrossRef] [PubMed]
111. Karra, S.; Sebii, H.; Yaich, H.; Bouaziz, M.A.; Blecker, C.; Danthine, S.; Attia, H.; Besbes, S. Effect of extraction methods on the physicochemical, structural, functional, and antioxidant properties of the dietary fiber concentrates from male date palm flowers. *J. Food Biochem.* **2020**, *44*, e13202. [CrossRef] [PubMed]
112. Walsh, S.K.; Lucey, A.; Walter, J.; Zannini, E.; Arendt, E.K. Resistant starch—An accessible fiber ingredient acceptable to the Western palate. *Compr. Rev. Food Sci. Food Saf.* **2022**, *21*, 2930–2955. [CrossRef]
113. Omerspahic, M.; Al-Jabri, H.; Siddiqui, S.A.; Saadaoui, I. Characteristics of desalination brine and its impacts on marine chemistry and health, with emphasis on the Persian/Arabian gulf: A review. *Front. Mar. Sci.* **2022**, *9*, 525. [CrossRef]
114. Alnawajha, M.M.; Kurniawan, S.B.; Imron, M.F.; Abdullah, S.R.S.; Hasan, H.A.; Othman, A.R. Plant-based coagulants/flocculants: Characteristics, mechanisms, and possible utilization in treating aquaculture effluent and benefiting from the recovered nutrients. *Environ. Sci. Pollut. Res. Int.* **2022**, *29*, 58430–58453. [CrossRef]
115. Rambabu, K.; Bharath, G.; Banat, F.; Show, P.L. Biosorption performance of date palm empty fruit bunch wastes for toxic hexavalent chromium removal. *Environ. Res.* **2020**, *187*, 109694. [CrossRef] [PubMed]
116. Rambabu, K.; AlYammahi, J.; Bharath, G.; Thanigaivelan, A.; Sivarajasekar, N.; Banat, F. Nano-activated carbon derived from date palm coir waste for efficient sequestration of noxious 2,4-dichlorophenoxyacetic acid herbicide. *Chemosphere* **2021**, *282*, 131103. [CrossRef]
117. Papaioannou, E.H.; Mazzei, R.; Bazzarelli, F.; Piacentini, E.; Giannakopoulos, V.; Roberts, M.R.; Giorno, L. Agri-Food Industry Waste as Resource of Chemicals: The Role of Membrane Technology in Their Sustainable Recycling. *Sustainability* **2022**, *14*, 1483. [CrossRef]
118. Mahmoodi, N.M.; Hayati, B.; Arami, M. Textile dye removal from single and ternary systems using date stones: Kinetic, isotherm, and thermodynamic studies. *J. Chem. Eng. Data* **2010**, *55*, 4638–4649. [CrossRef]
119. Banat, F.; Al-Asheh, S.; Al-Makhadmeh, L. Evaluation of the use of raw and activated date pits as potential adsorbents for dye containing waters. *Process Biochem.* **2003**, *39*, 193–202. [CrossRef]

MDPI AG
Grosspeteranlage 5
4052 Basel
Switzerland
Tel.: +41 61 683 77 34

*Applied Sciences* Editorial Office
E-mail: applsci@mdpi.com
www.mdpi.com/journal/applsci

www.ingramcontent.com/pod-product-compliance
Lightning Source LLC
Chambersburg PA
CBHW061631210926
51850CB00009B/2505
* 9 7 8 3 7 2 5 8 4 8 4 1 6 *